VOLUME ONE HUNDRED AND EIGHTY SIX

Progress in Molecular Biology and Translational Science

Micro/Nanofluidics and Lab-on-Chip Based Emerging Technologies for Biomedical and Translational Research Applications - Part A

VOLUME ONE HUNDRED AND EIGHTY SIX

PROGRESS IN MOLECULAR BIOLOGY AND TRANSLATIONAL SCIENCE

Micro/Nanofluidics and Lab-on-Chip Based Emerging Technologies for Biomedical and Translational Research Applications - Part A

Edited by

ALOK PANDYA

Department of Engineering and Physical Sciences, Institute of Advanced Research, Gandhinagar, Gujarat, India

VIJAI SINGH

Department of Biosciences, School of Science, Indrashil University, Rajpur, Mehsana, Gujarat, India

ELSEVIER

ACADEMIC PRESS
An imprint of Elsevier

Academic Press is an imprint of Elsevier
50 Hampshire Street, 5th Floor, Cambridge, MA 02139, United States
525 B Street, Suite 1650, San Diego, CA 92101, United States
The Boulevard, Langford Lane, Kidlington, Oxford OX5 1GB, United Kingdom
125 London Wall, London EC2Y 5AS, United Kingdom

First edition 2022

ISBN: 978-0-323-98899-5
ISSN: 1877-1173

For information on all Academic Press publications
visit our website at https://www.elsevier.com/books-and-journals

Publisher: Zoe Kruze
Acquisitions Editor: Ashlie M. Jackman
Developmental Editor: Jhon Michael Peñano
Production Project Manager: James Selvam
Cover Designer: Matthew Limbert

Typeset by STRAIVE, India

Contents

Contributors *ix*
Preface *xiii*

1. An introduction to microfluidics and their applications **1**
Nidhi Verma, Parth Prajapati, Vijai Singh, and Alok Pandya
1. Introduction 1
2. Type of micro/nanofluidics 3
3. Design and fabrication of microfluidics devices 4
4. Paper based microfluidics 4
5. Application of microfluidics devices 7
6. Conclusion and future remarks 11
Acknowledgments 12
References 12

2. Design and fabrication of micro/nanofluidics devices and systems **15**
Terry Ching, Yi-Chin Toh, and Michinao Hashimoto
1. Introduction 16
2. Theory of micro/nanofluidics 17
3. Fabrication of micro/nanofluidic devices 20
4. Design thinking approach for micro/nanofluidics 43
5. Outlook 50
References 51

3. Detection and separation of proteins using micro/nanofluidics devices **59**
Ashish Badiye, Neeti Kapoor, and Ritesh K. Shukla
1. Introduction 60
2. Protein separation using microfluidics system/devices 62
3. Conclusion 74
References 75

4. Micro/nanofluidic devices for DNA/RNA detection and separation **85**
Nidhi Verma, Sakshi Walia, and Alok Pandya
1. Introduction 86
2. Separation of nucleic acid using NF and MF 87

3. Detection of nucleic acid using NF and MF using different fabrication techniques 94
4. Future prospects 103
Acknowledgments 104
References 104

5. Paper based microfluidics: A forecast toward the most affordable and rapid point-of-care devices 109

Anirban Sinha, Mainak Basu, and Prerna Chandna

1. Introduction and historical perspective of paper-based microfluidic devices 110
2. Fabrication methods 117
3. Fluid flow control and liquid handling 125
4. Applications of paper-based microfluidic devices 133
5. Current limitations and future works 142
6. Conclusion and final thoughts 148
Acknowledgment 149
References 149

6. Paper based micro/nanofluidics devices for biomedical applications 159

P.E. Resmi, P.V. Suneesh, T. Ramachandran, and T.G. Satheesh Babu

1. Importance of paper as a diagnostic device 160
2. Paper as a sample pretreatment platform in clinical diagnosis 161
3. Paper microfluidic diagnostic devices 163
4. Dipstick paper biosensors 164
5. Lateral flow assay (LFA) 164
6. MicroPAD (μPAD) 167
7. Fabrications of paper microfluidics 167
8. Detection techniques 174
9. Readout devices 178
10. Biomedical applications 180
11. Paper nanofluidics 183
12. Challenges and research gaps 184
13. Conclusions 185
14. Future outlook 185
Acknowledgment 185
References 185

7. Advances in microfluidics devices and its applications in personalized medicines 191

Rupesh Maurya, Nisarg Gohil, Gargi Bhattacharjee, Khushal Khambhati, Khalid J. Alzahrani, Suresh Ramakrishna, Dinh-Toi Chu, and Vijai Singh

1. Introduction 192
2. Microfluidics for cancer theranostics 193
3. Microfluidics for detection of cardiovascular diseases 194
4. Microfluidics for detection of diabetes 195
5. Microfluidics for detection of pulmonary disease and therapy 196
6. Conclusion and future perspectives 197

Acknowledgments 197
References 198

8. Microfluidics for single cell analysis 203

Rupesh Maurya, Nisarg Gohil, Gargi Bhattacharjee, Khalid J. Alzahrani, Suresh Ramakrishna, and Vijai Singh

1. Introduction 203
2. Design and fabrication of microfluidics devices and chips 204
3. Droplets-based single cell analysis and their applications 207
4. Use of microfluidics devices for single cell analysis 208
5. Conclusion and future remarks 210

Acknowledgments 211
References 211

9. Fluorescence based miniaturized microfluidic and nanofluidic systems for biomedical applications 217

Aishwarya P. Waghchoure, J. Prakasha Reddy, and Rajesh S. Bhosale

1. Introduction 218
2. Microfluidic systems for biomedical applications 219
3. Nanofluidic systems for biomedical applications 232
4. Conclusion and future prospects 239

Acknowledgments 239
References 240

10. Active matter dynamics in confined microfluidic environments 245

Rik Chakraborty, Arnab Maiti, Nikita Sharma, and Krishna Kanti Dey

11. Paper-microfluidic signal-enhanced immunoassays **267**

N. Sathishkumar and Bhushan J. Toley

1. Introduction and the history of paper-based immunoassays 267
2. Classification of signal enhancement strategies used in paper-based immunoassays 271
3. SWOT analysis: Paper-based immunoassays 282

References 284

12. Challenges and opportunities in micro/nanofluidic and lab-on-a-chip **289**

Nidhi Verma and Alok Pandya

1. Understand micro/nanofluidic (MF/NF) 289
2. Understand lab-on-a-chip 292
3. Challenges in designing micro/nanofluidic 293
4. Challenges in designing lab-on-a-chip device 296
5. Opportunities in micro/nanofluidic and lab-on-a-chip devices 299

Acknowledgments 301

References 301

Index *303*

Contributors

Khalid J. Alzahrani
Department of Clinical Laboratories Sciences, College of Applied Medical Sciences, Taif University, Taif, Saudi Arabia

T.G. Satheesh Babu
Amrita Biosensor Research Lab; Department of Sciences, Amrita School of Engineering Coimbatore, Amrita Vishwa Vidyapeetham, Coimbatore, India

Ashish Badiye
Department of Forensic Science, Government Institute of Forensic Sciences, Nagpur, Maharashtra, India

Mainak Basu
MeFy Care Private Ltd., Pune, India

Gargi Bhattacharjee
Department of Biosciences, School of Science, Indrashil University, Rajpur, Mehsana, Gujarat, India

Rajesh S. Bhosale
Department of Chemistry, Indrashil University, Rajpur, Mehsana, Gujarat, India

Rik Chakraborty
Discipline of Physics, Indian Institute of Technology Gandhinagar, Palaj, Gandhinagar, Gujarat, India

Prerna Chandna
MeFy Care Private Ltd., Pune, India

Terry Ching
Pillar of Engineering Product Development; Digital Manufacturing and Design Centre, Singapore University of Technology and Design; Department of Biomedical Engineering, National University of Singapore, Singapore, Singapore

Dinh-Toi Chu
Center for Biomedicine and Community Health, International School, Vietnam National University, Hanoi, Vietnam

Krishna Kanti Dey
Discipline of Physics, Indian Institute of Technology Gandhinagar, Palaj, Gandhinagar, Gujarat, India

Nisarg Gohil
Department of Biosciences, School of Science, Indrashil University, Rajpur, Mehsana, Gujarat, India

Michinao Hashimoto
Pillar of Engineering Product Development; Digital Manufacturing and Design Centre, Singapore University of Technology and Design, Singapore, Singapore

Neeti Kapoor
Department of Forensic Science, Government Institute of Forensic Sciences, Nagpur, Maharashtra, India

Khushal Khambhati
Department of Biosciences, School of Science, Indrashil University, Rajpur, Mehsana, Gujarat, India

Arnab Maiti
Discipline of Physics, Indian Institute of Technology Gandhinagar, Palaj, Gandhinagar, Gujarat, India

Rupesh Maurya
Department of Biosciences, School of Science, Indrashil University, Rajpur, Mehsana, Gujarat, India

Alok Pandya
Department of Engineering and Physical Sciences, Institute of Advanced Research, Gandhinagar, Gujarat, India

Parth Prajapati
Department of Engineering and Physical Sciences, Institute of Advanced Research, Gandhinagar, Gujarat, India

T. Ramachandran
Department of Sciences, Amrita School of Engineering Coimbatore, Amrita Vishwa Vidyapeetham, Coimbatore, India

Suresh Ramakrishna
Graduate School of Biomedical Science and Engineering; College of Medicine, Hanyang University, Seoul, South Korea

J. Prakasha Reddy
Department of Chemistry, Indrashil University, Rajpur, Mehsana, Gujarat, India

P.E. Resmi
Amrita Biosensor Research Lab, Amrita School of Engineering Coimbatore, Amrita Vishwa Vidyapeetham, Coimbatore, India

N. Sathishkumar
Department of Chemical Engineering, Indian Institute of Science, Bengaluru, KA, India

Nikita Sharma
Discipline of Physics, Indian Institute of Technology Gandhinagar, Palaj, Gandhinagar, Gujarat, India

Ritesh K. Shukla
Biological and Life Sciences, School of Arts and Sciences, Ahmedabad University, Ahmedabad, Gujarat, India

Vijai Singh
Department of Biosciences, School of Science, Indrashil University, Rajpur, Mehsana, Gujarat, India

Anirban Sinha
MeFy Care Private Ltd., Pune, India

P.V. Suneesh
Amrita Biosensor Research Lab; Department of Sciences, Amrita School of Engineering Coimbatore, Amrita Vishwa Vidyapeetham, Coimbatore, India

Yi-Chin Toh
School of Mechanical, Medical and Process Engineering, Queensland University of Technology, Brisbane; Centre for Biomedical Technologies, Queensland University of Technology, Kelvin Grove, QLD, Australia

Bhushan J. Toley
Department of Chemical Engineering, Indian Institute of Science, Bengaluru, KA, India

Nidhi Verma
Department of Engineering and Physical Sciences, Institute of Advanced Research, Gandhinagar, Gujarat, India

Aishwarya P. Waghchoure
Department of Chemistry, Indrashil University, Rajpur, Mehsana, Gujarat, India

Sakshi Walia
Department of Biological Sciences and Biotechnology, Institute of Advanced Research, Gandhinagar, India

Preface

Microfluidics is an exponentially growing field of science and is employed in a wide range of areas. It refers to the science of a fluid handling platform that is designed and developed for manipulation of a small volume of fluids within a network of channels in the dimension range of 1–100 μm. Fundamentally, it is a subject of fluid mechanics that describes the physics of fluids at the microlength scale in microfluidic systems. Nowadays, fabricating microelectronic structures and photolithography are the most common methods used in microfluidic device design. Over the last 30 years, this technology has been established for printed circuit boards, integrated circuits, microscale gas chromatography, ink-jet printers, fluid engineering, analytical chemistry, synthetic chemistry, and life sciences. Most importantly, microfluidics enables the manipulation of small volumes of fluid in a precise and effective manner, which is crucial for the preparation and delivery of small synthetic molecules or biomedical applications.

Microfluidics and nanofluidics have numerous remarkable features, which has drawn the attention of scientific communities working in the field of medicine, biology, engineering, and chemistry. Pertaining to biological applications, a fluidic platform compromises with structures having length scales that are fundamentally analogous to the dimensions of prokaryotic and eukaryotic organisms, collections of cells, organelles, and those of small molecules and proteins. These features make microfluidics principally beneficial in studying biology and biomedicine. Besides, paper-based microfluidics is an excellent example of point-of-care (POC) diagnosis for accurate treatment of diseases. In context to the perceptive, the use of a fluidic platform in modern personalized health care and medicine is centered on patients, as in the case of diabetes, which is duly fulfilled after the development of POC blood glucose level-testing devices.

Microfluidics has led to the development of a series of new components and adoption of innovative techniques, such as pumping, mixing, and introducing, as well as storing fluids in the channels of microfluidic devices. Despite rapid progress in this area, there remain several unresolved issues, including suitable system integration, portability, and versatility. In this volume, we provide a comprehensive overview in terms of design, fabrication, system integration, and development, and its biomedical applications. This volume covers the state-of-the-art developments in the field of microfluidics/nanofluidics, including advanced system integration, wearable sensors, biomedical microfluidic devices, lab-on-a-chip (LOC), and micro-total analysis, which demonstrate capabilities to solve real-world problems of particular interest. This book is a key source of information available in a single volume and highlights a number of aspects of micro/nanofluidics-based miniaturized LOC devices in a manner that can help future investigators, researchers, students, and stakeholders perform their research with greater ease. This book will be an excellent basis from which scientific knowledge would grow and widen in the fields of micro/nanofluidics and LOC.

Alok Pandya
Vijai Singh

CHAPTER ONE

An introduction to microfluidics and their applications

Nidhi Verma[a], Parth Prajapati[a], Vijai Singh[b], and Alok Pandya[a,*]
[a]Department of Engineering and Physical Sciences, Institute of Advanced Research, Gandhinagar, Gujarat, India
[b]Department of Biosciences, School of Science, Indrashil University, Rajpur, Mehsana, Gujarat, India
*Corresponding author: e-mail address: alokpandya20@gmail.com

Contents

1. Introduction 1
2. Type of micro/nanofluidics 3
3. Design and fabrication of microfluidics devices 4
4. Paper based microfluidics 4
5. Application of microfluidics devices 7
 5.1 Detection and separation of proteins 7
 5.2 DNA/RNA detection and separation 8
 5.3 Microfluidics devices for drug discovery, screening and delivery 9
 5.4 Disease diagnostics and personalized medical applications 10
6. Conclusion and future remarks 11
Acknowledgments 12
References 12

Abstract

In the past decade, microfluidics and lab-on-chip are rapidly growing area. It has been used for number of biology to biotechnology and medicine applications. It has the potential to reduce cost of reagent and time of experiment. It has been used integrated with automation for extraction and detection of protein, nucleic acids, enzymes, metabolites and delivery of drug to target location. It has been used for drug discovery, high throughput screening of potent drug and used for delivery. Paper based microfluidics was used for point-of-care diagnosis for accurate treatment of diseases.

In this chapter, we highlight advances of microfluidics devices for number of biological and translational science applications.

1. Introduction

"Microfluidics" denotes to the science and technology of controlling fluids in channeled networks with measurements of 5–500 mm and its history has been the focus of a current review by Whiteside's. Microfluidic systems

Progress in Molecular Biology and Translational Science, Volume 186
ISSN 1877-1173
https://doi.org/10.1016/bs.pmbts.2021.07.006

help in transport of fluid with volumes varying from microliters to femtoliters in channels that are typically imprinted on the surface of a substrate and in certain cases may be fabricate directly on the substrate of glass. The type of substrate used is greatly varied based on the application of and depending on the type of analyte. Microfluidics and nanofluidic has numerous features that have drew attention in medicine, biology, engineering and chemistry. The application ranges from sample collect from bodily fluids like blood plasma, isolation of proteins and nucleic acid, on chip sample extraction, amplification and detection. The major reason for use of microfluidics in healthcare sector was its requirement of negligible quantity of sample as well as reagent requirement, causing minimum waste. The time required for analysis is quite lesser when compared to traditional techniques as well as in low cost and offer great deal of application in miniatured devices. Pertaining to the biological applications fluidic platform compromises with structures having length scales that are analogous to the fundamentally similar to dimensions of prokaryotic and eukaryotic cells, collections of cells, organelles, those of small molecules, cells and proteins. These features make microfluidics principally beneficial in studying biology and biomedicine. In context to the perceptive hovering around use of fluidic platform in modern personalized healthcare and medicine is centered around patients, wherein in case of diabetes, which much fulfilled after development of point of care blood glucose level testing device.

However, in accordance to the conventional approaches of detection of complex diseases, fluidic platforms are emerging as powerful methods which could tackle the challenges. Present time personalized healthcare relies on wide spectrum of patient history and diagnostic tests in order, to define the optimum medication option for each patient. It ranges began from simple classifications of patients to more precise stratifications of patients based on exclusive biomarker profiling details. Followed by individual response to treatment depending on the genetic makeup. The level of personalized healthcare is revolutionizing each day with microfluidic been at the center of the grid.

The number of applications of microfluidics varies from filed of biology to analytical and until chemistry to biochemistry as shown in Fig. 1. This have led to development of series of new components and adopting innovative techniques pumping, mixing, and introducing, as well as storing fluids in channels of microfluidic device. Despite rapid progress in this area, there remain several unresolved issues like portability and versatility. In this chapter we provide a comprehensive overview of recent advances in

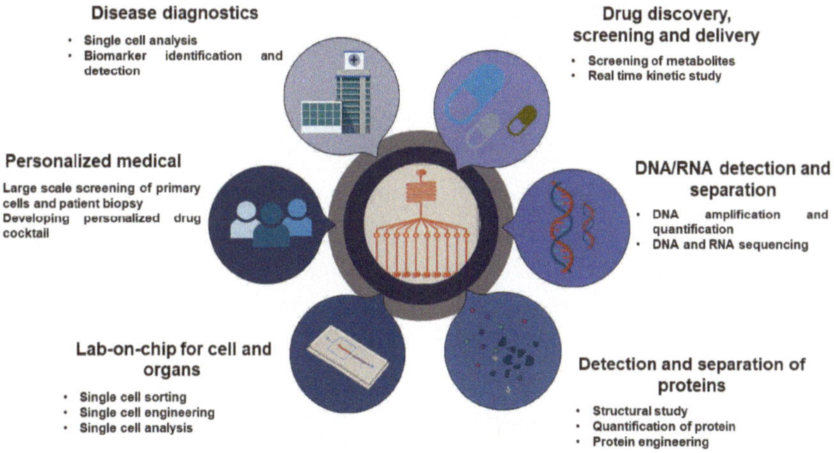

Fig. 1 Application of microfluidics in biology to personalized medicine.

nano/microfluidic technologies in terms of fabrication for personalized healthcare. We also underline the applications of paper-based microfluidics as a pioneering substrate for diagnosis. Furthermore, we provide insight into detection of proteins, nucleic acids and current trends of on chip cell and organ-based detection. Lastly it mentioned the application of microfluidics in personalized healthcare. This chapter also endeavors to communicate to the readers, from the authors' point of view the existing boundaries of microfluidics which require further research, and a few viewpoints in the directions for new analytical system may help in its development.

2. Type of micro/nanofluidics

The rigorous and exhaustive relationship of between designing and the substrate for the construction of the microfluidic devices. Materialistic attributed of the substrate have continuously been a vital point of development in establishment of microfluidic technologies and brings about with the range of applications. Until now several materials, including silicon, glass, quartz/fused silica, ceramics, metals and polymers, have been used in constructing microfluidic platforms orienting varied functionalities. Overall, a microfluidic device can be initially constructed by with various techniques and features the materials used, which serve as the ingredients for of various types of microfluidic devices the detail about the same is mentioned in detail in chapter "Micro/nanofluidic devices for DNA/RNA detection and separation" by Verma et al.

3. Design and fabrication of microfluidics devices

AutoCAD is a tool for 2D sketching and it has been widely used for design of microfluidics chips/devices. This final design can be printed on silicon wafer (with at least 1200 dpi). These chips have been fabricated using molding which is created by soft photolithography. It has been directly printed onto photomask which is a transparent film of silicon wafer. Silicon wafer is used for fabrication using Polydimethylsiloxane (PDMS) and curing agent. Both the chemicals are properly mixed followed by centrifuged at 4000 rpm for 10 min (it removes air bubbles). Subsequently, wafers are placed on a big sized petri plate and the mixture of PDMS and curing agent was poured on top of wafer and kept at 80 °C for 1 h for polymerization.[1]

After solidification, it was removed and using cut sharply. The hole was created using needle subsequently, and glass slide as well as polymer kept in oxygen plasma cleaner for 30 s then both were sticked together and kept for 24 h at 65 °C for proper bonding. Once chips are properly bonded then used for loaded cells and kept chip on microscope then Luria Bertani liquid media was provided to grow cell and images were taken at every 30 s for phase contrast and fluoresces image for 3 min time interval. A single cell was analyzed and behavior was monitored.[1]

4. Paper based microfluidics

Paper as a term is referred to fibrous porous mass-produced thin materials to be used in modern analytical substrate dated back to 1940s and 1950s, having major application in chromatography as well as electrophoresis. In the year 2007 "Paper-based microfluidics" or "lab on paper" evolved as a budding field of research, having a wide range of applications including health diagnostics, environmental monitoring as well as food quality testing. Over the time the field has evolved and now paper based microfluidic is referred to as μPADs. The major reasons that lead for making paper act as an attractive substrate to be used in microfluidic systems includes of properties of ubiquitous nature, an option of exceptionally cheap cellulosic based material. The paper as a substrate is highly compatible since inert in its nature making its way into all possible chemical, biochemical as well as medical applications. The next major reason for paper to make its way in microfluidics is because of its easy liquids transports via capillary forces without having requirement of any external equipment. Paper as a substrate not only help in microfluidic channels on paper, confined liquid flow within the channels but also help in liquid flow in guided and precise manner. A diverse nature of paper based microfluidic device has been proposed over time both 2D and even on predesigned pathways to transport liquids on paper. This decade is an explosion for research and development in paper-based microfluidics most work centered around developing low-cost and simple fabrication techniques and discovering innovative applications by integrating effective detection methods. This chapter directs to evaluate both the fabrication techniques and applications of microfluidics reported to date.[2]

The discovery paper-based microfluidics, which has been accredited to the Whitesides Group of Harvard University, but having a history long back initially reported by Müller et al. might be the origin of paper microfluidics.[3,4] In context to the fabricating, techniques for as per the literature major techniques includes of paper cutting, flexography printing, photolithography, ink jet printing, plotting with an analog plotter, plasma treatment ink jet etching, wax printing, screen printing and laser treatment. The principle adopted techniques for fabrication of the paper based microfluidic is usually hydrophilic zones as well as hydrophobic zones on a paper in order to create selective zones and flow channels. In order to pick the proper technique, wide range of factors needs to be included like cost of material, simplicity of fabrication, and the proposed applications of paper-based microfluidic devices. A simple paper based microfluidic device (μPAD) confirming the presence of disease visually are no longer adequate as diagnostic channel typically comprises multistep methods, including pretreatment, sample collection and analysis. Traditional methods of diagnostics processes made uses needles and syringes for collection of samples causing trouble in transport and storage of the samples. As well washing and transfer steps for sample analysis, making the overall procedure labor-demanding and incompatible for. Hopefully, the advancement of microfluidic technologies has helped in chip-based diagnosis fitting into point of care setting for diagnosis in order to overcome the drawbacks of traditional methods. However, traditional integrated chip-based diagnostic systems used substrates such as polymers, glass and silica confer deficiencies like requirement of pumps and valves as well as lengthy process of fabrication. Comparatively high expense, hinder their aptness for point of need diagnosis. In those terms paper microfluidics have merits of cost effectiveness, easy mass production and ease of use, arise as perfect tools for personal diagnosis, as sample collection/pretreatment, result outputs and signal transduction can be simplified into one μPAD with negligible need for complicated operation.

The application of paper-based microfluidics is eminence ranging from biomedical application, environmental as well as food industry applications. Considering the biomedical application, it covers a very broad range of section from sample extraction, and sample separation along with signal amplification and transduction. The sample type ranges from nucleic acids (RNA as well as DNA), protein molecules, disease causing important biomarkers. Sample extraction is of great significance to isolate targets of interest (especially DNA/RNA) from complicated samples. One such application of paper based microfluidic was seen in detection of hematocrit.[5] Berry et al. developed a paper based microfluidic device for the detection of hematocrit in whole blood. This paper-based microfluidic device enabled controlled transport of red blood cells and the measurement for detection of hematocrit was in terms of the ratio of RBC packed cell volume to total volume of whole blood. The device costed only inexpensive US$0.03 per device where the distance traveled by RBCs was proportional to the detection hematocrit.[5] Similarly, Alizadeh et al. utilized paper for developing paper microfluidic colorimetric immunosensor in detection of carcinoembryonic antigen. The immunosensor was developed using mixture of chitosan as well as ionic liquid functionalized with a primary antibody on the paper of detection. The detection was accompanied by a color change and have a limit of detection of 0.51 $pg\,mL^{-1}$.[6] One challenge however still that need to be overcome in using μPADs is achieving lower limits of detection with good selectivity due to a major paper-based microfluidics relied on colorimetric detection. Over the time integration of paper with other techniques has emerged as an effective option to overcome such problems.

Electrochemical detection coupled with μPADs is referred to as ePADs, offers a decent balance for low-cost detection while also providing high selectivity and sensitivity.[7] Noiphung et al. presented an ePAD device that incorporated blood separation where whole blood can be used for analysis. Serum was then passed from these regions to a zone of detection. The detection zone consisted of glucose oxidase and in contact surface modified electrode for the electrochemical detection of glucose. Apart from enzymatic detection, the detection of metabolites such as uric acid and ascorbic acid using amperometry was reported by Carvalhal et al. with the detection limits of 0.02 mM.[8] For increasing the overall performance of paper-based devices numerous challenges need to be addressed. Foremost, for better stability of the device easy storage of reagents at room temperature, thus removing the requirement for storage facilities, which may be helping in advancing the paper devices in remote locations. Besides, combined paper-based sample collection is still restricted, particularly for blood analysis. Although microneedle-based methods have been anticipated. Furthermore, the expansion of targets of interest and user-friendly interpretation approaches will unquestionably encourage the profound usage of μPADs in real life.[9]

5. Application of microfluidics devices

5.1 Detection and separation of proteins

Applications of micro and nanofluidic in genomics and proteomics is abundant where the detection of lower levels of nucleic acids such as DNA, RNA is usually achieved via target amplification using techniques such as PCR, as the detection and separation of proteins is challenging as they are present in low abundance in blood, body fluids, etc. This problem has prompted the development of both label-mediated and label-free detection strategies in microfluidic devices.[10] One such application for identification and detection of protein was carried out by Gao et al. Electrospray ionization mass spectrometry was used for the detection of the protein. The microfluidic platform was capable of simultaneous digestion as well as analysis of the protein. The microfluidic device was made of dimethyl siloxane and coupled to a poly(vinylidene fluoride) with the trypsin adsorbed on membrane. The extent of protein digestion was controlled by time of residency of protein inside the trypsin-adsorbed membrane, protein concentration and the reaction temperature. The resulting protein sample mixtures was straight investigated using electrospray ionization mass spectrometry and electrophoretic separations was used for resolving preceding to the analysis by mass spectrometric. The microfluidic system not only enabled rapid identification of proteins in minutes but also provides separation of protein for the analysis of complex protein mixtures.[11] Counting out on the most important human body fluids which accounts of plasma and blood in terms of proteome reputation. Profiling of plasma protein helps in analyzing the physiological and pathological conditions associated with enormous human diseases, still only limited plasma proteins are analyzed for routine analysis in clinical tests a more rigorous analysis can help in better diagnosis. One major reason that account for limited plasma proteome analysis is the complexity of the system. as well as degradation of proteins in sampled blood. In order to overcome this problem Fan et al. developed an integrated microfluidic system along with blood barcode chip for on-chip blood separation and rapid of the plasma proteins from whole blood. The device proved to be low-cost, non-invasive and predominantly important in point-of-care settings.[12] Fan et al. reported a sensitive droplet microfluidic for detection of membrane protein assisted with fluorescence signal amplifications. The membrane protein-triggered conformation alters into hairpin probe and to improve the accuracy of detection by eliminating steps of separation and washing. The developed microfluidic system provides a high-end platform for the analysis of a single cell along monodisperse droplet which as an independent micro-reactor for both aptamer in addition to nicking enzyme aided fluorescence signal amplification, enhanced rate of reaction and highly sensitive detection of membrane protein was achieved.[13] Another microfluidic platforms adopted for detection of proteins are affinity-based, that depend on selective capturing of analytes, usually proteins, functionalizing with an appropriate ligand molecules making microfluidics to be used for label-free biosensing. It usually is integrated with an acoustic based or surface plasmon resonance based sensors.[14–17] Finally as an affinity sensors the number of molecules binding depends on the target molecule transport, kinetics as well as thermodynamics of the binding of the analyte to the ligand.[18] Interventions of nanoparticles in microfluidics as well as nanofluidic has opened new avenues for detection of proteins. Nanopore sensing are one of the recent developments which detect the change in the ion current induced by analyte of interest passing through a nanopore separated electrolyte reservoirs and act as miniature version of coulter counting. This has been applied to a wide range of monomeric proteins as well as amyloid fibrils which help in revealing important information.[19–22] In the framework of protein studies, solid state nanopores helped in studying the translocation of lysozyme during the process of aggregation. Based on nanopore analysis it was concluded that protein fibril translocation time was linearly along with the length, helping retrieving information of the sizes of supra-molecular protein structures. Another nanopore for detecting digitally encoded DNA resulted in detection antibodies of four different types having the same isotype at nanomolar concentration levels. In certain cases it has been reported that proteins have been utilized as nanopores.[23,24]

5.2 DNA/RNA detection and separation

In molecular biology, extraction of nucleic acids is time taking and labor-intensive process. In the past decade, microfluidics devices integrated with automation have been used for high-throughput extraction of sample from number of samples. It has been reduced the cost and time for nucleic acids extraction from blood sample for further research.[25] In this study, Geissler et al. have designed microfluidics device for extraction of bacterial nucleic acids, along with lysis of cell and polymerase chain amplification of target gene.[26]

In more recently, Sullivan et al. designed a microfluidics device for purification of nucleic acids from blood samples for point-of-care diagnosis. The 33 μL blood sample was used for extraction of nucleic acid with 30 min followed by recombinase polymerase amplification (RPA) for rapid amplification of target gene.[27] In this study, Kim et al. have designed a continuous flow (CF) microchip using PDMS for detection of DNA using PCR. This chips have three different temperatures like Thermal cycler for maintaining denaturation, annealing and extension of reactions. PCR products were obtained with 8–30 min which is quite rapid for detection of gene.[28]

In addition, Hashimoto et al. also designed continuous-flow polymerase chain reaction (CF-PCR) devices for amplification of target gene. It has maintained temperatures and ran 20 cycles. DNA product (500 bp) was obtained within 1.7 min and 997 bp fragment was detected within 3.2 min.[29]

Similarly, Li et al. designed CF-PCR microfluidic chips that was integrated with electrophoresis and DNA hybridization all in one. It has been used for detection of *Porphyromonas gingivalis*, *Treponema denticola* (amplification time was 2 min and 31 s while PCR product detection time was 3 min 43 s) and *Tannerella forsythia*. Therefore, minimum 125 cfu/mL bacteria were also detected. This device has the potential used for POC for detection of DNA for disease diagnosis and accurate treatment of diseases.[30] In this study, Schabmueller et al. have designed silicon/Pyrex-based chip for detection of small quantities of DNA (1 μL). It has been used for detection of DNA and chip was controlled by sensor. This allows us to detection of DNA using small of quantity.[31]

5.3 Microfluidics devices for drug discovery, screening and delivery

Microfluidics devices offer drug discovery and their high throughput screening from large size library. In this study, Pang et al. designed microfluidics with multichannel that has ability to rapidly screen potent drugs from library. They have identified number of enzymes/proteins and metabolites.[32] Similarly, Dittrich and Manz used microfluidics for bioassay, separation and chemical synthesis of metabolites.[33] In this study, Vladisavljević et al.

was used in biomedicine, chemicals and nutraceuticals. It has been used for fine tuning of flow rate to changes the size, composition and dispersion of drug molecules.[34] In the past decade, microfluidics offers and changes the properties of drug, shape, size and pharmacokinetics nature. Tissue and cells were used for screening and testing of drug in vitro in a high throughput manner at low cost.[35]

Drug discovery is being progress and still number of challenges and issue has arisen. In this study, Sant et al. designed microfluidics devices for control of drug size and shape that can be used for easy drug delivery to target locations. It was used for biomimetic gastrointestinal (GI) tract models for drug delivery and testing effect of drug.[36] Similarly, Ghosh et al. have developed extended-release GI luminal delivery systems. It has been used for administration and therapeutic use of drugs. The sustainable and slow released of drug that can help to increase efficacy of drug therapy.[37]

Small interference RNAs (siRNAs) was used for cancer therapy. Microfluidics was used for polymer-lipid hybrid nanoparticles (P/LNPs) that was loaded with VEGF siRNA. It showed toxic effect on tumor cells that can help to easily treat it. This study can be expanded for delivery of drug or small RNA or aptamer for control of growth of cancer cells in an easy manner.[38]

5.4 Disease diagnostics and personalized medical applications

Nanofluidic and Microfluidic technologies are an evolving tools which are capable to address the challenges imposed by traditional diagnostic devices.[39] Microfluidic platform has the potential to significantly advance healthcare globally through the progress of devices for disease diagnosis and description. Such approaches enable on-chip point of care diagnosis and timely monitoring of infectious diseases from minuet samples of bodily fluids. These technologies have potential to be integrated into a single device and enable targeted delivery of samples to specific reaction chambers in an ordered manner. In context to its applications into point of care diagnostic devices, nanofluidic can improve disease detection with limited sample amount. Certain features, of nanofluidic and microfluidic devices have been used for preparation of sample, fractionation of blood flow purification of small molecules and nucleic acid extraction. Charmet et al. reported innovative procedures for fluid handling, and detection signal in microfluidics into a single, user-friendly point-of-need assay that duplicates all steps of ELISA, at much reasonable cost. It was referred to as "mChip" assay in the diagnosis of HIV in just 13 μL of blood and capability in performing simultaneously diagnose HIV and syphilis without compromise on the sensitivities and specificities Largely the device in an excellent miniatured device to diagnosis of multiple infectious diseases where lab facilities are limited.[10] Another exceptional disease diagnosis system can be developed by using on-chip detection combined with imaging techniques to detect diseases in a quick and precise manner. One such innovative imaging platform referred as "mobile-based clinical microscopy" has been integrated and designed into a portable mobile phone microscopy system for confronting global health.[40] Generally, most regions of developing countries have either scarce infrastructural resource and deficiency of healthcare facilities. The clinical viability of the technique was efficacious detection of tuberculosis and malaria by bright field and fluorescence imaging, respectively. This technique is possibly valuable in emerging as a portable platform that can be combined with disposable nanofluidic/microfluidic devices for global health applications.[40,41] Recently, a microfluidic device was established to achieve rapid, and inexpensive CD T-lymphocyte counts, especially for HIV supervising in resource-limited settings.[42] This device not only enabled imaging and capture of CD4 cells with a miniaturized flow chamber by digital camera technology and microscope optics. More recently, a label-free recognition method was also established for counting label-free CD4+ T-lymphocytes in under developed area.[43–45] Microfluidics-based immunoassays are another potential field for disease diagnosis which could prove to be of great importance where sensitivity requirements are high, fast and accurate results.[46] Yager et al. developed an on-chip diffusion immunoassay to assess the concentration of small molecules within microfluidic channels.[47]

6. Conclusion and future remarks

In spite of the exhilarating advancement in the field of nanofluids and microfluidics, there are still many obstacles for the application in personalized medicine as well as in routine diagnostics especially for the regions where lab facilities are limited mostly the case in the under developed countries. Nano/Microfluidic POC diagnostics have important advantages over conventional diagnostics such as reducing costs and increasing compactness and disposability. Many microfluidic devices still rely on detection methods which are complex and have requirement of external equipment, which restricts the use of these fluidic platform as point of care devices. Even though few colorimetric based devices have been developed for it is low on sensitivity and have poor limit of. Integration of electrochemical detection can prove to be one of the detector systems to be utilized for better quantitation of the analyte. Although optical detection continues to be an attractive option for microfluidic analysis and combining sensitive optical detectors might act as an inexpensive option for microfluidics-based devices for a potential candidate as point of care devices. Integration of smartphones might help in monitoring personal health status and monitor lab immunoassay. Therefore, we suppose that the amalgamation of smartphone technologies with microfluidic devices can influence on personalized health care and disease monitoring in the near future. The future trend in nanofluidic devices also includes innovative techniques for easy fabrication, collection of samples and preparation, reagent storage and fully combined lab-on-a-chip system. Microfluidic biochips that can right away analyses crude samples in real world are lacking. Aside from diseases monitoring, evaluation of drug monitoring explosives and foodborne contaminants detection has also attracted extensive attention. In the end, the accomplishment or disappointment of microfluidics will probably be governed by their use outside of academic research laboratories, much like traditional microfluidic devices. Also, early regulatory approval and clinical adoption is necessary for better adaptation of microfluidics platforms. While important attention should be duly be given for improved clinical diagnostics addressing disposability, cost-effectiveness, ease of use and portability centering the areas for future development.

Acknowledgments

A.P. and N.V. gratefully acknowledge Gujarat State Biotechnology Mission (GSBTM Project ID: AJJV48) and SHODH: Scheme of Developing High Quality Research Scholarship (Ref No: 201901440009) for financial support. The financial support from Gujarat State Biotechnology Mission (GSBTM) (Project ID: 5LY45F), Gujarat, India to V.S. is duly acknowledged.

References

1. Carrera J, Rodrigo G, Singh V, Kirov B, Jaramillo A. Empirical model and in vivo characterization of the bacterial response to synthetic gene expression show that ribosome allocation limits growth rate. *Biotechnol J*. 2011;6(7):773–783.
2. Li X, Ballerini DR, Shen W. A perspective on paper-based microfluidics: current status and future trends. *Biomicrofluidics*. 2012;6(1), 011301.
3. Whitesides GM. The origins and the future of microfluidics. *Nature*. 2006;442(7101): 368–373.
4. Whitesides GM. Cool, or simple and cheap? Why not both? *Lab Chip*. 2013;13(1): 11–13.

5. Berry SB, Fernandes SC, Rajaratnam A, DeChiara NS, Mace CR. Measurement of the hematocrit using paper-based microfluidic devices. *Lab Chip*. 2016;16(19):3689–3694.
6. Alizadeh N, Salimi A, Hallaj R. Mimicking peroxidase activity of Co2 (OH) 2CO3-CeO2 nanocomposite for smartphone based detection of tumor marker using paper-based microfluidic immunodevice. *Talanta*. 2018;189:100–110.
7. Adkins J, Boehle K, Henry C. Electrochemical paper-based microfluidic devices. *Electrophoresis*. 2015;36(16):1811–1824.
8. Carvalhal RF, Simão Kfouri M, de Oliveira Piazetta MH, Gobbi AL, Kubota LT. Electrochemical detection in a paper-based separation device. *Anal Chem*. 2010;82 (3):1162–1165.
9. Tian T, Bi Y, Xu X, Zhu Z, Yang C. Integrated paper-based microfluidic devices for point-of-care testing. *Anal Methods*. 2018;10(29):3567–3581.
10. Charmet J, Arosio P, Knowles TP. Microfluidics for protein biophysics. *J Mol Biol*. 2018;430(5):565–580.
11. Gao J, Xu J, Locascio LE, Lee CS. Integrated microfluidic system enabling protein digestion, peptide separation, and protein identification. *Anal Chem*. 2001;73(11):2648–2655.
12. Fan R, Vermesh O, Srivastava A, et al. Integrated barcode chips for rapid, multiplexed analysis of proteins in microliter quantities of blood. *Nat Biotechnol*. 2008;26(12): 1373–1378.
13. Li L, Wang Q, Feng J, Tong L, Tang B. Highly sensitive and homogeneous detection of membrane protein on a single living cell by aptamer and nicking enzyme assisted signal amplification based on microfluidic droplets. *Anal Chem*. 2014;86(10):5101–5107.
14. Luo X, Davis JJ. Electrical biosensors and the label free detection of protein disease biomarkers. *Chem Soc Rev*. 2013;42(13):5944–5962.
15. Burg TP, Mirza AR, Milovic N, et al. Vacuum-packaged suspended microchannel resonant mass sensor for biomolecular detection. *J Microelectromech Syst*. 2006;15 (6):1466–1476.
16. Papadakis G, Friedt J, Eck M, Rabus D, Jobst G, Gizeli E. Optimized acoustic biochip integrated with microfluidics for biomarkers detection in molecular diagnostics. *Biomed Microdevices*. 2017;19(3):16.
17. Mitsakakis K, Gizeli E. Multi-sample acoustic biosensing microsystem for protein interaction analysis. *Biosens Bioelectron*. 2011;26(11):4579–4584.
18. Squires TM, Messinger RJ, Manalis SR. Making it stick: convection, reaction and diffusion in surface-based biosensors. *Nat Biotechnol*. 2008;26(4):417–426.
19. Lanphere C, Offenbartl-Stiegert D, Dorey A, et al. Design, assembly, and characterization of membrane-spanning DNA nanopores. *Nat Protoc*. 2021;16(1):86–130.
20. Martyushenko N, Bell NA, Lamboll RD, Keyser UF. Nanopore analysis of amyloid fibrils formed by lysozyme aggregation. *Analyst*. 2015;140(14):4882–4886.
21. Yusko EC, Prangkio P, Sept D, Rollings RC, Li J, Mayer M. Single-particle characterization of Aβ oligomers in solution. *ACS Nano*. 2012;6(7):5909–5919.
22. Rodriguez-Larrea D, Bayley H. Multistep protein unfolding during nanopore translocation. *Nat Nanotechnol*. 2013;8(4):288–295.
23. Bell NA, Keyser UF. Digitally encoded DNA nanostructures for multiplexed, single-molecule protein sensing with nanopores. *Nat Nanotechnol*. 2016;11(7):645.
24. Wallace EV, Stoddart D, Heron AJ, et al. Identification of epigenetic DNA modifications with a protein nanopore. *Chem Commun*. 2010;46(43):8195–8197.
25. Brassard D, Geissler M, Descarreaux M, et al. Extraction of nucleic acids from blood: unveiling the potential of active pneumatic pumping in centrifugal microfluidics for integration and automation of sample preparation processes. *Lab Chip*. 2019;19(11): 1941–1952.
26. Geissler M, Brassard D, Clime L, et al. Centrifugal microfluidic lab-on-a-chip system with automated sample lysis, DNA amplification and microarray hybridization for

identification of enterohemorrhagic *Escherichia coli* culture isolates. *Analyst*. 2020;145 (21):6831–6845.
27. Sullivan BP, Bender AT, Ngyuen DN, Zhang JY, Posner JD. Nucleic acid sample preparation from whole blood in a paper microfluidic device using isotachophoresis. *J Chromatogr B*. 2021;1163:122494.
28. Kim JA, Lee JY, Seong S, et al. Fabrication and characterization of a PDMS–glass hybrid continuous-flow PCR chip. *Biochem Eng J*. 2006;29(1–2):91–97.
29. Hashimoto M, Chen P-C, Mitchell MW, Nikitopoulos DE, Soper SA, Murphy MC. Rapid PCR in a continuous flow device. *Lab Chip*. 2004;4(6):638–645.
30. Li Z, Ju R, Sekine S, Zhang D, Zhuang S, Yamaguchi Y. All-in-one microfluidic device for on-site diagnosis of pathogens based on an integrated continuous flow PCR and electrophoresis biochip. *Lab Chip*. 2019;19(16):2663–2668.
31. Schabmueller C, Lee M, Evans A, Brunnschweiler A, Ensell G, Leslie D. Closed chamber PCR chips for DNA amplification. *Eng Sci Educ J*. 2000;9(6):259–264.
32. Pang H-M, Kenseth J, Coldiron S. High-throughput multiplexed capillary electrophoresis in drug discovery. *Drug Discov Today*. 2004;9(24):1072–1080.
33. Dittrich PS, Manz A. Lab-on-a-chip: microfluidics in drug discovery. *Nat Rev Drug Discov*. 2006;5(3):210–218.
34. Vladisavljević GT, Khalid N, Neves MA, et al. Industrial lab-on-a-chip: design, applications and scale-up for drug discovery and delivery. *Adv Drug Deliv Rev*. 2013;65 (11 – 12):1626–1663.
35. Tsui JH, Lee W, Pun SH, Kim J, Kim D-H. Microfluidics-assisted in vitro drug screening and carrier production. *Adv Drug Deliv Rev*. 2013;65(11–12):1575–1588.
36. Sant S, Tao SL, Fisher OZ, Xu Q, Peppas NA, Khademhosseini A. Microfabrication technologies for oral drug delivery. *Adv Drug Deliv Rev*. 2012;64(6):496–507.
37. Ghosh A, Li L, Xu L, et al. Gastrointestinal-resident, shape-changing microdevices extend drug release in vivo. *Sci Adv*. 2020;6(44), eabb4133.
38. Huang X, Lee RJ, Qi Y, et al. Microfluidic hydrodynamic focusing synthesis of polymer-lipid nanoparticles for siRNA delivery. *Oncotarget*. 2017;8(57):96826.
39. Lee WG, Kim Y-G, Chung BG, Demirci U, Khademhosseini A. Nano/microfluidics for diagnosis of infectious diseases in developing countries. *Adv Drug Deliv Rev*. 2010;62(4–5):449–457.
40. Granot Y, Ivorra A, Rubinsky B. A new concept for medical imaging centered on cellular phone technology. *PLoS One*. 2008;3(4), e2075.
41. Breslauer DN, Maamari RN, Switz NA, Lam WA, Fletcher DA. Mobile phone based clinical microscopy for global health applications. *PLoS One*. 2009;4(7), e6320.
42. Rodriguez WR, Christodoulides N, Floriano PN, et al. A microchip CD4 counting method for HIV monitoring in resource-poor settings. *PLoS Med*. 2005;2(7), e182.
43. Cheng X, Irimia D, Dixon M, et al. A microfluidic device for practical label-free CD4 + T cell counting of HIV-infected subjects. *Lab Chip*. 2007;7(2):170–178.
44. Cheng X, Irimia D, Dixon M, et al. A microchip approach for practical label-free CD4+ T-cell counting of HIV-infected subjects in resource-poor settings. *J Acquir Immune Defic Syndr*. 2007;45(3):257–261.
45. Bhattacharyya A, Klapperich C. Design and testing of a disposable microfluidic chemiluminescent immunoassay for disease biomarkers in human serum samples. *Biomed Microdevices*. 2007;9(2):245–251.
46. Parsa H, Chin CD, Mongkolwisetwara P, Lee BW, Wang JJ, Sia SK. Effect of volume-and time-based constraints on capture of analytes in microfluidic heterogeneous immunoassays. *Lab Chip*. 2008;8(12):2062–2070.
47. Hatch A, Kamholz AE, Hawkins KR, et al. A rapid diffusion immunoassay in a T-sensor. *Nat Biotechnol*. 2001;19(5):461–465.

CHAPTER TWO

Design and fabrication of micro/nanofluidics devices and systems

Terry Ching[a,b,c], Yi-Chin Toh[d,e], and Michinao Hashimoto[a,b,*]

[a]Pillar of Engineering Product Development, Singapore University of Technology and Design, Singapore, Singapore
[b]Digital Manufacturing and Design Centre, Singapore University of Technology and Design, Singapore, Singapore
[c]Department of Biomedical Engineering, National University of Singapore, Singapore, Singapore
[d]School of Mechanical, Medical and Process Engineering, Queensland University of Technology, Brisbane, QLD, Australia
[e]Centre for Biomedical Technologies, Queensland University of Technology, Kelvin Grove, QLD, Australia
*Corresponding author: e-mail address: hashimoto@sutd.edu.sg

Contents

1. Introduction 16
2. Theory of micro/nanofluidics 17
3. Fabrication of micro/nanofluidic devices 20
   3.1 Subtractive manufacturing 22
   3.2 Formative manufacturing 28
   3.3 Additive manufacturing 33
   3.4 Overview of manufacturing techniques 39
4. Design thinking approach for micro/nanofluidics 43
   4.1 Why is design thinking relevant to the development of micro/nanofluidics? 43
   4.2 What is the design thinking approach? 44
5. Outlook 50
References 51

Abstract

This chapter provides an overview of the science, engineering, and design methods required in the development of micro/nanofluidic devices. Section 2 provides the scientific background of fluid mechanics and physical phenomena in micro/nanoscale. Section 3 gives a brief overview of the existing fabrication techniques employed in micro/nanofluidics. The techniques are grouped into three categories: (1) subtractive manufacturing, (2) formative manufacturing, and (3) additive manufacturing. The advantages and disadvantages of each manufacturing technique are also discussed. Implementation of the fluidic devices beyond laboratory demonstrations is not trivial, which requires a good understanding of the problems of interest and the end-users. To that end, Section 4 introduces the design thinking approach and its application to develop micro/nanofluidic devices. Finally, Section 5 concludes the chapter with future outlooks.

Progress in Molecular Biology and Translational Science, Volume 186

ISSN 1877-1173
https://doi.org/10.1016/bs.pmbts.2021.07.007

1. Introduction

In the early 1990s, Manz et al. introduced the concept of microfluidics to integrate a range of microscale components to perform chemical analysis, which was termed total chemical analysis system (TAS).[1] The fabrication of microfluidic devices was based on the processes and materials developed for the micro-electromechanical system (MEMS). However, due to the high manufacturing cost, the requirement for a cleanroom, and the difficulty in handling devices consisting of silicon and/or glass, microfluidics was not yet widely adopted. In the late 1990s, Whitesides et al. introduced the fabrication techniques that were collectively termed soft lithography.[2,3] Soft lithography refers to a series of techniques using micropatterned polydimethylsiloxane (PDMS) stamps. Replica molding with PDMS followed by plasma bonding served as a convenient route to prototype microfluidic systems, which ignited the field of microfluidics.[4] The simplicity of this microfabrication technique provided researchers with a versatile tool to study microscale fluid flows involving biological and chemical constituents. After two decades since the introduction of soft lithography, microfluidics has seen applications in multiple disciplines of science and engineering.[5] At present, the fabrication of microfluidic devices is also no longer limited to MEMS, cleanroom technologies, and soft lithography. Many other techniques such as embossing, injection molding, and 3D printing were explored and introduced. For example, recent advances in digital fabrication have enabled fabricating microfluidic devices with computer-aided design (CAD) files. Three-dimensional (3D) printing has enabled rapid prototyping of complex 3D channels embedded in thermoplastics, photoresists, and hydrogels.

The advancement in manufacturing techniques for micro/nanofluidics has enabled the field to make strides in the demonstrated application, including medical diagnostics, drug discovery, drug delivery, microscale chemical production, combinatorial synthesis biochemical assays, artificial organs, and micropropulsion. While these applications represented the potential of this field, many of these applications are yet to make the academic-to-industry transition required to secure widespread adoption and commercialization. There is, therefore, a dire need to bridge the gap between academic exploration and real-life applications. One solution to bridge the gap would require engagement and mutual understanding between researchers and the end-users. Fortunately, there already exists a tool that is widely adopted by multinational corporations and universities inculcating meaningful engagement—design thinking.[6–8] The field of micro/nanofluidics can benefit from the design thinking approach that ultimately helps bridge the gap between academic explorations and real-life applications.

This chapter provides an overview of the science, engineering, and design methods required to develop micro/nanofluidic devices. Section 2 introduces the background of fluid mechanics and physical phenomena in micro/nanoscale. Section 3 gives a brief overview of the existing fabrication techniques employed in micro/nanofluidics. Section 4 discusses the design thinking approach applicable to the development of micro/nanofluidic devices. Finally, Section 5 concludes the chapter with a discussion of the outlooks.

2. Theory of micro/nanofluidics

In Richard Feynman's 1959 talk that was titled "There's plenty of room at the bottom," he introduced his audience to a world of tiny technologies and possibilities it entailed. In his talk, he explained that when the size scale reduces, gravitational force becomes inconsequential and surface tension becomes dominant. The domination of certain forces due to scaling laws causes fluid to behave differently in micro/nanoscale.[9] One of the most prominent phenomena is laminar flow, which occurs in a regime where viscosity-related effects are more dominant than inertia-related effects. In the meter scale, if a submarine suddenly stops operating its propellers, the submarine does not immediately stop moving due to the dominant inertial-related effects. In the micrometer scale, however, when a bacterium stops swimming, it is brought to a halt quickly due to the dominating viscous effects. Some other important phenomena that take precedence in the minuscule scale include diffusion, fluidic resistance, and interfacial forces. Several review articles and books offer excellent discussions on these topics in great detail.[10–12] In this section, we briefly introduce the essential characteristics of microscale flows relevant to the design and fabrication of micro/nanofluidic devices.

Reynolds number (Re). Reynolds number is a dimensionless number indicating the relative magnitude of the inertial force to the viscous force, and it is used to predict flow patterns in differing flow situations. Laminar flow occurs when $Re < 2300$, and turbulent flow occurs when $Re > 2300$[9]

$$Re = \frac{\rho v D}{\mu}$$

where ρ is the fluid density, v is the fluid velocity, μ is the fluid viscosity, and D is the characteristic dimension of the flow (such as a hydraulic diameter of the pipe). Since D is small (10^{-6}–10^{-9}) in micro/nanofluidic channels, Reynolds number usually falls below 2300, and fluids undergo laminar flows. Laminar flows imply the absence of flow current perpendicular to the motion of the fluid particles. Consequently, multiple streams of fluids in contact with one another do not mix by convective flows, and only mixes by diffusion. Because of this constraint, mixing in microchannels has been extensively studied in the early 2000s[13]; physical interventions (e.g., herringbone structures on surfaces, 3D interconnections) are required to enhance the mixing.

Diffusion. Diffusion is the net movement of anything (e.g., atom, ions, molecules) from a region of higher concentration to a region of lower concentration. Diffusion happens in liquids and gases because their particles move randomly from place to place. The equation expresses a simplified Fickian mathematical model of diffusion over a distance:

$$d^2 = 2Dt$$

where d is the distance a particle moves at time t, and D is the diffusion coefficient of the particle. Diffusion of molecules over a distance is an important consideration in micro/nanofluidics because diffusion is the primary mode of mass transport (i.e., mixing) in micro/nanofluidics. For example, it takes ~5 s for saline molecules ($D = 10^{-5}\ cm^2\ s^{-1}$) to travel a distance of 100 μm by diffusion.[14] The required time (~5 s) must be considered when designing channel geometry if homogenous mixing is required.

Fluidic Resistance. When driving a fluid through a channel, drag forces act in the opposite direction to resist the flow. This phenomenon can be referred to as fluid resistance. Fluidic resistance for laminar flows in a microchannel is described by the equations:

$$R = \frac{12\eta L}{wh^3} \text{ (for rectangular channels)}$$

$$R = \frac{8\eta L}{\pi r^4} \text{ (for circular channels)}$$

where η is fluid viscosity, L is the channel length, w and h is the cross-sectional width and height of the channel, and r is the radius of a circular channel. Since the channel geometry (w, h, r) is on the order of 10^{-6}–10^{-9} m, fluidic resistance is inevitably high in microchannels. As such, high pressure is required to counter the high fluid resistance. In order to push fluid through a channel at a given flowrate (Q), the required pressure (ΔP) is given by:

$$\Delta P = QR$$

Fluid resistance is an important consideration when choosing the flow control system. In microfluidics, there are different modes to drive the fluids. In the volumetric flow, the flowrates (Q) are selected (using a syringe pump, for example), and the pressure (ΔP) is built up accordingly. Alternatively, in the pressure-driven flow, a chosen pressure difference (ΔP) is applied across the channel (using a compressed gas tank or a difference in the height of fluids, for example), and the flowrates (Q) are determined accordingly. It is important to realize the maximum pressure threshold of the devices. Operating the devices beyond that threshold typically leads to leakages, disconnection of tubes, or delamination of the bonding that forms the device.

Flow profile in a microchannel. The velocity profile in a fluidic channel tends to be parabolic in most simple cases of laminar flow. The fluid velocity is the fastest in the middle of the channel and zero near the channel wall (Fig. 1B). The simplified flow velocity profile of fluid flow is given by:

$$v(y) = \frac{3}{2}v_m\left[1 - \left(\frac{y}{0.5H}\right)^2\right] \textit{(flow between parallel plates)}$$

$$v(r) = v_m\left[1 - \frac{r^2}{R^2}\right] \textit{(flow in circular channel)}$$

where $v(y)$ is the velocity at the location y, $v(r)$ is the velocity at the location r, v_m is the maximum velocity, H is the distance between the parallel plate, and R is the radius of the tube. Average velocity (v_a) is $v_m/2$. The velocity profile is an important consideration because it implicates the distribution of molecules transported within a fluidic channel.

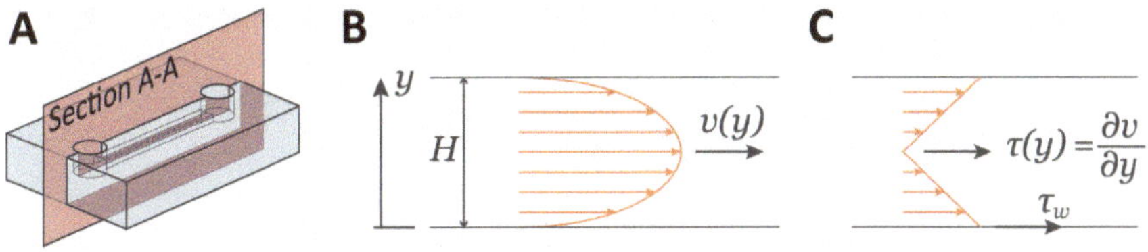

Fig. 1 (A) Illustration of a fluidic device highlighted with a red plane (Section A-A). (B) Section A-A view of the parabolic velocity profile of fluid flow in a microfluidic channel. (C) Section A-A view of highlighting the distribution of the linear shear stress of fluidic flow in a microfluidic channel.

Shear stress. Shear is caused by the friction between fluid particles, and the magnitude of the shear varies across the channel. The highest shear stress is found at the wall interface and zero in the middle of the channel (Fig. 1C). Simplified expressions of wall shear stress, τ_w is given by[15,16]:

$$\tau_w = \frac{12\eta Q}{h^2 w}\ (\textit{flow between parallel plates})$$

$$\tau_w = \frac{4\eta Q}{R^3 \pi}\ (\textit{flow in circular channels})$$

where η is fluid viscosity, Q is the fluid velocity, R is the radius of the circular channel, and w and h is the cross-sectional width and height of the channel, respectively. Wall shear stress is an important consideration when dealing with biological samples. For example, when dealing with adherent cells on the fluidic wall surfaces, it is important to avoid shear stress detrimental effects on cells.[17,18] The shear stress may affect the property of the cells on the surface and compromise their viability.

Interfacial forces. Interfacial forces are the cohesive forces between liquid molecules that cause the liquid to resist separation. Some examples of interfacial forces include surface tension, electrostatic, and Van der Waals forces. To illustrate, gravitational forces become inconsequential with respect to interfacial forces in micro/nanoscale. Interfacial forces are sufficient to drive fluid through a microchannel even in opposition to forces like gravity (i.e., capillary action).[9] Several microfluidic technologies take advantage of interfacial forces, including: paper fluidics,[19,20] droplet microfluidics,[21–23] digital microfluidics.[24,25] Interfacial forces are also an important phenomenon in nanofluidics. As demonstrated by Conlisk et al., a pressure-driven flow is usually not feasible for nanofluidic channels as pressure drop can increase up to ~50,000% when the channel diameter was decreased from 80 nm to 10 nm.[26] As such, it is common to employ electrokinetic flow that takes advantage of the interfacial forces.[27,28]

Overall, we discussed several important physical phenomena and the behavior of the fluids in the micro/nanofluidics. Such knowledge aids in the development of micro/nanofluidics, which is discussed in Sections 3 and 4.

3. Fabrication of micro/nanofluidic devices

In this section, we introduce existing fabrication techniques currently employed to fabricate fluidic devices. The fabrication of micro/nanofluidic devices can be divided into three manufacturing processes: (1) subtractive, (2) formative, and (3) additive. Understanding the advantages and disadvantages of existing fabrication techniques guides us to choose fabrication techniques to satisfy the identified requirements. We discuss how these manufacturing processes are applied to fabricate micro/nanofluidic devices. We also discuss their advantages and disadvantages.

Fabrication of a fluidic device by subtractive manufacturing is illustrated (Fig. 2A). Subtractive manufacturing is a process where the successive removal of material forms shapes from a solid block of material. Subtractive manufacturing is the earliest discovered form of manufacturing. As early as the stone age, archeologists found evidence of humans used lithic reduction to fashion tools out of stones or rocks.[29,30] Lithic reduction is a process where a 3D object is constructed by successively removing material away from a solid block of material. For the fabrication of microfluidic channels, channel features are created by selectively removing materials from a solid block of material, as illustrated by the rotating mill bit (Fig. 2A). Typically, grooves are created

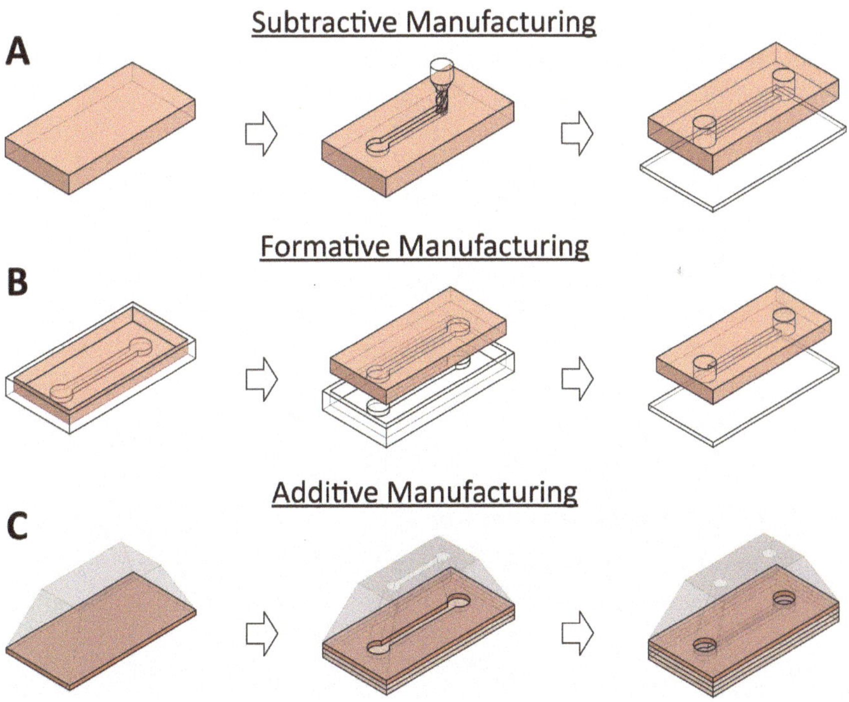

Fig. 2 Illustration highlighting the steps involved in fabrication of a fluidic device by (A) subtractive manufacturing (e.g., micro-milling), (B) formative manufacturing (e.g., PDMS replica molding), and (C) additive manufacturing (e.g., stereolithography 3D printing).

from flat substrates. After creating the features, a secondary step of bonding is required to form the fluidic channels surrounded by the matrix (Fig. 2A).

Fabrication of fluidic devices by formative manufacturing is illustrated (Fig. 2B). Formative manufacturing is a process where a material is formed into shapes as defined by the mold. With the discovery of metallic elements, humans developed new processes to shape metals into utensils, jewelry, and weapons. Evidence of formative manufacturing was discovered as early as the copper age, where humans poured molten metal into molds with cavities carrying the desired shape to fashion tools.[31] Fig. 2B illustrates the fabrication of a fluidic device formed by a commonly used formative manufacturing process—soft lithography. In soft lithography, uncured polydimethylsiloxane (PDMS) is poured in a prefabricated mold with protruding features that define the fluidic channels. After allowing the PDMS to cure, the PDMS replica can be removed from the mold and bonded to a separate substrate to form the channels.

Fabrication of fluidic devices by additive manufacturing is illustrated (Fig. 2C). Additive manufacturing is a process where a shape is formed by the successive addition of material, usually in a layer-by-layer manner. The concept of additive manufacturing, where a 3D object is built layer-by-layer, is not a recent invention. The earliest evidence of additive manufacturing is dated back to 10,000 years ago, where homes were built brick by brick.[32] Computer-controlled automation of fabrication *via* additive manufacturing is a relatively recent invention.[33] Fig. 2C illustrates the fabrication of a fluidic device by 3D printing using a stereolithography (SL) printer. Each layer is built by exposing a photopattern to a vat of uncured resin. Only the areas exposed to light are cured. Repeated exposure to the pattern of light over multiple layers enabled fabricating a 3D fluidic device.

3.1 Subtractive manufacturing

Fluidic devices were fabricated using subtractive manufacturing processes during the infancy of the dissemination of microfluidics. The manufacturing processes were borrowed from the fields of semiconductor and microelectromechanical systems (MEMS).[34–36] As early as the 1990s, bulk micromachining has been applied to the fabrication of microfluidic devices. Bulk micromachining mainly relies on etching to selectively remove materials from a substrate.[37,38] Fig. 3A illustrates the steps required in bulk micromachining to fabricate microfluidic devices.[35] Briefly, bulk micromachining

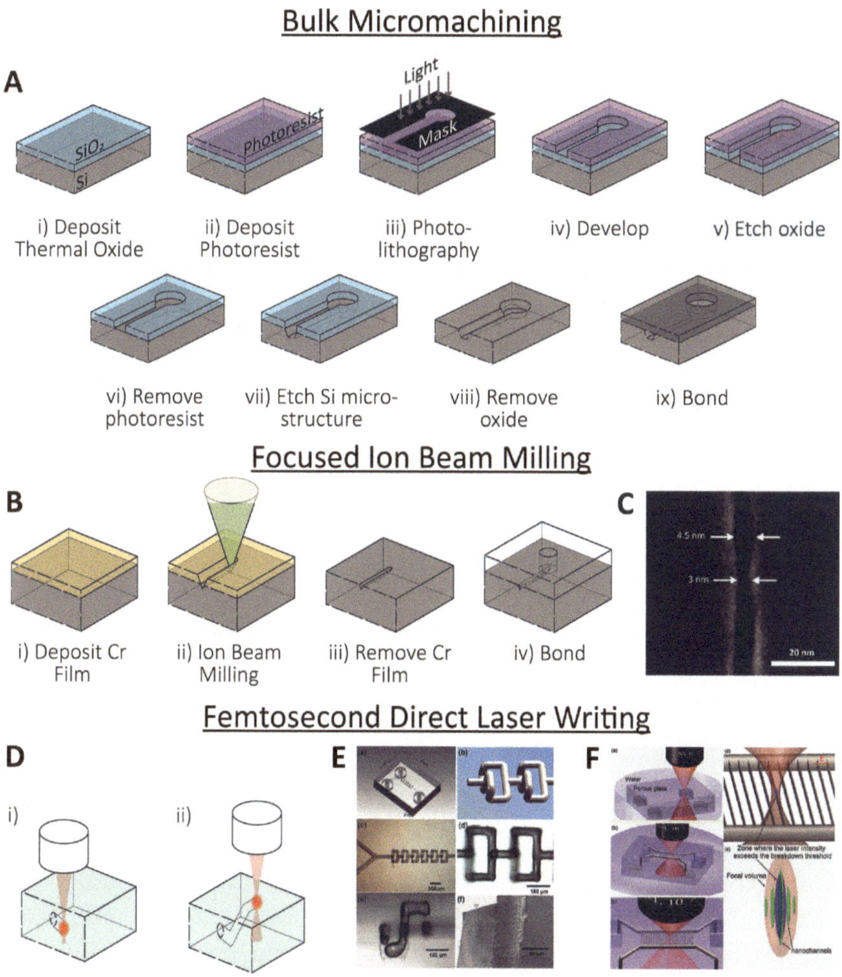

Fig. 3 Illustration highlighting the steps involved in the fabrication of fluidic devices by bulk micromachining, focused ion beam (FIB) milling and femtosecond direct laser writing (DLW). (A) Schematic illustration of the process of bulk micromachining. (B) Schematic illustration of the process of FIB milling. (C) Micrograph showing the attainable feature size by FIB milling. (D) Schematic illustration of the process of femtosecond DLW. (E) Micrograph showing the attainable three-dimension channels by femtosecond DLW. (F) Micrograph of sub-50 nm channels by femtosecond DLW. *Panel (C): Reproduced from Menard LD, Ramsey JM. Fabrication of sub-5 nm nanochannels in insulating substrates using focused ion beam milling. Nano Lett. 2011;11:512–517 with permission. Panel (E): Reproduced from Ke K, Hasselbrink EF, Hunt AJ. Rapidly prototyped three-dimensional nanofluidic channel networks in glass substrates. Anal Chem. 2005;77:5083–5088 with permission. Panel (F): Reproduced from Liao Y, Cheng Y, Liu C, Song J, He F, Shen Y, et al. Direct laser writing of sub-50 nm nanofluidic channels buried in glass for three-dimensional micro-nanofluidic integration. Lab Chip. 2013;13:1626 with permission.*

consisted of photolithography (Fig. 3A, i–iv) followed by etching (Fig. 3A, v–viii) and bonding (Fig. 3A, ix). First, a layer of thermal oxide was deposited on a silicon substrate, followed by a layer of photoresist. Photolithography was then performed by positioning a photomask over the deposited photoresist, followed by exposure to light. Here, we illustrated the use of a positive photoresist, where the exposure to light made the photoresist removable by baking the chemical bonds during the development. The exposed regions of the silicon oxide layer (i.e., the regions not covered by the photoresist) were then removed by etching. Next, the photoresist layer was removed, leaving only the silicon oxide with the channel pattern. The silicon substrate was then etched, where only the exposed areas were removed by the etchant. The silicon oxide layer was finally removed, followed by bonding to form the channels (e.g., using anodic or fusion bonding[39,40]). The devices manufactured by bulk micromachining on silicon substrates were applied to electrophoresis[41,42] and polymerase chain reaction (PCR) amplification.[43] Silicon is a well-established material in the semiconductor industry, but it is not a favorable material for microfluidics due to the lack of transparency. Glass and quartz were thus attractive alternatives. To this end, glass microfluidic devices have been fabricated using photolithography and etching. Both dry etching and wet etching have been used to fabricate microchannels.[44] The limiting factor of the bulk micromachining was the attainable feature size, which was defined by the resolution of photolithography and etching. In general, the narrowest feature size that was achieved by photolithography is ~0.3 μm,[45] which is constrained by the diffraction limit of the light used. As such, standard photolithography is not amenable for the fabrication of nanochannels (<150 nm). Sub-150 nm features have been successfully fabricated using electron-beam lithography (EBL) and nanoimprint lithography (NIL) to pattern the photoresist prior to etching.[46,47]

Focused ion beam (FIB) milling has been used to subtract materials from silicon and glass substrates.[48,49] In FIB milling, a finely focused beam of ions (usually gallium ions) is fired on the substrate to selectively remove a small amount of material (Fig. 3B). This method is particularly useful in creating sub-10 nm channels; such channel dimensions are difficult to achieve by the conventional bulk micromachining processes. Nanochannels are milled into the quartz substrate through a chromium film using FIB milling (Fig. 3B). Devices are then sealed using fusion bonding. Sub-5 nm nanochannels have been demonstrated using FIB (Fig. 3C).[50–52] A similar process, helium ion-beam writing, was also demonstrated to fabricate nanochannels.[53]

Femtosecond direct laser writing (DLW) is another technique used to create fluidic devices.[54] In bulk micromachining and FIB milling, channels are first created on the surface of the substrate, followed by bonding to seal the channels. In contrast, channels are directly created within a high-silica porous glass substrate in femtosecond DLW, eliminating the need to seal the device (Fig. 3D).[54] Femtosecond DLW also permits fabricating 3D channel networks, unlike bulk micromachining and FIB milling (Fig. 3E).[55–58] While femtosecond DLW still does not match the feature size created by FIB milling (i.e., sub-5 nm), Liao et al. has demonstrated the fabrication of sub-50 nm nanochannels (Fig. 3F).[59]

Bulk micromachining, FIB milling, and femtosecond DLW are well-established methods to fabricate fluidic devices in silicon and glass substrates. However, these fabrication techniques are expensive and require dedicated facilities and equipment (e.g., cleanroom, thin-film deposition devices, mask aligners, and FIB machines), which may not be available in all research settings. These limitations motivated the development of other subtractive manufacturing techniques that can be readily performed. For example, the laser ablation of polymeric sheets or films was demonstrated (Fig. 4A). Laser ablation is a cost-effective technique for prototyping fluidic devices. Patko et al. used a laser cutter to pattern the outline of the channel on double-sided tape that is later sandwiched between two rigid substrates (Fig. 4B).[60] Similarly, Thompson et al. used a laser to ablate channel outlines on polyester films and sealing the channels by laminating with other polymer films.[61] Other than using a laser to cut through polymer films, microchannels can be engraved on polymer sheets by controlling the laser power and writing speed (Fig. 4C). Klank et al. employed a laser-cutter to engrave fluidic channels of varying depth on poly(methyl methacrylate) sheets and investigated multiple bonding methods to seal the channels (Fig. 4D and E).[62]

Besides laser ablation, a cutting plotter can be used to cut polymer thin-films, and later bonded with other substrates to form microfluidic channels (Fig. 5A).[63–65] This technique is called xurography. Xurography is a useful alternative for laser ablation because some polymers may emit toxic gases or produce poor surface finishing (e.g., polyvinylchloride, polycarbonate, polytetrafluoroethylene, Teflon). Hizawa et al. used a cutting plotter to cut fluoropolymer films which are not suitable for laser ablation. After cutting the channel outline with a cutting plotter, heat and pressure were applied to laminate and seal the channels (Fig. 5B).[66] The same technique was also extended to fabricate multilayer channels. Currently, the best cutting plotters are capable of fabricating features down to 20 μm.[67]

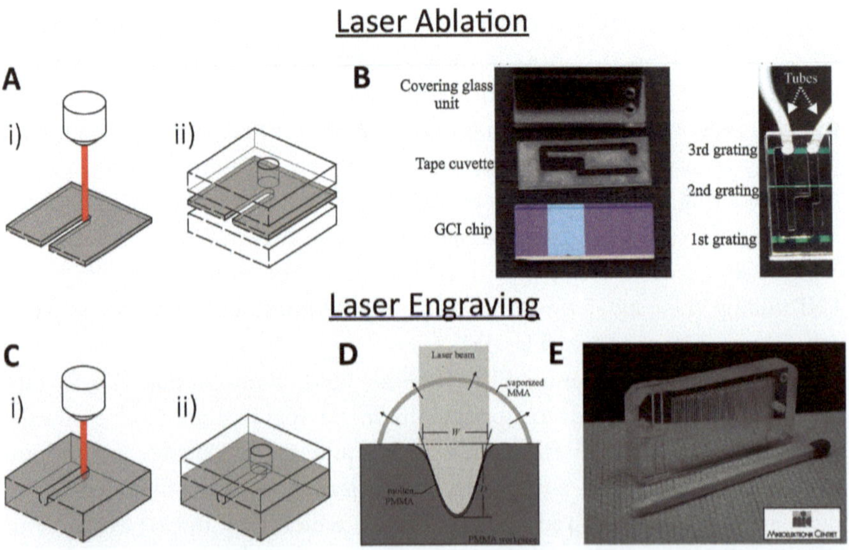

Fig. 4 Illustrations highlighting the steps involved in the fabrication of fluidic devices by laser ablation and laser engraving. (A) Schematic illustration of the process of laser ablation. (B) Photographs demonstrating a fluidic device fabricated by laser ablation of tape sandwiched between rigid sheets. (C) Schematic illustration of the process of laser engraving. (D, E) Schematic and photograph of a microfluidic device fabricated by laser engraving. *Panel (B): Reproduced from Patko D, Mártonfalvi Z, Kovacs B, Vonderviszt F, Kellermayer M, Horvath R. Microfluidic channels laser-cut in thin double-sided tapes: cost-effective biocompatible fluidics in minutes from design to final integration with optical biochips. Sensors Actuators B Chem. 2014;196:352–356 with permission. Panels (D) and (E): Reproduced from Klank H, Kutter JP, Geschke O, CO_2-laser micromachining and back-end processing for rapid production of PMMA-based microfluidic systems. Lab Chip. 2002;2:242–246 with permission.*

Lastly, micromilling can subtract materials to fabricate fluidic devices (Fig. 5C).[68] For example, Owen et al. employed micromilling to machine microchannels on the faces of Lego bricks; the surfaces of the bricks were subsequently sealed with transparent adhesive polyethylene films to form microchannels (Fig. 5D).[69] They demonstrated the fabrication of microfluidic modules with varying functions that can be assembled to form a microfluidic system.

3.1.1 Advantages and disadvantages of subtractive manufacturing

Bulk micromachining is a well-established method used in the industry. Bulk micromachining is applied to manufacture glass or quartz devices with good resolution (~1 μm) and precision. However, bulk micromachining

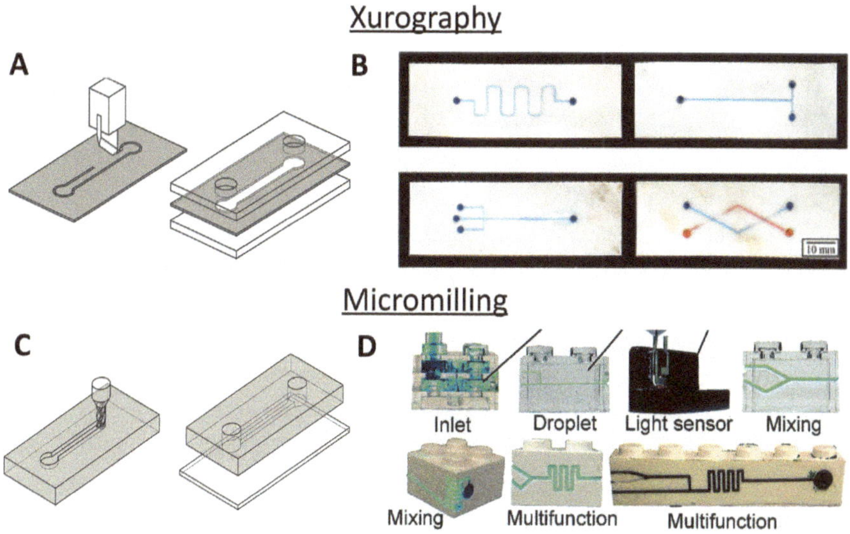

Fig. 5 Illustration highlighting the steps involved in the fabrication of fluidic devices by xurography and micromilling. (A) Schematic illustration of xurography for the fabrication of microchannels. (B) Photograph of various microfluidic devices fabricated by xurography. (C) Schematic illustration of micromilling for the fabrication of microchannels. (D) Photography of various microfluidic devices fabricated by micromilling features on Lego® bricks. *Panel (B): Reproduced from Hizawa T, Takano A, Parthiban P, Doyle PS, Iwase E, Hashimoto M. Rapid prototyping of fluoropolymer microchannels by xurography for improved solvent resistance. Biomicrofluidics. 2018;12:064105 with permission. Panel (D): Reproduced from Owens CE, Hart AJ. High-precision modular microfluidics by micromilling of interlocking injection-molded blocks. Lab Chip. 2018;18:890–901 with permission.*

is a highly specialized manufacturing process that requires a substantial understanding of photolithography, etching, and bonding techniques. The fabrication also requires highly sophisticated and costly equipment used in a cleanroom. Furthermore, etching also requires strong acids and solvents that are hazardous and toxic to humans. While such setups are not available to all researchers, some commercial companies offer services to fabricate glass and quartz fluidic devices, including Elveflow, Dolomite, Chipshop, and Darwin Microfluidics. These companies also provide services to custom-build fluidic devices based on the specifications defined by the customers.

For the small channel features in the nanometer scale (<150 nm), FIB milling is the most suitable candidate. Femtosecond DLW also produces nanometer-scale features (~50 nm). Fluidic devices fabricated using femtosecond DLW also do not require an additional bonding step. Similar to bulk micromachining, the need for sophisticated and expensive equipment is the drawback of FIB milling and Femtosecond DLW. Laser ablation, xurography, and micromilling are low-cost fabrication techniques compared to bulk micromachining, FIB milling, and Femtosecond DLW. For example, the setup cost (i.e., cost of a laser cutter, cutting plotter, or micromilling machine) is usually $< 10{,}000$ USD. The advantage of these techniques is also the fast design-to-prototyping cycle. Generally, users can turn a design into a prototype within hours. The main limitation, however, is the attainable feature resolution; typically, the attainable minimum channel width is on the order of 100 μm. Lastly, these techniques are developed for polymeric materials that may not offer characteristics desirable for micro/nanofluidic devices (i.e., transparency, chemical resistance).

Overall, subtractive manufacturing applied in micro/nanofluidics has seen substantial development in the last three decades. It is no longer limited to MEMS-based fabrication techniques. Many low-cost, fast prototyping techniques were introduced in the past decade, and we expect innovations in subtractive manufacturing techniques in the years to come.

3.2 Formative manufacturing

Formative manufacturing is the process where the material is formed into the desired shape as defined by a mold. Whitesides and Quake introduced methods that are collectively termed soft lithography, which paved an avenue for PDMS microfluidics. Soft lithography offers rapid and relatively low-cost methods to fabricate microfluidic devices using PDMS stamps.[2,3] This method was a favorable replacement for the bulk micromachining that was slow and costly. Over the last 20 years, soft lithography remains the most widely adopted method to fabricate microfluidic devices in the academic setting. We note that soft lithography refers to a collection of techniques that include: replica molding, microcontact printing (μCP), microtransfer molding, micromolding in capillary, solvent-assisted micromolding (SAMIM), phase-shifting edge lithography, nanotransfer printing, decal transfer lithography, and nanoskiving.[70] This section discusses the process involved in PDMS replica molding.

A representative workflow to fabricate PDMS devices by replica molding is illustrated (Fig. 6A). Fig. 6A, i–iv illustrates the fabrication of the master mold, and Fig. 6A, v–vii illustrates the casting and bonding of elastomeric polymer. First, SU-8 (i.e., a negative photoresist) was spin-coated on the

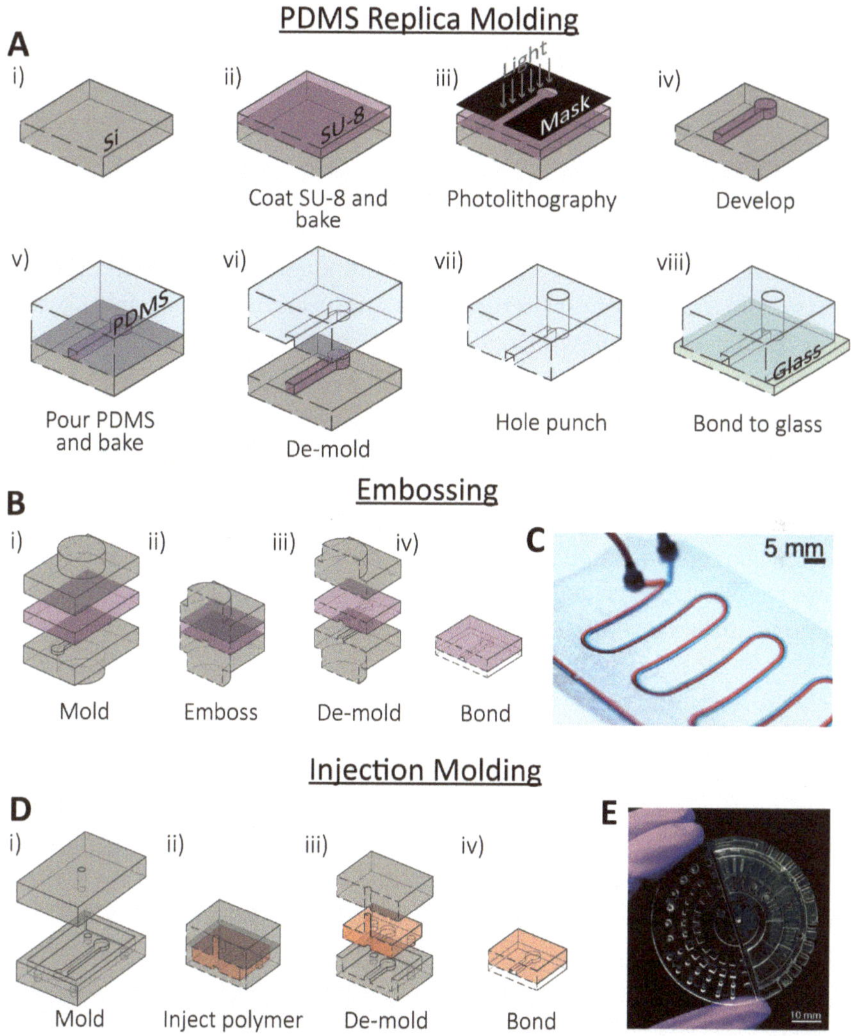

Fig. 6 Illustration of steps involved in the fabrication of fluidic devices by PDMS replica molding, hot embossing and injection molding. (A) Schematic illustration of the process of PDMS replica molding. (B) Schematic illustration of the process of hot embossing. (C) Photography of a microfluidic device fabricated by cellophane films using 3D printed molds. (D) Schematic illustration of the process of injection molding. (E) Photograph of an injection molded microfluidic device. *Panel (C): Reproduced from Hamedi MM, Ünal B, Kerr E, Glavan AC, Fernandez-Abedul MT, Whitesides GM. Coated and uncoated cellophane as materials for microplates and open-channel microfluidics devices. Lab Chip. 2016;16:3885–3897 with permission. Panel (E): Reproduced from Lee UN, Su X, Guckenberger DJ, Dostie AM, Zhang T, Berthier E, et al. Fundamentals of rapid injection molding for microfluidic cell-based assays. Lab Chip. 2018;18:496–504 with permission.*

silicon substrate. The thickness of the SU-8 coating was precisely controlled by varying the rotation speed, acceleration, and viscosity of SU-8.[71] The thickness of the SU-8 coating determined the height of the fluidic channel. After coating, a soft-baking process (5–10 min at 65 °C) was carried out to partially remove the solvent and solidify the photoresist before exposure to UV light. Next, a photomask was positioned over the photoresist coating and exposed to UV light (350–400 nm). Two types of photomasks were demonstrated for the use: (1) hard photolithography mask made from a glass or quartz substrate patterned with a thin chrome layer, and (2) soft photolithography mask with a plastic film printed with black ink with high resolution (20,000 dot-per-inch (DPI)). A hard mask, being more costly, generally gives higher resolutions (~1 μm) than a soft mask (~10 μm). SU-8 is a negative photoresist where the parts exposed to the UV light are crosslinked; the parts not exposed to the light remain soluble to the developer solution. After the exposure to UV light, the photoresist was heated for post-baking to accelerate the SU-8 polymerization. The unpolymerized parts of the SU-8 were finally removed by soaking the whole substrate in a developer solvent (i.e., 1-methoxy-2-propyl acetate). An optional hard baking step was carried out after this step to conclude the fabrication of the mold. Prior to PDMS replication, the mold usually underwent an additional step of salinization to ensure easy demolding of PDMS. Salinization assisted in the demolding of cured PDMS.

After salinization, uncured PDMS was poured over the mold, followed by degassing to remove bubbles in a chamber under reduced pressure. PDMS was then heated to accelerate the curing. After curing, the PDMS was readily removed from the mold to obtain a micropatterned PDMS stamp. Access ports to connect tubing were created to the PDMS replica by a biopsy punch. Finally, the PDMS stamp was bonded to a substrate to form the channels. Both glass and PDMS were suitable substrates for bonding, where their surfaces were activated in oxygen plasma (or air plasma) that converted Si-CH_3 to Si-OH. A covalent Si-O-Si bond formed between the PDMS stamp and the activated surface. The sealed device was baked to ensure the formation of Si-O-Si bonds to complete the sealing.

SU-8 master molds have been the most common type of master mold for replica molding of PDMS. However, replica molding of PDMS can be carried out with other surfaces. Hwang et al. published a comprehensive review of unconventional molds employed for PDMS casting.[72] Replica molding was also demonstrated using PDMS as a mold to cast rigid polymers (including SU-8).[73] Dissolvable or removable structures also served as molds to form micro/nanochannels. For example, sacrificial nanofiber or nanowires was demonstrated for the use of templates for the fabrication of

PDMS nanofluidic devices.[74,75] 3D printed master molds or sacrificial templates have also been demonstrated to cast a secondary material.[76,77] Water-soluble polyvinyl alcohol (PVA) has been widely used for the fabrication of microchannels in PDMS and other substrates.[78,79]

Next, we discuss other formative techniques employed in the fabrication of microfluidics and nanofluidics, which are classified as embossing. Formative manufacturing *via* embossing is illustrated (Fig. 6B). Channel features are embossed on polymers using rigid stamps at elevated temperatures, which was called hot embossing. For hot embossing, rigid polymers are usually employed instead of elastomers; elastomeric materials can pose problems for the fabrication of nanochannels and nanoslits because of the deformability and tendency to collapse.[80] In the formative manufacturing of nanochannels or nanoslits, rigid polymers are thus preferred.[81] To this end, methods to imprint nanostructures defined by the mold have been developed. Pioneering work by Chou et al. demonstrated fabricating nanochannels *via* imprinting, which they termed nanoimprint lithography (NIL).[82,83] In NIL, a master mold carrying the channel features was stamped on a heated polymer substrate to create the feature on the polymer surface. Fabrication of the master mold used for NIL was carried out by the combination of standard photolithography, EBL, and FIB milling. Alternatively, periodic nanopatterns were also used to create nanostructures on the master mold.[84] Other types of embossing techniques for the fabrication of microfluidic devices include stamp embossing, roller embossing, UV embossing, and thin-film laminate.[85–89] Researchers also demonstrated that embossing was applied for the rapid prototyping of microfluidic devices using a 3D-printed mold. Hamedi et al. demonstrated the use of 3D-printed embossing molds to create fluidic devices with cellophane films (Fig. 6C).[90] Embossing is largely employed to produce devices in large quantities in the industry.

Injection molding is yet another formative manufacturing technique used to fabricate fluidic devices. Injection molding is a fabrication technique used to obtain molded products by injecting molten polymeric into a mold and then cooling and solidifying them. Microinjection molding occurs in four major steps. The two halves of the mold were compressed to create a mold cavity (Fig. 6D, i). The heated thermoplastic in a liquid state was then injected at a specific rate to fill the mold cavity (Fig. 6D, ii). The mold was cooled, and the casted object was removed from the mold (Fig. 6D, iii). The sprue was removed from the casted object. Finally, the channel was formed by sealing the micropatterned surface to a separate substrate (e.g., thermal fusion bonding or solvent bonding[91]). Injection molding is a manufacturing technique used for the mass production of plastic parts. It is not feasible for small-scale production in academic research laboratories where rapid iterative development of the devices is required. Because biocompatible polymers can easily be employed for injection molding, injection molding is favorable to fabricate fluidic devices for biomedical applications. Thermoplastic polymers such as polystyrene and cyclic olefin copolymer have been shown advantageous in microfluidics due to biocompatibility and optical property. Lee et al. demonstrated the compatibility of injection-molded microfluidic devices for cell culture and optical microscopy (Fig. 6E).[92]

3.2.1 Advantages and disadvantages of formative manufacturing

The steps required in formative manufacturing are conceptually different from subtractive manufacturing. Subtractive manufacturing involves processes that directly modify the fluidic device. Unlike subtractive manufacturing, an extra process of fabricating the mold is required in formative manufacturing. These master molds can be reused multiple times with little variability to the replica. Formative manufacturing may offer high repeatability and high precision, which are dependent on the quality of the master mold. Therefore, the fabrication of the master mold is crucial to determine the final quality of the fabricated devices. Formative manufacturing, except for soft lithography, is the most suitable for high-volume manufacturing in the industry. The main limitation of formative manufacturing is (1) the cost of fabricating the molds and (2) the availability of materials and instruments. Fabrication of the master mold is generally expensive; fine feature size requires sophisticated fabrication processes such as EBL and FIB milling. Chemically inert and rigid materials (such as silicon and glass) are typically not suitable for formative manufacturing as they cannot be readily molded or molten-injected. Formative manufacturing is the most compatible with curable polymer resins and thermoplastics, limiting the chemical compatibility of the fabricated devices.

Among the methods of formative manufacturing, soft lithography is the most cost-effective method, considering the cost required for the facilities and the manufacturing of the mold. Once the master mold is obtained, replica molding is carried out without a cleanroom environment. Soft lithography is arguably the most adopted method to prototype microfluidic devices. Several features of soft lithography make it suitable for research use in academic laboratories. Firstly, the process is suitable for small-scale prototyping requiring multiple iterations. Secondly, PDMS is a well-characterized material amenable to biomedical applications. Thirdly, the elasticity of PDMS allows extending functions (such as valves, pumps, and actuators), making it possible to create devices incorporating unique dynamic actuation combined with cell cultures.[93] The drawback of the soft lithography also comes from the property of PDMS. For example, nonpolar solvents (e.g., benzene and pentane) are not compatible with PDMS devices. PDMS is also known to absorb small molecules such as drugs.[12] The absorption of small molecules should be accounted for when PDMS devices are applied in research such as drug testing.[94] Different approaches have been developed to create microchannels using castable polymers other than PDMS.[73,95] Besides the property of the materials, it is noted that the fabrication of the mold for soft lithography still requires multistep cleanroom-based technologies.

Embossing can generate devices at high throughput with great precision, making it an excellent option for mass production of the devices. The dimension of the channel is determined by the dimension of the feature in the mold, and micro/nanochannels can be fabricated. Conventionally, these molds are fabricated with metal (e.g., steel, silicon). Fabrication of metal molds is both expensive and time-consuming, and the use of metal molds is economical only for high-volume production. Researchers recently demonstrated the use of polymer molds for embossing, making it more suitable for small scale-production that required iterative development.[81,90] Similar to hot embossing, injection molding is an excellent option for fabricating commercial microfluidic devices due to its high throughput with good precision. This technique also utilizes a wide range of thermoplastics. Unfortunately, injection molding is only economical for high volume production (>10,000 devices) because of the high cost of the mold (10,000–20,000 USD).[92] Fabrication of injection mold requires dedicated machinery, which is often outsourced to specialists with dedicated facilities.

Overall, formative manufacturing techniques such as embossing and injection molding offers advantages in high volume production that are unparalleled by other manufacturing techniques, making them the ideal choice of fabrication for the commercialization of micro/nanofluidic chips. On the other hand, soft lithography is a popular prototyping method suitable for small-scale production in the academic setting. In general, formative manufacturing offers good repeatability and precision; therefore, we foresee its continued relevance and critical contribution in the future development of micro/nanofluidics.

3.3 Additive manufacturing

Subtractive manufacturing and formative manufacturing required multiple steps in fabrication, such as molding and bonding. Recent advances in additive manufacturing, especially 3D printing, have offered alternative routes to fabricate microfluidic devices in automated manners. A computer-aided design (CAD) file is used in 3D printing, and the printer builds the object in a layer-by-layer fashion. Post-processing in additive manufacturing is relatively simple (such as cleaning the printed models), and they are not as labor-intensive as the fabrication processes described in subtractive and formative manufacturing. We discuss three mechanisms of 3D printing relevant to the fabrication of microfluidic devices: (a) light-assisted 3D printing, (b) inkjet 3D printing, and (c) extrusion-based 3D printing.

Light-assisted 3D printing. In light-assisted 3D printing, a 3D object is built layer-by-layer using light to polymerize a photocurable resin. Fig. 7A illustrates the working principle of a digital light projection printer (DLP), a subset of light-assisted 3D printers. The light source exposes a pattern of illumination (with a fixed wavelength of light) to the photoresin, and

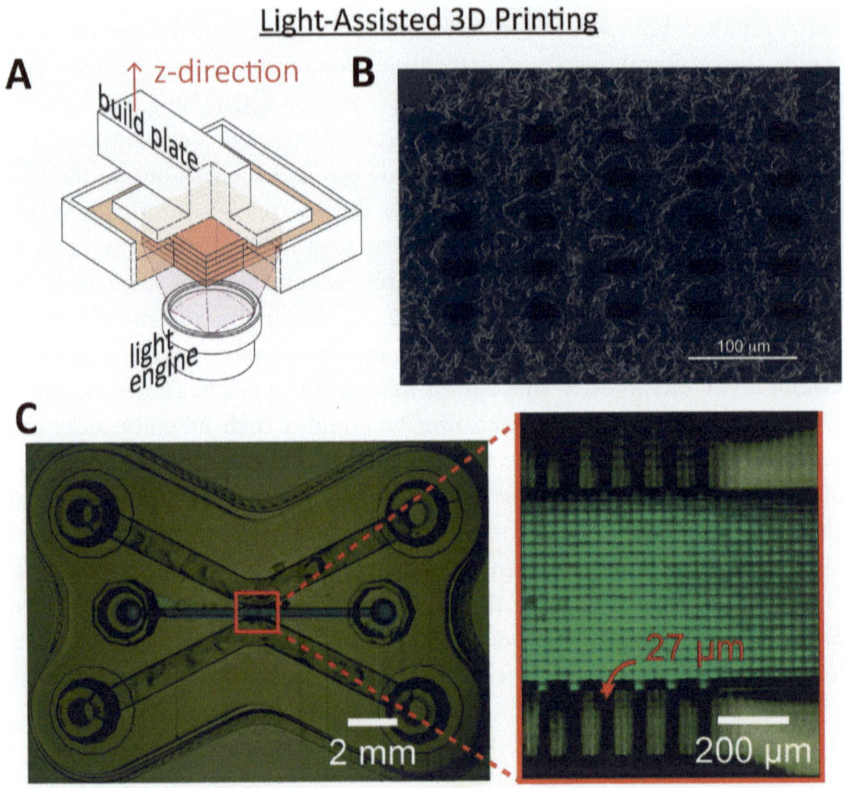

Fig. 7 (A) Illustration highlighting the working principle of a DLP printer. (B) SEM image of 3D serpentine channel cross-section fabricated with a custom DLP printer. (C) Photograph of a microfluidic device printed with PEGDA resin with a commercial DLP printer. *Panel (B): Reproduced from Gong H, Bickham BP, Woolley AT, Nordin GP. Custom 3D printer and resin for 18 μm × 20 μm microfluidic flow channels. Lab Chip. 2017;17:2899–2909 with permission. Panel (C): Reproduced from Kim YT, Bohjanen S, Bhattacharjee N, Folch A. Partitioning of hydrogels in 3D-printed microchannels. Lab Chip. 2019;19:3086–3093 with permission.*

the areas exposed to the light polymerize. After curing one layer, the build stage moves upwards (*z*-direction) to free up space for the polymerization of the subsequent layer. Other light-assisted printers exist in the market, including laser-scanning printers and liquid crystal display (LCD) printers.[96] 3D printing of microscale fluidic devices with a light-assisted 3D printer requires optimizing materials and operating parameters. Commercially available printers and photoresin are not yet optimized to print fluidic devices in the microscale ($<100\ \mu m$). Owing to the working principle of light-assisted 3D printing, clogging of channels may occur. Clogging of channels is usually attributed to undesired polymerization within the channel void or the inability to evacuate the uncured photoresin.[97] To prevent undesired clogging of channels, fine-tuning of photoresin and light-source is required. For example, by carefully tuning the properties of the resin and light source, Gong et al. successfully fabricated fluidic devices with a cross-sectional dimension below 20 μm (Fig. 7B).[98,99]

An advantage of light-assisted 3D printing is the ability to fabricate 3D architectures with multiple layers. For example, researchers printed multilayer fluidic devices to function as valves and pumps.[100,101] Motivated by the desire to use 3D-printed fluidic devices for biological applications, the development of biocompatible resins and hydrogels to print fluidic devices is under active investigation. One commonly used crosslinker is poly(ethylene glycol) diacrylate (PEGDA). Kuo et al. formulated a PEGDA-based precursor (258 MW) to 3D-print microfluidic devices with channel width as small as $\sim 27\ \mu m$ (Fig. 7C).[102,103] Using customized PEGDA hydrogel, Grigoryan et al. used a DLP printer to fabricate a complex branching network to mimic native vasculature, showcasing the truly free-forming capabilities of 3D printing.[104]

Most light-assisted printers, including DLP printers and LCD printers, are limited to the fabrication of microscale and milliscale fluidic channels. For nanoscale printing, advanced light-assisted printing technologies such as two-photon polymerization (2PP) direct laser writing (DLW) are used. DLW offers excellent voxel resolution at the nanometer scale.[105] With the trade-off of improved resolution, DLW usually requires a long print time with limited print volume.[106] Due to these limitations of DLW, there are only a few demonstrations to apply DLW to fabricate entire microfluidic devices.[107–109]

Inkjet 3D printing. In inkjet-assisted printing, a photocurable resin is jetted through an inkjet nozzle and subsequently cured by exposing the resin to UV light (Fig. 8A). They are usually known as multiple-jet (MJ) printers because they can accommodate multiple nozzles for multi-material printing.

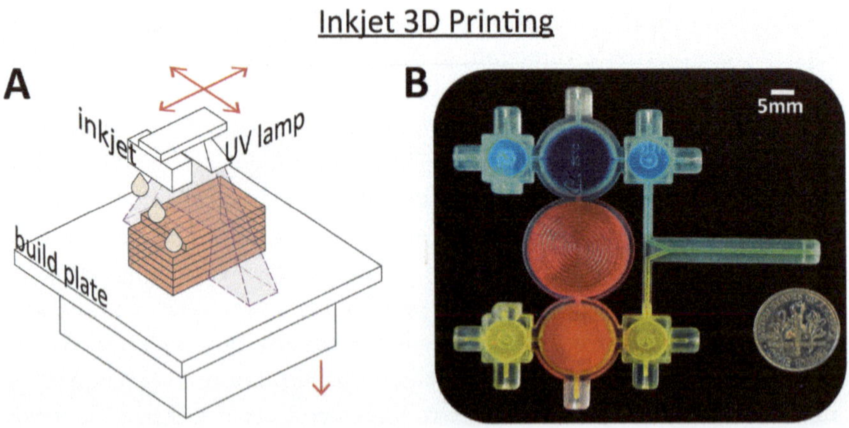

Fig. 8 (A) Illustration highlighting the working principles of an inkjet 3D printer. (B) Photograph of a finger-powered fluidic device printed with a MJ 3D printer. *Reproduced from Sweet EC, Mehta R, Xu Y, Jew R, Lin R, Lin L. Finger-powered fluidic actuation and mixing via MultiJet 3D printing. Lab Chip. 2020;20 with permission.*

Multi-material printing is important for constructing microfluidic devices. To print fluidic devices, the channel void is first printed with a sacrificial material that can be dissolved in post-processing. Several variations of MJ printing techniques are currently under patent (e.g., Stratasys, 3D Systems). At present, low-cost variants of MJ printing are not yet available, and most users rely on proprietary printers and materials. Fabrication of fluidic devices using MJ printers has been demonstrated, including a finger-powered fluidic system (Fig. 8B).[110–113]

Extrusion-based 3D printing. In extrusion-based printing, materials are extruded through a nozzle over multiple layers to create 3D objects (Fig. 9A). Extrusion-based methods are the most commercialized and widely used technology in 3D printing. The expiration of the original fused-deposition modeling (FDM) patent led to an open-source development community,[114] leading to the introduction of a wide array of inexpensive FDM printers accessible to the general public. In FDM 3D printing, a solid thermoplastic filament is fed into the print head, where the filament is heated to a semi-molten state before being forced through the nozzle. The molten material hardens spontaneously upon extrusion when exposed to the ambient temperature. Direct ink writing (DIW) is another extrusion-assisted 3D printing technology where liquid precursors loaded in syringes are pneumatically extruded through a nozzle. DIW 3D printing may be combined with UV irradiation and heating to accommodate different mechanisms to solidify liquid precursors for 3D modeling.

Multiple types of thermoplastics have been demonstrated for hot embossing and injection molding (e.g., acrylonitrile butadiene styrene

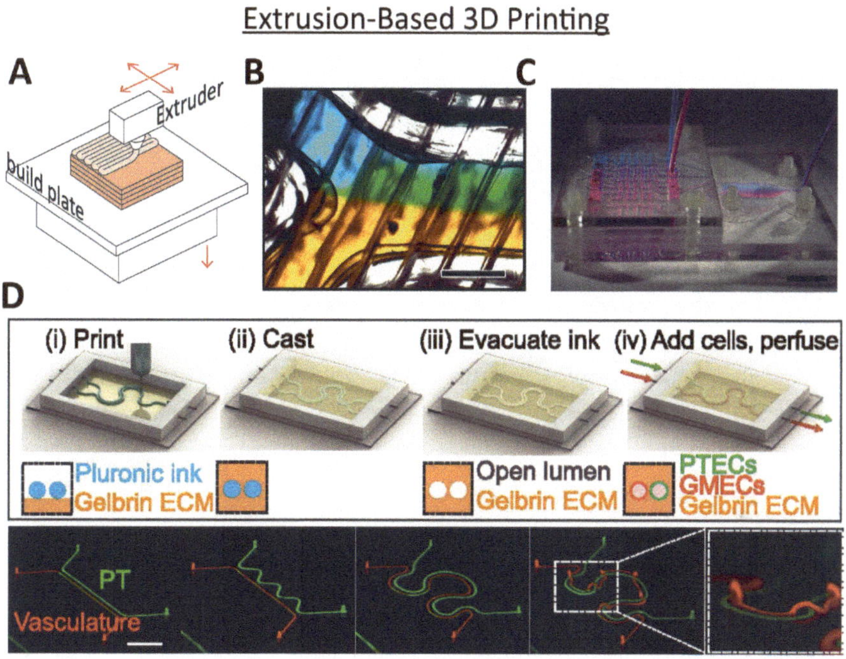

Fig. 9 (A) Illustration highlighting the working principles of an extrusion-based 3D printing. (B) Micrograph of a Y-channel printed by an FDM 3D printer, highlighting the surface roughness inherent to FDM printed devices. (C) Photograph of a fluidic device fabricated by DIW of silicone sealant on PMMA sheets. (D) Schematic and photographs showcasing the fabrication of complex fluidic networks by fugitive ink strategy. *Panel (B): Reproduced from Macdonald NP, Cabot JM, Smejkal P, Guijt RM, Paull B, Breadmore MC. Comparing microfluidic performance of three-dimensional (3D) printing platforms. Anal Chem. 2017;89:3858–3866 with permission. Panel (C): Reproduced from Ching T, Li Y, Karyappa R, Ohno A, Toh Y-C, Hashimoto M. Fabrication of integrated microfluidic devices by direct ink writing (DIW) 3D printing. Sensors Actuators B Chem. 2019;297:126609 with permission. Panel (D): Reproduced from Lin NY, Homan KA, Robinson SS, Kolesky DB, Duarte N, Moisan A, et al. Renal reabsorption in 3D vascularized proximal tubule models. Proc Natl Acad Sci U S A. 2019;116:5399–5404 with permission.*

(ABS), polystyrene (PS), polycarbonate (PC), polycaprolactone (PCL), polylactic acid (PLA), polybutylene terephthalate (PBT), and polyglycolic acid (PGA)). These thermoplastics also apply to FDM printing.[115] However, FDM is not suitable for printing microfluidic devices, crucially because the dimension of the printed filaments is usually larger (>150 μm) than the size of microchannels. Furthermore, owing to the working mechanism of extrusion-based printers, where strands of filaments are laid sequentially next to one another, the printed filaments tend to fail to form a uniform bond with its partnering filaments and may result in leakages.[116] Macdonald et al. also found that FDM-printed microfluidic devices produced the highest surface roughness relative to devices printed with DLP and MJ printers.[117]

The successive printing of filaments creates undulating surfaces that compromise the smoothness of the surface and the optical transparency (Fig. 9B).

To overcome these limitations of extrusion-based 3D printing to fabricate microchannels, our group developed a hybrid method that combines DIW printing with laser-ablated PMMA sheets.[118] Instead of printing the entire device as a monolithic entity, we printed the outline of the fluidic channels on a PMMA sheet. The channels are then sealed by securing another PMMA sheet over the printed outlines (Fig. 9C). This hybrid method allowed for the creation of fine channel dimensions (width = 30 μm) without compromising the optical transparency of the device.

Extrusion-based printers are also used to print sacrificial structures to form microchannels.[78,119] Lin et al. developed a method to apply DIW 3D printing to print fugitive inks that are evacuated after casting a secondary ECM (Fig. 9D).[120,121] Parekh et al. developed a method to apply DIW to print liquid metal that was evacuated after casting PDMS over the printed features.[122]

3.3.1 Advantages and disadvantages of additive manufacturing

Additive manufacturing is a promising manufacturing technique with the potential to bring in a new paradigm of microfluidics. There are several advantages of 3D printing in the fabrication of microfluidic devices: (1) fast design-to-prototype cycle, (2) free-forming capabilities (i.e., 3D channel architecture), (3) applicable materials, and (4) cost-effective. Firstly, a fast design-to-prototype cycle is advantageous to perform multiple iterative development cycles demanded to develop fluidic devices or systems.[123] Most subtractive and formative manufacturing techniques do not have the design-to-prototype cycle comparable to additive manufacturing. For example, the fabrication of PDMS microchannels *via* replica molding takes several days from design to prototype. In comparison, 3D printing requires several hours to obtain a prototype. Secondly, additive manufacturing is capable of truly 3D channel architecture of microchannels.[124] Fluidic channels fabricated by subtractive and formative manufacturing are mostly planar. The 3D capabilities of additive manufacturing have the potential to fabricate biomimetic vascular constructs to recapitulate their native counterparts *in vivo*. Thirdly, additive manufacturing allows the patterning of unique materials not compatible with subtractive and formative manufacturing (e.g., hydrogels embedded with cells, food).[125,126] For example, the 3D patterning of cell-laden inks enabled bioprinting that cannot be achieved in other processes. Lastly, additive manufacturing is becoming more cost-effective than subtractive and formative manufacturing, mainly due to the reduced cost of 3D printers. Low-cost SL printers (based on LCD) with a planar (x–y) resolution of 47 μm are currently available at 300 USD.

Several technical challenges need to be yet addressed for additive manufacturing to be a standard method in the field of microfluidics. Namely, (1) attainable channel dimensions, (2) surface roughness, and (3) availability of materials. The attainable channel dimensions are still larger than those attainable in subtractive and formative manufacturing. It should be noted that the minimal resolution as specified by the manufacturer does not usually translate to the attainable channel dimensions.[106] Most 3D-printed fluidic devices are millifluidic devices, where the cross-sectional dimension of channels is larger than 100 μm.[127] For instance, inkjet-assisted and extrusion-based techniques are only capable of fabricating millifluidic devices. Currently, the most promising technique is light-assisted printing, where the channel dimensions of <20 μm have been readily fabricated. Surface roughness is another limitation of additive manufacturing. Surface roughness inherent to the mechanisms of printing compromises the optical transparency of the devices. Extrusion-based printing and inkjet printing result in greater surface roughness than light-assisted printing.[117] Lastly, the availability of materials suitable for additive manufacturing is still waiting to be expanded. For example, the biocompatibility of the materials needs to be investigated, envisaging the biomedical application of the 3D printed devices. Extrusion-based 3D printing accommodates a wide selection of materials, including polymers, hydrogels, and metals. As such, extrusion-based methods are the most accustomed to biocompatibility. Unfortunately, material availability for inkjet- and light-assisted 3D printing is still limited. Most commercially available resins are yet to be assessed for their biosafety,[105] and these resins may well contain leachates toxic to cells.[77] At present, many research groups formulate their resin and/or hydrogel for biomedical applications.[102–104,128]

Overall, the application of additive manufacturing to fabricate micro/nanofluidics is still in its infancy. There are still challenges that need to be addressed before it can compete with subtractive and formative manufacturing in the fabrication of micro/nanofluidic devices. Nonetheless, we foresee that additive manufacturing holds promise and has the potential to play a critical role in the future development of micro/nanofluidics if the limitations are adequately addressed.

3.4 Overview of manufacturing techniques

Table 1. summarizes the manufacturing process, highlighting the advantages and limitations of each technique. Fig. 10 summarizes the attainable channel dimensions by different manufacturing processes discussed in this section.

Table 1 Advantages and limitations of various manufacturing processes.

Manufacturing methods		Materials	Advantages	Limitations	References
Subtraction	Bulk micromachining (photolithography + etching)	Glass; quartz; silicon	High precision Cost-effective	Requires clean-room facilities Slow design-to-prototype cycle Material availability Planar channels Requires bonding	35,36,129,130
	Focused ion beam (FIB) milling	Glass; quartz; silicon; PDMS	High precision	Requires expensive equipment Requires multiple passes to create large features	48,49,52,131
	Femtosecond direct laser writing (DLW)	High-silica porous glass	High precision Can fabricate 3D microfeatures Do not require sealing of devices	Requires expensive equipment Requires multiple passes to create large features	54,59,132
	Laser ablation/ engraving	PMMA; paper	Low-cost Quick design-to-prototype process	Limited resolution (>100 μm)	133–135
	Xurography	Most polymer films (e.g., PTFE, Teflon)	Low-cost Quick design-to-prototype process	Limited resolution (>100 μm)	66,67
	Micromilling	Most rigid polymers	Quick design-to-prototype process	Surface roughness may affect optical transparency Limited resolution (>100 μm)	68,69

Formative	Soft lithography	PDMS; elastomers	Applicable to rigid and flexible substrates Master mold can be outsourced to external vendors Cost-effective	Requires cleanroom facilities for the fabrication of molds Labor intensive	2,136
	Embossing	Most polymer (e.g., polycarbonate, polymethylmethacrylate, cyclic olefin copolymer, and polyethylene terephthalate)	Fine channel features (<5 nm, NIL) Suitable for both small-scale and large-scale production Suitable for high-aspect-ratio features	Requires fabrication of mold	90,137
	Injection molding	Most polymers	Effective for large-scale manufacturing Material availability	Requires fabrication of injection mold Not suitable for small-scale production Not suitable for high-aspect ratio features	92,138
Additive	Light-assisted	Photocurable resins; photocurable hydrogels	Quick design-to-prototype process Good attainable channel resolution	Requires post-processing to evacuate uncured resin Limited selection of materials	99,103

Continued

Table 1 Advantages and limitations of various manufacturing processes.—cont'd

Manufacturing methods	Materials	Advantages	Limitations	References
Inkjet	Photocurable resins (mostly proprietary)	Quick design-to-prototype cycle Multiple material printing	Surface roughness Difficulty in evacuating sacrificial material Proprietary equipment and resin required	110–113
Extrusion-based	Thermoplastics (e.g., PLA, ABS, PET, TPU); metal; composites; ceramics; hydrogels	Quick design-to-prototype cycle Wide selection of suitable materials Low setup cost	Poor feature size (>100 μm) High surface roughness	78,117,118,120

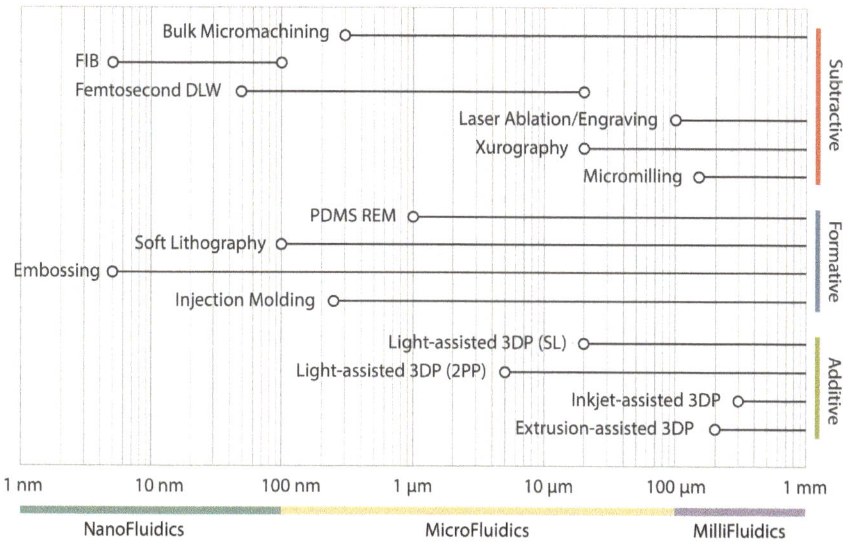

Fig. 10 Chart showing the range of attainable feature size (approximate) by the various manufacturing techniques.

4. Design thinking approach for micro/nanofluidics

This section introduces the design thinking approach as a possible toolkit that researchers can apply to design and develop micro/nanofluidic devices as solutions for the problems they handle. First, we describe why design thinking is relevant to the development of micro/nanofluidics. Next, we introduce the design thinking approach by highlighting the five stages in the design thinking approach.

4.1 Why is design thinking relevant to the development of micro/nanofluidics?

In a letter published in Trends in Biotechnology, Caicedo et al. mentioned that while microfluidics has made significant strides in academic research, the field is flooded with academic proofs-of-concept, with little translational research that tackles real-world scientific and technological problems.[139] They gave several reasons to explain the poor adoption of microfluidics in mainstream biomedical research and the biotech industry. The reasons can be summarized as such: the failure of the academics to empathize with the end-users and their needs. We end this chapter with a brief section on design thinking because design thinking is an approach that emphasizes understanding the users and their needs. Today, design thinking is a tested and proven approach that is adopted by professional firms and universities alike for the development of innovative solutions to complex problems.[6] We hope that academics working in micro/nanofluidics can adopt the design thinking approach to aid them in taking a user-centric approach in the innovation of micro/nanofluidics. Ultimately, we hope that the application of design thinking can bridge the gap between the exploration and the translation of the research.

4.2 What is the design thinking approach?

Design thinking is a philosophy and a set of tools aimed to help designers solve problems creatively. It is an iterative process centered around (1) understanding the users, (2) challenging assumptions, and (3) redefining the problem to identify solutions that might not be apparent initially. It is a user-centric framework that encourages designers and engineers to focus on the end-users for whom products are designed. The ideology of design thinking asserts that a hands-on, user-centric approach to problem-solving can lead to innovation. The design thinking framework consists of five stages: *empathize*, *define*, *ideate*, *prototype*, and *test*.[140] While we introduce these five stages sequentially, design thinking should not be understood as a recipe where all the stages must be followed sequentially. Importantly, design thinking is a mindset, and these five steps merely serve as a framework to guide the designer. Besides, these five steps do not necessarily follow a linear path in practice. Instead, they may toggle between stages in iterative loops (Fig. 11).[140] This section does not aim to give a comprehensive review of design thinking, which is provided elsewhere.[141–143] Instead, we discuss the relevance of each stage to the design and development of micro/nanofluidic devices.

4.2.1 *Empathize*

The first stage of design thinking is to gain an empathetic understanding of the users and their needs. The primary goal in this stage is gathering information about (1) who I am building the product for, and (2) what their problems are. An empathetic understanding of the user may require being immersed in the physical environment of the users and gaining insight into the users and the constraints they have. Consider the available resources (e.g., fabrication facilities and materials). For example, is there a cleanroom facility available? If no, some fabrication techniques may not be accessible. Consider the environment of the users (e.g., laboratory) where the device is fabricated, assembled, and operated. When immersing into the physical

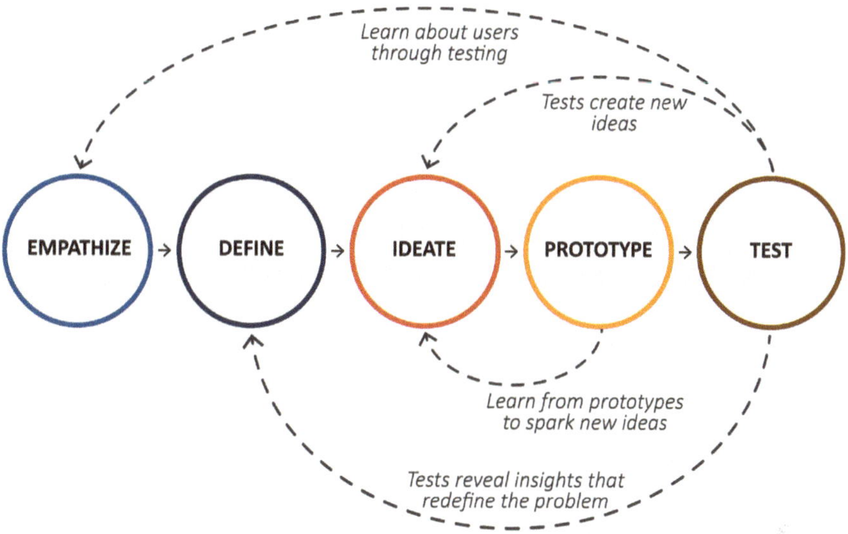

Fig. 11 Diagram illustrating the five-stage design thinking framework. *Adapted from Plattner H, Meinel C, Weinberg U.* Design-Thinking. *Springer; 2009.*

environment of the users, alternative considerations that were not apparent may surface. Considerations can be as trivial as transporting the device from an incubator to a biosafety cabinet without compromising sterility. It can be the accessibility of electrical sockets near the incubator for the connection of fluidic pumps or the method to connect electrical cables to the pumps located in the incubator. These considerations are information that can help guide the development of the fluidic device. Immersing in the physical environment of the users can also help designers and engineers gain new ideas on repurposing existing components for other applications. Components found in a fluidic device (i.e., fluidic reservoir) sometimes do not require complete redesigning. Existing elements that are readily found in most laboratories can be adapted for a different purpose. One good example is the repurposing of the pipette tip as a fluidic reservoir for microfluidic devices,[144,145] and disposal tips as droplet generators.[146]

Other than understanding the users, it is also of paramount importance to understand the needs of the users. For example, what is the research question of interest? Or, how can implementing a fluidic device help address the problem? Before these questions are adequately addressed, one cannot adopt the use of fluidic devices as an approach. Design and fabrication of fluidic devices is not a trivial task, and it should only be pursued if the use of a fluidic device can help address the problem.

4.2.2 Define

The second stage, *define*, aims to define the problem and requirement by drawing insights gathered in the stage of *empathize*. This stage aids the designer and engineer in articulating the problem clearly in the form of a problem statement. Specific to the development of a fluidic device requires clearly defining the engineering requirements when identifying the problem statement. This step aims not merely to list all the good-to-have requirements but also to identify the requirements of paramount importance.

Define is similar to the process of abstraction in software engineering; abstraction aims to remove the details and attributes in a system to focus on those that are of greater importance. Abstraction is vital in the development of fluidic devices because integrating too many elements in a single fluidic device is technically challenging. Through the process, we differentiate essential requirements from merely good-to-have requirements. Here, we briefly highlight common requirements that need to be defined to develop fluidic devices for particular applications.

Materials. The choice of material is crucial in the development of fluidic devices. Materials determine the property and function of the devices (e.g., chemical and biological compatibility, integration of functional components, optical observation). Materials are specific to methods of manufacturing. For example, silicon or glass-based materials are generally suited to subtractive manufacturing processes; PDMS is suited for formative manufacturing processes. Materials should be selected to fulfill the needs of the end-users. For instance, designing a device for biology applications, biocompatibility, and sterilization tolerance must be considered. Other important criteria include physicochemical properties, design, fabrication possibilities, and cost.[147]

Channel architecture. The channel architecture is designed to meet the needs of the end-users. One prominent example is the lung-on-a-chip, where the needs of the end-users (i.e., mimicking a tissue barrier between the alveolar and the capillaries) influenced the channel architecture of the fluidic device.[93] Droplet generators are another canonical example in microfluidics where the function informs the channel architecture of the device. Specific channel architecture is necessary for the generation of droplets (i.e., T-junction, flow-focusing, or co-flow). Furthermore, channel geometry directly affects the function of the device, determining the size and frequency of the formation of the droplets.[148] Importantly, the requirements in channel architecture limit the selection of the fabrication technique. For example, formative manufacturing techniques are generally applied to fabricate 2D planar channel networks and not suitable in the fabrication of 3D channel networks.

Form factor. Form factor refers to the size, shape, and other physical specifications of the device. The form factor influences how the user interacts with the device. For example, a small footprint makes the operation of the device cumbersome. Consideration of form factors is also essential to use the device with other existing devices (e.g., a microscope, an incubator for cells and bacteria). For the observation, considerations to incorporate an optical access window (e.g., glass slide, coverslip) are essential.

Fluid Control. Fluid control is paramount in a fluidic system. By definition, micro/nanofluidics aim to study the control and manipulation of fluid at a minuscule scale. Considerations on the selection of flow control should be performed at the onset of the device development. Choosing a specific flow control method (i.e., gravity-driven, pressure-driven, or peristaltic pump) may warrant a complete redesign of the device. The type of flow control method also affects the flow profile. Byun et al. have written a comprehensive review summarizing the various flow control techniques used for microfluidic devices.[149]

Functional components. Functional components add unique roles for the device, such as sensing, switching, and stimulating. Surfaces that are functionalized with metals and hydrogels also offer functions. For example, the integration of surface-patterned electrodes permits the development of on-chip sensors that can be used to detect flow, pressure, and temperature.[150] Mechanical stimulations can be integrated into the fluidic device through fluid flow-induced shear forces[17,151] and stretching and compression.[93,152–155] Integrating functional components with fluidic systems is technically challenging and should be considered only if it is essential in addressing the needs of the users.

4.2.3 Ideate

The third stage of the design thinking framework is *ideate*. *Ideate* is the process of generating, developing, and communicating new ideas to address the problem. This stage aims to consolidate the knowledge in the first two stages to identify narrowed-down solutions that meet the needs and requirements. Ideation can be carried out by generating as many ideas and solutions at the beginning and narrowing them down to choose potential solutions to be investigated in the subsequent stages (i.e., *prototyping* and *testing*). Design thinking does not provide a fixed method to ideate the solutions. In general, it is suggested to take an interdisciplinary approach to generate ideas.

Micro/nanofluidics, by nature, is very interdisciplinary. It combines expertise in the fields of biology, chemistry, physics, and engineering. Many of the fluidic-based emerging technologies for biomedical and translational research applications involves collaboration between multiple disciplines.[94,156] In light of this, the collaboration between multiple disciplines and involving the end-users is recommended when generating ideas. Fabrication of fluidic devices is also a highly specialized discipline. It is vital to consult and work with fabrication experts to discuss possible solutions.

4.2.4 Prototype

Prototype follows once the potential solutions are ideated. This stage aims to get a conceptual idea into a prototype to validate the possible solutions proposed in the previous stages (*emphasize*, *define*, and *ideate*). The goal of prototyping is not to finish the final product but to learn about the strengths and weaknesses of the ideated solutions and to revise them (Fig. 11).

The development of prototypes employs a common practice in the engineering industry: design for manufacturing and assembly (DFMA). DFMA aims to design a product that can be manufactured and assembled easily and economically. DFMA is primarily practiced in the industry with the motivation of cost-saving. The concept of DFMA is applicable for academic researchers interested in the small-scale production of fluidic devices in streamlined manners. Some considerations for DFMA includes:

Minimize the number of parts and steps. It is generally a good practice to minimize the number of parts in a system and the steps to fabricate them. An increase in the number of parts inevitably leads to an increase in assembly time. An increased number of parts tends to result in high error rates during assembly. The same argument applies to the number of steps required for the fabrication of the device. For example, the fabrication of the mold for PDMS replica molding becomes cumbersome once the mold consists of multiple heights and 3D features, which can be prone to errors (i.e., misalignment of masks, misalignment of PDMS replicas). Wherever possible, such designs of channels should be avoided.

Design parts with self-locating features. Examples of implementing parts with self-locating features can be as simple as incorporating inlet and outlet features in the SU-8 mold to assist in the subsequent hole punching (Fig. 12A). The step to create inlets and outlets on PDMS slabs for tubing is done by hand. The addition of the gutter in the design of the channel helps align the biopsy punch. Some fluidic devices inevitably require multiple layers of channels; such channels require bonding multiple layers of polymeric

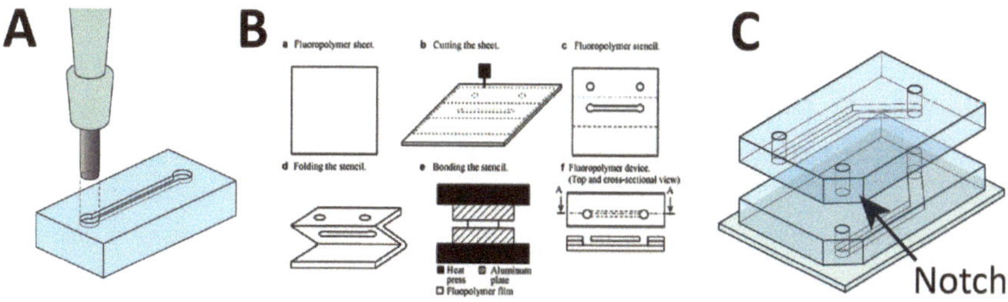

Fig. 12 (A) Illustration of the incorporation of self-locating feature on the PDMS slab (blue) to assist in the alignment of the biopsy punch. (B) Illustration demonstrating the fabrication of film-based microfluidic devices by xurography. Dotted incisions were included to assist in the folding and alignment of the multiple layers. (C) Illustration demonstrating the incorporation of a notch to help prompt the correct alignment during assembly. *Panel (B): Reproduced from Hizawa T, Takano A, Parthiban P, Doyle PS, Iwase E, Hashimoto M. Rapid prototyping of fluoropolymer microchannels by xurography for improved solvent resistance. Biomicrofluidics. 2018;12:064105 with permission.*

or elastomeric sheets together with precise alignment. Adding perforation is one strategy employed to assist in the alignment of film-based devices. Hizawa et al. programmed the cutting plotter to include perforation (or dotted incision) on the film-based microfluidic devices to assist the folding and alignment of the multilayer channels (Fig. 12B).[66] Another example of self-alignment was achieved with the pre-fabricated holes to assemble microfluidic devices with multiple layers of polymeric sheets.[64]

Design component asymmetry and symmetry. To eliminate the chances to assemble parts in incorrect orientations, the components with asymmetry may be used. For example, a simple notch on components prompts the assembly in the correct orientation (Fig. 12C). Alternatively, component symmetry can be considered so that components can be assembled regardless of their orientations.

Use standardized parts. Luer-lock is a good example of a standardized system of small-scale fluid fittings. It is the standardized interface used in commercially available syringes and needles. The benefit of implementing Luer fittings in a fluidic device is to provide the ease of connecting and disconnecting between devices. Several groups have implemented Luer fittings in their device design.[138,157] Another simple practice is to avoid implementing tubing with different sizes on the same device. This precaution simplifies the assembly and operation of the device.

Modularize design. Modular design refers to a design approach that subdivides a system into subsystems called modules that can be independently created, modified, and replaced. The main benefit of modular design is the modification of a specific module without the need to modify the entire system. More researchers implement the modular design approach in the development of microfluidic devices over the past decade. In such devices, modules are readily changed and reoriented to accommodate specific user needs.[69,158–161] A recent demonstration includes a modular body-on-a-chip platform, where different organ-inspired chips are linked together to predict changes in drug levels and organ-specific toxicities quantitatively over time.[94]

4.2.5 Test

The last stage, *test*, involves operating the prototype in a scenario relevant to the actual users. The strengths and weaknesses of an idea or prototype can only be understood when the product is tested in real-world scenarios. The goal of testing is to evaluate and gather feedback from the users as they use the product.

Test is also considered an extension of *empathize* conducted with a tangible prototype because testing involves observing and gathering feedback from the users as they operate the prototypes. When undertaken correctly, this stage can feed additional information to other stages of the design thinking framework that can lead to the iterative development of the fluidic device (Fig. 11). *Test* is often conducted with the end-users who can provide feedback to the prototypes. Ultimately, with a fast feedback loop, we expect ideas to evolve fast, resulting in thoughtful and effective solutions that address the needs of the users.

5. Outlook

Microfluidics and nanofluidics have found a wide range of adoption from basic science to applied engineering since the appearance of the concept of μTAS in the early 1990s. In the early days, the methods of manufacturing were reliant on MEMS-based micromachining. The introduction of soft lithography, arguably, ignited the development of micro/nanofluidics, followed by the adaptation in diverse disciplines. MEMS-based micromachining requires a high level of expertise at the cost of high resolution and precision. Soft lithography offered simplicity in the manufacturing procedure and empowered researchers from diverse fields in science and engineering to build, study, and solve innovative problems that were previously impossible without microfluidics.[4] Recently, the introduction of 3D printing technologies brought in yet another manufacturing paradigm, offering a fast design-to-prototype cycle beneficial in device development, where prototypes can be quickly validated to identify potential problems and perform iterative development. Just as soft lithography substantially improved the design-to-prototype time relative to the traditional MEMS-based micromachining, 3D printing has vastly shortened the design-to-prototype cycle down to several hours. The short design-to-prototype cycle is also advantageous for the implementation of design thinking. Furthermore, the introduction of 3D printing helped in the democratizing of manufacturing processes to individual labs all across the world. Before the introduction of 3D printing, the manufacturing of fluidic devices by subtractive and formative processes usually require specialized facilities and equipment that are not accessible to all research settings. Today, largely due to the introduction of affordable printers, 3D printers have become commodities found in most academic laboratories. Crucially, the digital nature of 3D printing allows researchers to share digital designs with the research community all around the world.

While the continual introduction of new manufacturing technologies has enabled substantial progress in the field of micro/nanofluidics, most of the publications are merely proof-of-concept with little translational value to secure successful widespread adoption and commercialization.[139] After several decades since the appearance of the concept of μTAS, the search for *killer applications* in the field of microfluidics is still ongoing.[162] The device developers in the field of micro/nanofluidics may lack an empathetic understanding of their users (i.e., biologists, clinicians, industry research scientists), and they are not adequately addressing the needs of the end-users. Perhaps, academic engineers working on biomedical investigations should engage in thoughtful partnerships with biologists, clinicians, and industry research scientists to understand their needs. Perhaps, inculcating a design thinking mindset is a good place to start.

References

1. Manz A, Graber N, Widmer Há. Miniaturized total chemical analysis systems: a novel concept for chemical sensing. *Sensors Actuators B Chem*. 1990;1:244–248.
2. Duffy DC, Mcdonald JC, Schueller OJA, Whitesides GM. Rapid prototyping of microfluidic systems in poly(dimethylsiloxane). *Anal Chem*. 1998;70:4974–4984.
3. Unger MA, Chou H-P, Thorsen T, Scherer A, Quake SR. Monolithic microfabricated valves and pumps by multilayer soft lithography. *Science*. 2000;288:113–116.
4. Lee A. The third decade of microfluidics. *Lab Chip*. 2013;13:1660–1661.
5. Convery N, Gadegaard N. 30 years of microfluidics. *Micro Nano Eng*. 2019;2:76–91.
6. Iskander N. Design thinking is fundamentally conservative and preserves the Status Quo. *Harv Bus Rev*. 2018;5.

7. Kolko J. Design thinking comes of age. *Harv Bus Rev*. 2015;66–71.
8. Tschimmel K. Design thinking as an effective toolkit for Innovation. In: *ISPIM Conference Proceedings, The International Society for Professional Innovation Management (ISPIM)*; 2012:1.
9. Beebe DJ, Mensing GA, Walker GM. Physics and applications of microfluidics in biology. *Annu Rev Biomed Eng*. 2002;4:261–286.
10. Song Y, Zhao X, Tian Q, Liang H. Fundamental concepts and physics in microfluidics. In: *Microfluidics: Fundamentals, Devices, and Applications*. Wiley; 2018.
11. Squires TM, Quake SR. Microfluidics: fluid physics at the nanoliter scale. *Rev Mod Phys*. 2005;77:977.
12. Van Meer B, De Vries H, Firth K, et al. Small molecule absorption by PDMS in the context of drug response bioassays. *Biochem Biophys Res Commun*. 2017;482:323–328.
13. Mansur EA, Mingxing Y, Yundong W, Youyuan D. A state-of-the-art review of mixing in microfluidic mixers. *Chin J Chem Eng*. 2008;16:503–516.
14. Dietzel A. *A Brief Introduction to Microfluidics, Microsystems for Pharmatechnology*. Springer; 20161–21.
15. Buchanan CF, Verbridge SS, Vlachos PP, Rylander MN. Flow shear stress regulates endothelial barrier function and expression of angiogenic factors in a 3D microfluidic tumor vascular model. *Cell Adhes Migr*. 2014;8:517–524.
16. Lu H, Koo LY, Wang WM, Lauffenburger DA, Griffith LG, Jensen KF. Microfluidic shear devices for quantitative analysis of cell adhesion. *Anal Chem*. 2004;76:5257–5264.
17. Ingber D. Mechanobiology and diseases of mechanotransduction. *Ann Med*. 2003;35:564–577.
18. Bhatia SN, Ingber DE. Microfluidic organs-on-chips. *Nat Biotechnol*. 2014;32:760–772.
19. Akyazi T, Basabe-Desmonts L, Benito-Lopez F. Review on microfluidic paper-based analytical devices towards commercialisation. *Anal Chim Acta*. 2018;1001:1–17.
20. Carrell C, Kava A, Nguyen M, et al. Beyond the lateral flow assay: a review of paper-based microfluidics. *Microelectron Eng*. 2019;206:45–54.
21. Hashimoto M, Shevkoplyas SS, Zasońska B, Szymborski T, Garstecki P, Whitesides GM. Formation of bubbles and droplets in parallel, coupled flow-focusing geometries. *Small*. 2008;4:1795–1805.
22. Hashimoto M, Mayers B, Garstecki P, Whitesides GM. Flowing lattices of bubbles as tunable, self-assembled diffraction gratings. *Small*. 2006;2:1292–1298.
23. Zhu P, Wang L. Passive and active droplet generation with microfluidics: a review. *Lab Chip*. 2017;17:34–75.
24. Samiei E, Tabrizian M, Hoorfar M. A review of digital microfluidics as portable platforms for lab-on a-chip applications. *Lab Chip*. 2016;16:2376–2396.
25. Wang H, Chen L, Sun L. Digital microfluidics: a promising technique for biochemical applications. *Front Mech Eng*. 2017;12:510–525.
26. Conlisk AT. The Debye-Hückel approximation: its use in describing electroosmotic flow in micro-and nanochannels. *Electrophoresis*. 2005;26:1896–1912.
27. Weerakoon-Ratnayake KM, O'Neil CE, Uba FI, Soper SA. Thermoplastic nanofluidic devices for biomedical applications. *Lab Chip*. 2017;17:362–381.
28. Schoch RB, Han J, Renaud P. Transport phenomena in nanofluidics. *Rev Mod Phys*. 2008;80:839.
29. Bradley BA. *Lithic reduction sequences: a glossary and discussion*. In: *Lithic Technology*. 1975;5–13.
30. Conard NJ, Adler DS. Lithic reduction and hominid behavior in the Middle Paleolithic of the Rhineland. *J Anthropol Res*. 1997;53:147–175.
31. Artioli G, Angelini I, Kaufmann G, Canovaro C, Dal Sasso G, Villa IM. Long-distance connections in the copper age: new evidence from the Alpine Iceman's copper axe. *PLoS One*. 2017;12:e0179263.

32. Kumar S. Introduction. In: Kumar S, ed. *Additive Manufacturing Processes*. Cham: Springer International Publishing; 2020:1–19.
33. Hull CW. *Method of and Apparatus for Production of Three Dimensional Objects by Stereolithography*. U.S. Patent No. 5,236,637, 1993.
34. Abgrall P, Gué AM. Lab-on-chip technologies: making a microfluidic network and coupling it into a complete microsystem—a review. *J Micromech Microeng*. 2007;17: R15–R49.
35. Manz A, Fettinger JC, Verpoorte E, Lüdi H, Widmer HM, Harrison DJ. Micromachining of monocrystalline silicon and glass for chemical analysis systems A look into next century's technology or just a fashionable craze? *Trends Anal Chem*. 1991;10:144–149.
36. Kopp MU. Chemical amplification: continuous-flow PCR on a chip. *Science*. 1998;280:1046–1048.
37. Fiorini GS, Chiu DT. Disposable microfluidic devices: fabrication, function, and application. *BioTechniques*. 2005;38:429–446.
38. Moreau WM. *Semiconductor Lithography: Principles, Practices, and Materials*. Springer Science & Business Media; 2012.
39. Kricka L, Wilding P, Kost G, Welsh J. *Micromechanics and Nanotechnology*. New York: John Wiley and Sons; 1996:45.
40. Kovacs GTA. *Micromachined Transducers Sourcebook*. Kovacs, WCB McGraw-Hill; 1998:19–154.
41. Volkmuth WD, Austin RH. DNA electrophoresis in microlithographic arrays. *Nature*. 1992;358:600–602.
42. Harrison DJ, Fluri K, Seiler K, Fan Z, Effenhauser CS, Manz A. Micromachining a miniaturized capillary electrophoresis-based chemical analysis system on a chip. *Science*. 1993;261:895–897.
43. Kopp MU, Mello AJ, Manz A. Chemical amplification: continuous-flow PCR on a chip. *Science*. 1998;280:1046–1048.
44. Iliescu C, Taylor H, Avram M, Miao J, Franssila S. A practical guide for the fabrication of microfluidic devices using glass and silicon. *Biomicrofluidics*. 2012;6:016505.
45. Madou MJ. *Fundamentals of Microfabrication: The Science of Miniaturization*. CRC Press; 2018.
46. Tegenfeldt JO, Prinz C, Cao H, et al. The dynamics of genomic-length DNA molecules in 100-nm channels. *Proc Natl Acad Sci U S A*. 2004;101:10979–10983.
47. Cabodi M, Turner SWP, Craighead HG. Entropic recoil separation of long DNA molecules. *Anal Chem*. 2002;74:5169–5174.
48. Tseng AA. Recent developments in micromilling using focused ion beam technology. *J Micromech Microeng*. 2004;14:R15.
49. Tseng AA. Recent developments in nanofabrication using focused ion beams. *Small*. 2005;1:924–939.
50. Menard LD, Mair CE, Woodson ME, Alarie JP, Ramsey JM. A device for performing lateral conductance measurements on individual double-stranded DNA molecules. *ACS Nano*. 2012;6:9087–9094.
51. Menard LD, Ramsey JM. Electrokinetically-driven transport of DNA through focused ion beam milled nanofluidic channels. *Anal Chem*. 2013;85:1146–1153.
52. Menard LD, Ramsey JM. Fabrication of sub-5 nm nanochannels in insulating substrates using focused ion beam milling. *Nano Lett*. 2011;11:512–517.
53. Yang J, Ferranti DC, Stern LA, et al. Rapid and precise scanning helium ion microscope milling of solid-state nanopores for biomolecule detection. *Nanotechnology*. 2011;22:285310.
54. Liao Y, Cheng Y. Femtosecond laser 3D fabrication in porous glass for micro- and nanofluidic applications. *Micromachines*. 2014;5:1106–1134.

55. Sugioka K, Hanada Y, Midorikawa K. Three-dimensional femtosecond laser micromachining of photosensitive glass for biomicrochips. *Laser Photonics Rev*. 2010;4:386–400.
56. Sugioka K, Cheng Y, Midorikawa K. Three-dimensional micromachining of glass using femtosecond laser for lab-on-a-chip device manufacture. *Appl Phys A Mater Sci Process*. 2005;81:1–10.
57. Liao Y, Song J, Li E, et al. Rapid prototyping of three-dimensional microfluidic mixers in glass by femtosecond laser direct writing. *Lab Chip*. 2012;12:746–749.
58. Ke K, Hasselbrink EF, Hunt AJ. Rapidly prototyped three-dimensional nanofluidic channel networks in glass substrates. *Anal Chem*. 2005;77:5083–5088.
59. Liao Y, Cheng Y, Liu C, et al. Direct laser writing of sub-50 nm nanofluidic channels buried in glass for three-dimensional micro-nanofluidic integration. *Lab Chip*. 2013;13:1626.
60. Patko D, Mártonfalvi Z, Kovacs B, Vonderviszt F, Kellermayer M, Horvath R. Microfluidic channels laser-cut in thin double-sided tapes: cost-effective biocompatible fluidics in minutes from design to final integration with optical biochips. *Sensors Actuators B Chem*. 2014;196:352–356.
61. Thompson BL, Ouyang Y, Duarte GR, Carrilho E, Krauss ST, Landers JP. Inexpensive, rapid prototyping of microfluidic devices using overhead transparencies and a laser print, cut and laminate fabrication method. *Nat Protoc*. 2015;10:875–886.
62. Klank H, Kutter JP, Geschke O. CO 2-laser micromachining and back-end processing for rapid production of PMMA-based microfluidic systems. *Lab Chip*. 2002;2:242–246.
63. Kim J, Surapaneni R, Gale BK. Rapid prototyping of microfluidic systems using a PDMS/polymer tape composite. *Lab Chip*. 2009;9:1290–1293.
64. Bartholomeusz DA, Boutté RW, Andrade JD. Xurography: rapid prototyping of microstructures using a cutting plotter. *J Microelectromech Syst*. 2005;14:1364–1374.
65. Yuen PK, Goral VN. Low-cost rapid prototyping of flexible microfluidic devices using a desktop digital craft cutter. *Lab Chip*. 2010;10:384–387.
66. Hizawa T, Takano A, Parthiban P, Doyle PS, Iwase E, Hashimoto M. Rapid prototyping of fluoropolymer microchannels by xurography for improved solvent resistance. *Biomicrofluidics*. 2018;12:064105.
67. Cosson S, Aeberli LG, Brandenberg N, Lutolf MP. Ultra-rapid prototyping of flexible, multi-layered microfluidic devices via razor writing. *Lab Chip*. 2015;15:72–76.
68. Guckenberger DJ, Groot TED, Wan AMD, Beebe DJ, Young EWK. Micromilling: a method for ultra-rapid prototyping of plastic microfluidic devices. *Lab Chip*. 2015;15:2364–2378.
69. Owens CE, Hart AJ. High-precision modular microfluidics by micromilling of interlocking injection-molded blocks. *Lab Chip*. 2018;18:890–901.
70. Qin D, Xia Y, Whitesides GM. Soft lithography for micro- and nanoscale patterning. *Nat Protoc*. 2010;5:491–502.
71. Martinez-Duarte R, Madou M. SU-8 photolithography and its impact on microfluidics. In: *Microfluidics and Nanofluidics Handbook*. CRC Press; 2011:231–268.
72. Hwang Y, Candler RN. Non-planar PDMS microfluidic channels and actuators: a review. *Lab Chip*. 2017;17:3948–3959.
73. Hashimoto M, Langer R, Kohane DS. Benchtop fabrication of microfluidic systems based on curable polymers with improved solvent compatibility. *Lab Chip*. 2013;13:252–259.
74. Bellan LM, Strychalski EA, Craighead HG. Nanochannels fabricated in polydimethylsiloxane using sacrificial electrospun polyethylene oxide nanofibers. *J Vac Sci Technol*. 2008;26:1728.
75. Zhang L, Gu F, Tong L, Yin X. Simple and cost-effective fabrication of two-dimensional plastic nanochannels from silica nanowire templates. *Microfluid Nanofluid*. 2008;5:727–732.

76. Glick CC, Srimongkol MT, Schwartz AJ, et al. Rapid assembly of multilayer microfluidic structures via 3D-printed transfer molding and bonding. *Microsyst Nanoeng*. 2016;2:16063.
77. Ferraz MAMM, Nagashima JB, Venzac B, Le Gac S, Songsasen N. 3D printed mold leachates in PDMS microfluidic devices. *Sci Rep*. 2020;10:1–9.
78. Goh WH, Hashimoto M. Fabrication of 3D microfluidic channels and in-channel features using 3D printed, water-soluble sacrificial mold. *Macromol Mater Eng*. 2018;303:1700484.
79. Goh WH, Hashimoto M. Dual sacrificial molding: fabricating 3D microchannels with overhang and helical features. *Micromachines*. 2018;9:523.
80. Haque AU, Kumar A. *Nanochannels for Nanofluidics Fabrication Aspects*. Oak Ridge, TN, United States: Oak Ridge National Lab. (ORNL); 2012.
81. Chantiwas R, Park S, Soper SA, et al. Flexible fabrication and applications of polymer nanochannels and nanoslits. *Chem Soc Rev*. 2011;40:3677–3702.
82. Chou SY, Krauss PR, Renstrom PJ. Imprint of sub-25 nm vias and trenches in polymers. *Appl Phys Lett*. 1995;67:3114–3116.
83. Chou SY, Krauss PR, Renstrom PJ. Imprint lithography with 25-nanometer resolution. *Science*. 1996;272:85–87.
84. Levy SL, Craighead HG. DNA manipulation, sorting, and mapping in nanofluidic systems. *Chem Soc Rev*. 2010;39:1133.
85. Peng L, Deng Y, Yi P, Lai X. Micro hot embossing of thermoplastic polymers: a review. *J Micromech Microeng*. 2013;24:013001.
86. Wang X, Liedert C, Liedert R, Papautsky I. A disposable, roll-to-roll hot-embossed inertial microfluidic device for size-based sorting of microbeads and cells. *Lab Chip*. 2016;16:1821–1830.
87. Lin T-Y, Do T, Kwon P, Lillehoj PB. 3D printed metal molds for hot embossing plastic microfluidic devices. *Lab Chip*. 2017;17:241–247.
88. Liedert R, Amundsen LK, Hokkanen A, et al. Disposable roll-to-roll hot embossed electrophoresis chip for detection of antibiotic resistance gene mecA in bacteria. *Lab Chip*. 2012;12:333–339.
89. Weigl BH, Bardell R, Schulte T, Battrell F, Hayenga J. Design and rapid prototyping of thin-film laminate-based microfluidic devices. *Biomed Microdevices*. 2001;3:267–274.
90. Hamedi MM, Ünal B, Kerr E, Glavan AC, Fernandez-Abedul MT, Whitesides GM. Coated and uncoated cellophane as materials for microplates and open-channel microfluidics devices. *Lab Chip*. 2016;16:3885–3897.
91. Tsao C-W, DeVoe DL. Bonding of thermoplastic polymer microfluidics. *Microfluid Nanofluid*. 2009;6:1–16.
92. Lee UN, Su X, Guckenberger DJ, et al. Fundamentals of rapid injection molding for microfluidic cell-based assays. *Lab Chip*. 2018;18:496–504.
93. Huh D, Matthews BD, Mammoto A, Montoya-Zavala M, Hsin HY, Ingber DE. Reconstituting organ-level lung functions on a chip. *Science*. 2010;328:1662–1668.
94. Novak R, Ingram M, Marquez S, et al. Robotic fluidic coupling and interrogation of multiple vascularized organ chips. *Nat Biomed Eng*. 2020;4:407–420.
95. Mukhopadhyay R. *When PDMS Isn't the Best*. ACS Publications; 2007.
96. Quan H, Zhang T, Xu H, Luo S, Nie J, Zhu X. Photo-curing 3D printing technique and its challenges. *Bioact Mater*. 2020;5:110–115.
97. Ching T, Toh Y-C, Hashimoto M. Fabrication of complex 3D fluidic networks via modularized stereolithography. *Adv Eng Mater*. 2019;22(3):1901109.
98. Gong H, Beauchamp M, Perry S, Woolley AT, Nordin GP. Optical approach to resin formulation for 3D printed microfluidics. *RSC Adv*. 2015;5:106621–106632.
99. Gong H, Bickham BP, Woolley AT, Nordin GP. Custom 3D printer and resin for 18 μm × 20 μm microfluidic flow channels. *Lab Chip*. 2017;17:2899–2909.

100. Lee Y-S, Bhattacharjee N, Folch A. 3D-printed Quake-style microvalves and micropumps. *Lab Chip*. 2018;18:1207–1214.
101. Gong H, Woolley AT, Nordin GP. High density 3D printed microfluidic valves, pumps, and multiplexers. *Lab Chip*. 2016;16:2450–2458.
102. Kuo AP, Bhattacharjee N, Lee YS, Castro K, Kim YT, Folch A. High-precision stereolithography of biomicrofluidic devices. *Adv Mater Technol*. 2019;4:1800395.
103. Kim YT, Bohjanen S, Bhattacharjee N, Folch A. Partitioning of hydrogels in 3D-printed microchannels. *Lab Chip*. 2019;19:3086–3093.
104. Grigoryan B, Paulsen SJ, Corbett DC, et al. Multivascular networks and functional intravascular topologies within biocompatible hydrogels. *Science*. 2019;364:458–464.
105. Bhattacharjee N, Urrios A, Kang S, Folch A. The upcoming 3D-printing revolution in microfluidics. *Lab Chip*. 2016;16:1720–1742.
106. Nielsen AV, Beauchamp MJ, Nordin GP, Woolley AT. 3D printed microfluidics. *Annu Rev Anal Chem*. 2020;13:45–65.
107. Mayer F, Richter S, Westhauser J, Blasco E, Barner-Kowollik C, Wegener M. Multimaterial 3D laser microprinting using an integrated microfluidic system. *Sci Adv*. 2019;5:eaau9160.
108. Di Giacomo R, Krödel S, Maresca B, et al. Deployable micro-traps to sequester motile bacteria. *Sci Rep*. 2017;7:1–8.
109. Son AI, Opfermann JD, McCue C, et al. An implantable micro-caged device for direct local delivery of agents. *Sci Rep*. 2017;7:1–16.
110. Anderson KB, Lockwood SY, Martin RS, Spence DM. A 3D printed fluidic device that enables integrated features. *Anal Chem*. 2013;85:5622–5626.
111. Sochol R, Sweet E, Glick C, et al. 3D printed microfluidic circuitry via multijet-based additive manufacturing. *Lab Chip*. 2016;16:668–678.
112. Causier A, Carret G, Boutin C, Berthelot T, Berthault P. 3D-printed system optimizing dissolution of hyperpolarized gaseous species for micro-sized NMR. *Lab Chip*. 2015;15:2049–2054.
113. Sweet EC, Mehta R, Xu Y, Jew R, Lin R, Lin L. Finger-powered fluidic actuation and mixing via MultiJet 3D printing. *Lab Chip*. 2020;20:3375–3385.
114. Crump SS. *Apparatus and Method for Creating Three-Dimensional Objects*. U.S. Patent No. 5,121,329, 1992.
115. Waheed S, Cabot JM, Macdonald NP, et al. 3D printed microfluidic devices: enablers and barriers. *Lab Chip*. 2016;16:1993–2013.
116. Au AK, Huynh W, Horowitz LF, Folch A. 3D-printed microfluidics. *Angew Chem Int*. 2016;55:3862–3881.
117. Macdonald NP, Cabot JM, Smejkal P, Guijt RM, Paull B, Breadmore MC. Comparing microfluidic performance of three-dimensional (3D) printing platforms. *Anal Chem*. 2017;89:3858–3866.
118. Ching T, Li Y, Karyappa R, Ohno A, Toh Y-C, Hashimoto M. Fabrication of integrated microfluidic devices by direct ink writing (DIW) 3D printing. *Sensors Actuators B Chem*. 2019;297:126609.
119. Shimizu A, Goh WH, Itai S, Karyappa R, Hashimoto M, Onoe H. ECM-based microfluidic gradient generator for tunable surface environment by interstitial flow. *Biomicrofluidics*. 2020;14:044106.
120. Lin NY, Homan KA, Robinson SS, et al. Renal reabsorption in 3D vascularized proximal tubule models. *Proc Natl Acad Sci U S A*. 2019;116:5399–5404.
121. Kolesky DB, Homan KA, Skylar-Scott MA, Lewis JA. Three-dimensional bioprinting of thick vascularized tissues. *Proc Natl Acad Sci U S A*. 2016;113:3179–3184.
122. Parekh DP, Ladd C, Panich L, Moussa K, Dickey MD. 3D printing of liquid metals as fugitive inks for fabrication of 3D microfluidic channels. *Lab Chip*. 2016;16:1812–1820.

123. Chan HN, Tan MJA, Wu H. Point-of-care testing: applications of 3D printing. *Lab Chip*. 2017;17:2713–2739.
124. Kinstlinger IS, Miller JS. 3D-printed fluidic networks as vasculature for engineered tissue. *Lab Chip*. 2016;16:2025–2043.
125. Mantihal S, Kobun R, Lee B-B. 3D food printing of as the new way of preparing food: a review. *Int J Gastron Food Sci*. 2020;22:100260.
126. Ng WL, Chua CK, Shen Y-F. Print me an organ! why we are not there yet. *Prog Polym Sci*. 2019;97:101145.
127. Beauchamp MJ, Nordin GP, Woolley AT. Moving from millifluidic to truly microfluidic sub-100-μm cross-section 3D printed devices. *Anal Bioanal Chem*. 2017;409:4311–4319.
128. Urrios A, Parra-Cabrera C, Bhattacharjee N, et al. 3D-printing of transparent bio-microfluidic devices in PEG-DA. *Lab Chip*. 2016;16:2287–2294.
129. Terry SC, Jerman JH, Angell JB. A gas chromatographic air analyzer fabricated on a silicon wafer. *IEEE Trans Electron Devices*. 1979;26:1880–1886.
130. Manz A, Miyahara Y, Miura J, Watanabe Y, Miyagi H, Sato K. Design of an open-tubular column liquid chromatograph using silicon chip technology. *Sensors Actuators B Chem*. 1990;1:249–255.
131. Esmek FM, Bayat P, Pérez-Willard F, Volkenandt T, Blick RH, Fernandez-Cuesta I. Sculpturing wafer-scale nanofluidic devices for DNA single molecule analysis. *Nanoscale*. 2019;11:13620–13631.
132. Cao Y, Zeng X, Cai Z, Duan J. Laser micro/nano-fabrication techniques and their applications in electronics. In: *Advances in Laser Materials Processing*. Elsevier; 2010:629–670.
133. Nie J, Liang Y, Zhang Y, Le S, Li D, Zhang S. One-step patterning of hollow microstructures in paper by laser cutting to create microfluidic analytical devices. *Analyst*. 2013;138:671–676.
134. Cheng J-Y, Yen M-H, Wei C-W, Chuang Y-C, Young T-H. Crack-free direct-writing on glass using a low-power UV laser in the manufacture of a microfluidic chip. *J Micromech Microeng*. 2005;15:1147.
135. Hanada Y, Sugioka K, Kawano H, Ishikawa IS, Miyawaki A, Midorikawa K. Nano-aquarium for dynamic observation of living cells fabricated by femtosecond laser direct writing of photostructurable glass. *Biomed Microdevices*. 2008;10:403–410.
136. Anderson JR, Chiu DT, Jackman RJ, et al. Fabrication of topologically complex three-dimensional microfluidic systems in PDMS by rapid prototyping. *Anal Chem*. 2000;72:3158–3164.
137. Gale BK, Jafek AR, Lambert CJ, et al. A review of current methods in microfluidic device fabrication and future commercialization prospects. *Inventions*. 2018;3:60.
138. Utko P, Persson F, Kristensen A, Larsen NB. Injection molded nanofluidic chips: fabrication method and functional tests using single-molecule DNA experiments. *Lab Chip*. 2011;11:303–308.
139. Caicedo HH, Brady ST. Microfluidics: the challenge is to bridge the gap instead of looking for a 'killer app'. *Trends Biotechnol*. 2016;34:1–3.
140. Dam R, Siang T. *Stages in the Design Thinking Process*. Interaction Design Foundation; 2021.
141. Plattner H, Meinel C, Weinberg U. *Design-Thinking*. Springer; 2009.
142. Lewrick M, Link P, Leifer L. *The Design Thinking Toolbox: A Guide to Mastering the Most Popular and Valuable Innovation Methods*. John Wiley & Sons; 2020.
143. Lewrick M, Link P, Leifer L. *The Design Thinking Playbook: Mindful Digital Transformation of Teams, Products, Services, Businesses and Ecosystems*. John Wiley & Sons; 2018.

144. Kim KM, Choi YJ, Hwang J-H, et al. Shear stress induced by an interstitial level of slow flow increases the osteogenic differentiation of mesenchymal stem cells through TAZ activation. *PLoS One*. 2014;9:e92427.
145. Park JY, White JB, Walker N, et al. Responses of endothelial cells to extremely slow flows. *Biomicrofluidics*. 2011;5:022211.
146. Vijayan S, Hashimoto M. 3D printed fittings and fluidic modules for customizable droplet generators. *RSC Adv*. 2019;9:2822–2828.
147. Kurth F, Györvary E, Heub S, et al. *Engineered Microenvironments for Safety and Efficacy Testing, Organs-on-a-Chip Engineering*. Elsevier; 202047–130.
148. Chong ZZ, Tan SH, Gañán-Calvo AM, Tor SB, Loh NH, Nguyen N-T. Active droplet generation in microfluidics. *Lab Chip*. 2016;16:35–58.
149. Byun CK, Abi-Samra K, Cho Y-K, Takayama S. Pumps for microfluidic cell culture. *Electrophoresis*. 2014;35:245–257.
150. Kieninger J, Weltin A, Flamm H, Urban GA. Microsensor systems for cell metabolism–from 2D culture to organ-on-chip. *Lab Chip*. 2018;18:1274–1291.
151. Sebastian B, Dittrich PS. Microfluidics to mimic blood flow in health and disease. *Annu Rev Fluid Mech*. 2018;50:483–504.
152. Naruse K, Yamada T, Sai X-R, Hamaguchi M, Sokabe M. Pp125 FAK is required for stretch dependent morphological response of endothelial cells. *Oncogene*. 1998;17:455–463.
153. Shimizu A, Goh WH, Itai S, Hashimoto M, Miura S, Onoe H. ECM-based microchannel for culturing in vitro vascular tissues with simultaneous perfusion and stretch. *Lab Chip*. 2020;20:1917–1927.
154. Huh D, Kim HJ, Fraser JP, et al. Microfabrication of human organs-on-chips. *Nat Protoc*. 2013;8:2135–2157.
155. Park S-H, Sim WY, Min B-H, Yang SS, Khademhosseini A, Kaplan DL. Chip-based comparison of the osteogenesis of human bone marrow-and adipose tissue-derived mesenchymal stem cells under mechanical stimulation. *PLoS One*. 2012;7:e46689.
156. Herland A, Maoz BM, Das D, et al. Quantitative prediction of human pharmacokinetic responses to drugs via fluidically coupled vascularized organ chips. *Nat Biomed Eng*. 2020;4:421–436.
157. Au AK, Lee W, Folch A. Mail-order microfluidics: evaluation of stereolithography for the production of microfluidic devices. *Lab Chip*. 2014;14:1294–1301.
158. Ong LJY, Ching TTH, Chong LH, et al. Self-aligning tetris-like (TILE) modular microfluidic platform for mimicking multi-organ interactions. *Lab Chip*. 2019;19(13):2178–2191.
159. Bhargava KC, Thompson B, Malmstadt N. Discrete elements for 3D microfluidics. *Proc Natl Acad Sci U S A*. 2014;111:15013–15018.
160. Yuen PK. SmartBuild–a truly plug-n-play modular microfluidic system. *Lab Chip*. 2008;8:1374–1378.
161. Yuen PK. A reconfigurable stick-n-play modular microfluidic system using magnetic interconnects. *Lab Chip*. 2016;16:3700–3707.
162. Blow N. Microfluidics: in search of a killer application. *Nat Methods*. 2007;4:665–670.

CHAPTER THREE

Detection and separation of proteins using micro/nanofluidics devices

Ashish Badiye[a], Neeti Kapoor[a], and Ritesh K. Shukla[b,*]
[a]Department of Forensic Science, Government Institute of Forensic Sciences, Nagpur, Maharashtra, India
[b]Biological and Life Sciences, School of Arts and Sciences, Ahmedabad University, Ahmedabad, Gujarat, India
*Corresponding author: e-mail address: ritesh.shukla@ahduni.edu.in

Contents

1. Introduction 60
2. Protein separation using microfluidics system/devices 62
   2.1 Separation systems 63
   2.2 Nanofluidics 68
   2.3 Detection method 70
3. Conclusion 74
References 75

Abstract

Microfluidics is the technology or system wherein the behavior of fluids' is studied onto a miniaturized device composed of chambers and tunnels. In biological and biomedical sciences, microfluidic technology/system or device serves as an ultra-high-output approach capable of detecting and separating the biomolecules present even in trace quantities. Given the essential role of protein, the identification and quantification of proteins help understand the various living systems' biological function regulation. Microfluidics has enormous potential to enable biological investigation at the cellular and molecular level and maybe a fair substitution of the sophisticated instruments/equipment used for proteomics, genomics, and metabolomics analysis. The current advancement in microfluidic systems' development is achieving momentum and opening new avenues in developing innovative and hybrid methodologies/technologies. This chapter attempts to expound the micro/nanofluidic systems/devices for their wide-ranging application to detect and separate protein. It covers microfluidic chip electrophoresis, microchip gel electrophoresis, and nanofluidic systems as protein separation systems, while methods such as spectrophotometric, mass spectrometry, electrochemical detection, magneto-resistive sensors and dynamic light scattering (DLS) are discussed as proteins' detection system.

Progress in Molecular Biology and Translational Science, Volume 186
ISSN 1877-1173
https://doi.org/10.1016/bs.pmbts.2021.07.008

1. Introduction

The 21st century has witnessed several technological advancements and an increased knowledge base in scientific research & development.[1] Various innovative technologies are evolving faster, sensitive, portable, smaller in shape & size, and integrated on a single device. Beginning around three decades ago, this miniaturization has continued to integrate more and more laboratory protocols. Microfluidics is one of the emerging technologies of miniaturization that evolved and shifted the technological paradigm of scientific research and development.[2–4]

Microfluidics is defined as a miniatured device or system in which fluids flow through the chambers and tunnels within a microscale range. This technology requires a minimal sample volume to experiment compared to its counterpart sophisticated equipment/instruments.[4] Many other advantages include: quicker reaction time, portability, easy automation, cost-effectiveness, and the ability to perform multiple experiments parallelly onto a single device, shown in Fig. 1.

Microfluidics has enormous potential to enable biological investigation at the cellular and molecular level and maybe a fair substitution of the sophisticated instruments/equipment used for proteomics, genomics, and metabolomics analysis.[1] Currently, microfluidics technology/device or system

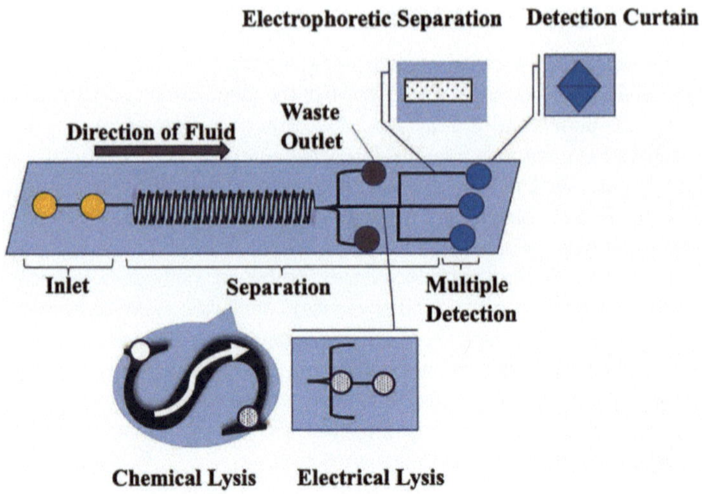

Fig. 1 Schematic representation (Top view) of the simple microfluidic device (with one inlet/multiple outlets) for multiplex protein detection.

plays a vital role in detecting various drugs/biomolecules in chemical and biomedical research. This technology is the best example of an interdisciplinary approach that is not limited to basic disciplines (physical, chemical, or biological/biomedical sciences) but is also interconnecting applied fields such as engineering, material science, and nanotechnology with the primary discipline.[5] The main motives for the evolution of microfluidics technology/system or device are to study the behavior of fluids at the microscale and develop innovative and highly sensitive applications in the diversified field of a scientific discipline that opens up new avenues and research opportunities.

In biological and biomedical sciences, microfluidic technology/system or device serves as an ultra-high-output approach capable of detecting and separating the biomolecules present even in trace quantities. Meanwhile, sophisticated analytical instruments are often unable to do so.[1,5]

Protein is one of the essential biomolecules that play a vital role in every living system's physiological process. Their identification and quantification help to understand the regulation of the biological function of the various living systems. Although proteins are found in abundance, yet they vary in amounts. The analysis of less copious proteins is inherently tricky.[6]

To date, different methods like Biuret, Lowery, BCA (bicinchoninic acid), Bradford, and fluorescence assays have been developed to detect and quantify the proteins.[7–13] These methods are easy and sensitive to detect biomolecules when available and specific and subtle for protein detection. However, due to contaminants' presence, the accuracy and sensitivity of these methods are sometimes affected. Moreover, the proteins in the sample cannot be detected using these methods if present below a certain threshold. This indicates that the detection of rare and trace amounts of proteins by these methods is a significant challenge.

Therefore, it is a need of the hour to develop a simple, exact, and sensitive method to detect trace quantities of protein. Various schemes/approaches have been developed to enhance the sensitivity and efficacy of protein detection and measurement. In this order, in a previous study, gold nanoparticles (Au NP)-based click-chemistry has been used to detect and quantify the protein.[14] Click-chemistry is the chemical scheme of methodology that usually refers to rapid reactivity, sensitivity, specificity, versatility, and ease of use. In this study, the protein was detected within 10 min using this method, which was at a low limit of about 0.2 μg/mL.[14] A nanoparticle (NPs) based fluorescent polymer sensor to detect and identify the proteins has been developed. Here proteins were detected and quantified using particle adsorption and fluorescence quenching.[15] In another study, the protein was quantified using a competitive method of particle adsorption to the protein. This is a very sensitive method in which bovine serum albumin (BSA) was detected at 7.0 μg/mL.[16] The methods above are significant and can play a key role in advancing detection and quantification of protein.

Over the last two decades, the reason behind the extensive use of microfluidic devices/systems for biochemical and clinical diagnostics include the advantages over conventional methods like small size/dimension of the device that leads to a significant reduction in consumption of costly reagent/chemicals, the rapid rate of reaction and ease of use.[17–19] Despite this, miniaturization of microfluidics devices/systems has led to new challenges such as laminar fluid flow and reagent mixing in a microchannel/tunnel due to low Reynolds number and slow diffusion process respectively.[6] Sample's concentration was still a challenge even in the microfluidics platform below the limit; therefore, micro/nanofluidics device/systems have evolved.[6]

Various types of nanostructures, nanoporous materials, and networks have been introduced into the device, wherein they act as pre-concentrators, accelerators, and molecular gates to counter the difficulties present in microfluidics.[20–26] These micro/nanofluidic devices/systems have been applied to quantify protein, enzyme kinetics, immunoassays, cell assays, etc.[24–31] This chapter attempted to elucidate the micro/nanofluidic systems/devices for their extensive application in detecting and separating protein.

2. Protein separation using microfluidics system/devices

The concept of microfluid technology was introduced by Manz et al. which later became a revolutionary approach to analytical chemistry.[32] In its initial year of development, the technology focused only on microfluidic chip electrophoresis (MCE), which exhibited extensive application in diverse fields such as biological, biomedical, chemical, and environmental sciences.[33–39] This technique is the best integration of miniaturization and automation that can be easily coupled with different types of detectors for sensitive detection, shown in Fig. 2. Microfluidic technology has further explored its application from sensing to the separation of biomolecules and has been an excellent platform for separating and analyzing proteins.

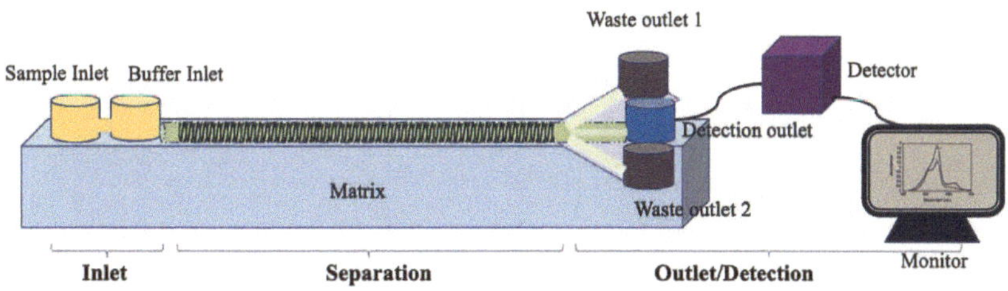

Fig. 2 Schematic representation (3-dimensional view) of automation of microfluidic device (with single inlet/single outlet) coupled with a specific detector for protein detection.

2.1 Separation systems

Chromatography, with some of its variants such as ion exchange, affinity, gel filtration, and electrophoresis, are the most effective and commonly used methods for separating and identifying proteins.[40] Both these approaches laid the foundation for microfluidic devices to separate proteins using various types of chromatographic resins. In this order, various methods have been used to incorporate chromatographic resin such as silica, magnetic beads into microchannel on microfluidic devices/systems.[41,42] This approach has developed new avenues for protein separation via microfluidic devices/systems. This approach is derived from the concept of nanofabricated size exclusion chromatography, in which nanomaterials were combined with various chromatographic microclimate/media,[43] shown in Table 1.[44–66] This concept has shown efficient results for protein separation when it is applied to "on-Chip RP-HPLC" (Reverse-phase high-performance liquid chromatography).[67] Some other SDS-PAGE (Sodium dodecyl sulfate-polyacrylamide gel electrophoresis) microfluidic devices are also commercially available for protein separation.

2.1.1 Microfluidic chip electrophoresis (MCE)

Since its inception in the 1980s, capillary electrophoresis (CE) is a valuable analytical technique that requires less amount of sample, has low electrolyte consumption, is rapid in analysis with higher resolution power of sample separation, and maybe effortlessly coupled with other sensitive detection systems.[68,69] In this order, various separation modes like capillary zone electrophoresis (CZE), micellar electrokinetic capillary chromatography (MEKC), capillary isoelectric focusing (CIEF), capillary isotachophoresis (CITP), and non-gel sieving-CE have evolved to separate and detect various biochemical substances.[70,71]

Table 1 Different microfluidic approaches for chromatographic applications for the pre-concentration/purification of polypeptides and proteins.

Material used	Analysis	Separation method	References
Bio-Gel P-6, Sephadex G-25, TSK-Gel SW, Toyopearl (HW-40S, HW-40F, HW-40C, HW-65S)	BSA and GlyGly peptide	Size-Exclusion Chromatography	Chirica et al.[44]
Sephadex G-25/DEAE-sepharose	Recombinant eGFP from *E. coli* lysate	Size-Exclusion Chromatography/ Ionic Exchange	Millet et al.[45]
Polycrystalline diamond surfaces	Myoglobin, conalbumin, ovalbumin trypsin inhibitor Hemoglobin/β-lactoglobulin mixture	Ionic Exchange Chromatography	Björkman et al.[46]
SP sepharose fast flow	FITC-labeled lysozyme	Ionic Exchange Chromatography	Shapiro et al.[47]
Ethylenediamine coated methacrylate monoliths	BSA and ovalbumin	Ionic Exchange Chromatography	Chan et al.[48]
Monoliths	Recombinant GFP	Ionic Exchange Chromatography/ Hydrophobic Interactions (Solid Phase Extraction)	Yu et al. [49]
Poly[hydroxyethyl acrylate-*co*-poly(ethylene glycol) diacrylate] Monolithic column	Cytochrome c, myoglobin, lysozyme, ribonuclease A, alpha-chymotrypsinogen	Hydrophobic Interactions	Li et al.[50]
Anti-FITC antibodies immobilized on monoliths with epoxy groups	FITC-labeled proteins (human serum albumin and IgG)	Immunoaffinity Chromatography	Gao et al.[51]
Anti-goat IgG antibodies immobilized onto streptavidin-coated glass beads	Alexa Fluor 647-labeled goat IgG	Immunoaffinity Chromatography	Peoples et al.[52]
Antibody-coated silica beads	C-reactive protein (human serum and cerebrospinal fluid)	Immunoaffinity Chromatography	People and Karnes[53]

Antibody modified magnetic core-shell nanoparticles	α-lactalbumin (α-lac) and anti-α-lac IgG	Immunoaffinity Chromatography	Teste et al.[54]
IDA-functionalized polymer matrix (Cu2+)	Histidine-tagged proteins: (LacI-His(6)-GFP and α-Synuclein-His(8)-YFP)	Immobilized Metal Affinity Chromatography	Deshpande et al.[55]
Hydrazide-derivatized PMMA surface	Carbonylated proteins Oxidized cytochrome *c* and TRITC-BSA/with cytochrome *c*	Affinity Chromatography	Hollins et al.[56]
Immobilized lectins (ConA, WGA) onto silica monoliths	Glycoproteins	Affinity Chromatography	Zhong and El Rasi[57]
Magnetite-doped polydimethylsiloxane	Phosphopeptides	Affinity Chromatography	Sandison et al.[58]
Different magnetic nanocomposites	Phosphopeptides	Affinity Chromatography	Batalha and Roque [59]
Oligonucleotide coated monoliths	Thrombin (human serum)	Aptamers Affinity Chromatography	Gao et al.[60]
Aptamer-functionalized magnetic beads	Thrombin	Aptamers Affinity Chromatography	Chen et al.[61]
Self-assembled gold nanoparticles	Exclusion-enrichment effect tested with FITC-labeled BSA	Exclusion-Enrichment Effect	Jen et al.[62], Chen et al.[63]
C18-functionalized silica-based monolith	Pre-concentration of proteins (insulin, cytochrome C, lysozyme, myoglobin, β-lactoglobulin, and hemoglobin)	Solid Phase Extraction	Alzahrani and Welham [64]
C18-functionalized silica-based monolith	Pre-concentration of milk proteins	Solid Phase Extraction	Alzahrani and Welham [65]
C8-functionalized mesoporous magnetic microspheres	Tryptic digestions of BSA/monoclonal antibodies	Solid Phase Extraction	Gasilova et al.[66]

Table adapted from Rodríguez-Ruiz I, Babenko V, Martínez-Rodríguez S, Gavira JA. Protein separation under microfluidic regime. *Analyst*, 2018;143:606–619, Rodríguez-Ruiz I, Conejero-Muriel M, Ackermann TN, Gavira JA, Llobera A. A multiple path photonic lab on a chip for parallel protein concentration measurements. *Lab Chip*, 2015;15:1133–1139. https://doi.org/10.1039/C4LC01332H.

CE is simple in instrumental design, due to which it can be easily miniaturized and automated to produce disposable microfluidic chips. CZE from the CE family is one of the most commonly used and extensively applied methods for separating microfluidic device/system. This automated device could analyze the amino acids up to the range of ~16 nM. Additionally, in this system, the integrated sample preparation method makes it suitable for various applications like exobiology and environmental monitoring.[72] CZE can easily separate charged species, whereas an electrophoretic mobility-based method, such as MEKC, can separate both charged or neutral biomolecules. Besides, MEKC can efficiently separate fluorescent-labeled standard proteins using PDMS serpentine microfluidic chip/device.[73] CIEF is another powerful and popular method for protein separation. When CIEF is immobilized with monolithic pH gradient materials into microchannels, then myoglobin, ribonuclease B, and a-casein can be separated.[74]

Advancement in technology for protein separation has exploited 2D separation mode on a microfluidic device. In this advanced technique, MEKC is coupled with CE, which rapidly separated peptides digested from tryptase.[75] Since 2D electrophoresis is capable of characterizing proteins precisely; therefore, when CIEF and CZE are coupled to a 2D microfluidic device, it can easily predict protein aggregation.[76] With this integration's help, protein aggregation can be seen in the early stage, which was a challenge before this technology was developed. Also, this technology can understand the mechanism of protein aggregation and assist in its prevention.

MCE can also detect and separate proteins, just like the SDS-PAGE method. However, in MCE, the loading and separation of protein samples can be precisely controlled on the microfluidic device. Moreover, highly reproducible data can also be acquired rapidly compared to the conventional SDS-PAGE method.[40] MCE can also easily be integrated with conventional methods like 2D electrophoresis, which offer information about a specific type of protein and Western blotting that enables multiplexed protein detection.[77–81] The significant advantage of MCE is the simple design of the device that does not have any pumps and valves. The only disadvantage of MCE is the requirement of an external power supply for conducting the separation process.

Pressure driven liquid chromatography microfluidic device is another version of the protein separation system, which is an alternative to high-performance liquid chromatography (HPLC). This variant has many advantages like lesser sample requirement, less consumption of costly solvents, and portability. This system has been developed using different preparatory methods for the stationary phase, such as filling with standard HPLC packing,[82] monolithic phases,[83,84] and wall coating.[85] Here, the monolithic phase and wall coating is the most successful approach for column preparation.

2.1.2 Microchip Gel Electrophoresis (MGE)

No doubt, MCE can most efficiently separate ions and micro molecules, but in bio-macromolecules such as proteins, DNA charge to mass ratios are similar. However, their sizes vary, and in this condition, Gel electrophoresis (GE) is a more appropriate method for detecting and separating macromolecules. On the GE platform, macromolecules are retarded and migrate slowly because of the bigger size.[86,87] Moreover, GE's traditional method requires large amounts of samples, and consumption of other reagents is also high, while bio-macromolecules are always precious and rare. Therefore, reducing the cost of this process has always been a significant challenge. Microfluidic/microchip GE (MGE) has thus been developed, wherein all processing steps become automated, and sample and reagent requirement is significantly reduced, and the speed is increased along with the affordability compared to its traditional counterpart.[88]

In this direction, Herr and co-workers researched on PGE for single-cell, proteins and riboswitches analysis using MGE technology.[89–93] Earlier, they developed an accurate microfluidic mobility shift assay that rapidly increases riboswitch's ligand binding and, subsequently, obtains protein structure data within a short period compared to conventional assay.[92] They also developed an open microfluidic electrophoretic cytometry in PGE to analyze the protein from single cells.[91]

Affinity chromatography is another method for protein separation that can be applied to a microfluidic device. In this improvised system, affinity beads or binding groups might be packed into a microchamber or bind on the microchannel's surface. In this order, avidin functionalized poly(methyl methacrylate) [PMMA] microchip has been developed by Xia et al. for the detection of biotinylated proteins.[94] This device allowed the separation of "post-transitionally modified (PTM) proteins," which are sparsely present in the samples. This type of microfluidic device can help detect any PTM protein after some relevant modification.[84]

In another study, Sarkar et al. have proposed a unique method for separating less frequent or rare proteins.[95] The proteins bind affinity capture beads with a definite antigen on their surface are resolved based on the flow fractionation principle. According to this principle, particles at or above the specified threshold size aligned in separate streams when flowing through a spiral channel. With this approach's help, three HIV-specific antibodies/antigen can be captured simultaneously, which indicate that microchip based on this approach might help detect specific markers for certain diseases. This microchip is cost-effective and provide high throughput result in terms of separation of selective proteins. Another method is used to separate protein that binds with magnetic beads. In this approach, an external magnet conjugated with streptavidin-coated beads binds with a flow platform, which helps separate biotinylated proteins, and this separation process (reaction and wash steps) can be performed within a minute.[96] This approach became popular because of its simple separation process and had scope for various binding proteins with the beads.

In another study, Hu et al. demonstrated a simple two-phase continuous partitioning system that can efficiently separate target proteins from the crude lysate. This approach can reduce emulsification, and as a result, 90% of protein content can be obtained.[97] In another study, a similar approach has been employed with genetically tagged recombinant proteins to drive the PEG-rich partitioning phase, which provides a high rate of retrieval of the autofluorescent protein and glutathione S-transferase.[98]

Field fractionation is another method for protein separation.[99] In this method, protein molecules that run near the wall are transported slowly compared to those in the center. A field perpendicularly employed onto the channel through which particles are migrated into congeries can be distinctly eluted. Currently, this method is not extensively used for protein separation because of its complexity.

2.2 Nanofluidics

After the microfluidics platform's successful attempts for the separation of biomolecules, the nanofluidics platform was also developed in 1995 for the separation and transportation of proteins.[100] Nanofluidics is defined as "the study of fluid behavior inside or around the device, which has at least one dimension lesser than 100 nm."[101] Various theoretical models have been developed that clearly explain the behavior of charged ions in nanochannels.[102–105]

In the nanofluidic system, two approaches have been most significantly involved in protein transport and separation via silicon membrane nanochannels.[106,107] In the first approach, Pricl et al. through molecular modeling depicted the specific interaction among the proteins and the silicon surface and derived structural and molecular parameters.[106] While in the second approach, protein transportation and separation were interpreted based on a continuum modeling scheme. Both modeling schemes combine and form the multiscale simulation tool, which helps to depict proteins' behavior in terms of interaction, transportation, and separation on the nanofluidic system. This tool itself helps to understand the specific interaction between protein molecules and the surface of nanochannels. This modeling scheme also allows us to design and optimize microfluidic and chemical sensing devices.[107]

Meanwhile, a different approach for the broad set of proteins has also been simulated, which was based on the Brownian dynamics method. This model includes both an atomic description of the protein molecules and their electrostatic and hydrophobic interactions. This model was also calibrated via experimental data as well as simulations that reproduce experimental trends well. As a result, this model interprets the structural and dynamical behavior of the molecules.[108]

In another study, Nelissen et al. demonstrated Langevin dynamics simulations that analyze single-file diffusion of colloidal particles, which calculated that if molecular-surface interaction was intense, it could lead to a slowdown of the sub-diffusive behavior of bio-molecules than $t_{1/2}$. Although this model was not initially derived for them, yet it can be extended toward proteins.[109]

Although proteins are prone to adsorb at the walls of micro/nanofluidics devices readily, yet nanofluidics has significantly extended toward protein separation and transportation via this method. In this regard, a previous study demonstrated that diffusion kinetics could be controlled by solute size tailoring the width of microfabricated nanochannels.[110] This study's experiment was conducted in silicon-based membranes of uniform channels, which exhibited that the nanopore width approach can display a non-Fickian behavior of diffusion kinetics. These findings became the basis for developing a prototype for a drug delivery system that was capable of providing a prolonged level of bovine serum albumin (BSA) in the blood. In another study, a nano-biosensor was proposed to monitor the binding affinity of the two proteins.[111] This study demonstrated that when two bound proteins were run through the nanochannels, their diffusion profiles were significantly more comprehensive than a single one, which a theoretical model supported. It indicated that the measurement of the diffusion length could provide the rapid detection of protein interactions. The same research group later developed a label-free sensor measuring adsorption/desorption coefficients of proteins on surfaces.[112] This sensor can measure the changes in conductivity in a nano-slit, and the result highly coincides with the theoretical scheme about the high and low ionic strength regimes.

Typically, nanofluidic devices can operate when the sample is low, and analytes are very diluted in concentration. Hence, Kim et al. proposed a mechanism that before protein separation, protein concentration should be measured. In this study, two microchannels were fabricated in Polydimethylsiloxane (PDMS), which are weakly bound to a glass surface, and in this way, a nanochannel is created in the area where these two microchannels are adjacent. Finally, in this study, it was shown that two different proteins could be separated with the help of switching the direction of the electric field.[24]

2.3 Detection method

At present, various methods are being used for the detection of protein via the microfluidic platform. Based on signal detection and transduction, these methods are divided into three groups. Several detection methods and their potential application in the microfluidic platform are shown in Fig. 3.

2.3.1 Spectrophotometric method

This method is based on the interaction of light with the sample's analytes. These interactions are mainly in the form of absorption of light in the UV–Vis range, and the form of fluorescence wherein photons are emitted from the sample when excited by an external light source or in the form of chemiluminescence where the chemical reaction emits photons. The major

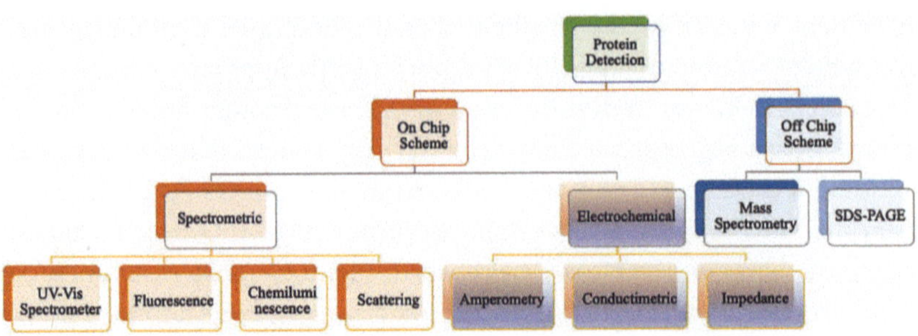

Fig. 3 Classification of the protein detection techniques integrated with the microfluidic system, and their potential configuration for on-chip and off-chip detection scheme.

constraint for this method is its conversion into the microfluidic platform where sample volume significantly decreases. Simultaneously, the optical path for sample and light interaction can also come down to a micron level[113] and require a sensitive detector for detection purposes.

In the spectrophotometric method, despite limited sensitivity, UV–Vis spectrophotometry is widely used detection purposes given the rapid and reproducible results.[113–128] It is known that proteins have an intrinsic property to absorb UV light. However, absorption depends mainly on amino acids with aromatic rings (tryptophan, tyrosine) and disulfide bonds from cysteine. The molar absorptivity of these amino acids provides an empirical equation at 280 nm (wavelength specific for protein absorption), making this method accessible and easy to apply.[129] This method relies on absorption properties in the UV–Vis range but is not based on absorbance schemes that have been reported for different macromolecular systems. These methods are generally used in a whole imaging configuration[122–127,130–134] and can also work for single-point detection systems. This system helps to detect and quantify the protein.[135] The fluorescence-based protein detection method is the most extensively used in the microfluidic platform.[77,78,99,136] Two different laser-induced fluorescence schemes can be applied. Single point fluorescence measurement[137–140] is the first scheme that allows online monitoring for continuous processing. In contrast, fluorescence imaging/scanning[78,141,142] is the second scheme to monitor the whole microfluidic channel. Despite high sensitivity, a significant constraint is that the sample should be fluorescent. It can either be tagged with a fluorescent-labeled probe or employed with functionalized fluorescent nano-beads for the separation process.[95,143,144] Functionalized fluorescent nano-beads is an approach that can provide lower detection limits of the analytes, while it can also be tagged with antibody-conjugated magnetic beads and an ex-situ fluorescent detector.[145] However, this method is costly and requires additional setup for fluorescence measurements, another major constraint for its extensive usage. Cost-effective LED light sources and photodetectors have also been proposed for the microfluidic platform through fluorescence[146,147] and absorbance-based detection,[148] which leads to a reduction in the cost of the method. Furthermore, image processing software and image stacking processes can improve the detection system's quality while reducing the analysis cost.[149]

Another spectrophotometric method for protein detection is based on chemiluminescence.[150–153] This method is developed with a simple design of the device, leading to a highly sensitive detection system compared to indirect UV detection and indirect fluorescence-based measurement.[153] Despite the highly sensitive detection system, this method is not used extensively because a sufficient amount of the optimization process is necessary to detect proteins in this system.

2.3.2 Mass spectrometry

Integration of mass spectrometry (MS) with Ligand Foot-printing (LiF) has become one of the most extensively applied methods for protein detection and visualization through a microfluidic system/device. Simultaneously, integration of MS with electrospray ionization or matrix-assisted laser desorption/ionization offers the most sensitive approach for protein like macromolecules detection, even at low concentration, within a minimal amount.[154–158] In the alternative method, multiple microfluidic channels are parallelly integrated with ICP-MS and significantly improved detection limits. With improved sensitivity, no adverse effects were seen on the microfluidic platform, and efficiency and resolution were also not affected because of this ultimate integration.[159] No doubt, this integration was excellent in reproducing results, but the high cost and larger size of MS again fall in conflict with the concept of miniaturization.

2.3.3 Electrochemical detection

Electrochemical detection provides high sensitivity and simple operation conditions as compared to spectrophotometric methods. This approach can be employed for protein detection using three strategies, which are amperometry-based detection,[160–164] conductivity-based detection,[165–167] and impedance-based detection systems.[168]

Among these three strategies, the conductivity-based detection system is a simple, precise, and cost-effective system that requires no reference electrodes compared to the amperometry based detection systems. Both conductivity and amperometry-based detection systems are non-selective approaches; therefore, conductive electrodes for direct sensing of proteins can act to increase their selectivity and activity. However, the impedance-based detection system works on the principle of change in impedance,[169] wherein the target molecules bind with receptors. Due to the higher frequency, the impedance-based detection systems are highly sensitive and responds even on minor changes, but electrode functionalization is needed. Interfacial engineering is generally required to reduce non-specific adsorption. Required needed.

In miniaturization, detection electrodes impose some constraints, such as periodical cleaning of the surface or attracting and detecting a low quantity of specific analytes by modifying or functionalizing the surface on their usage. To overcome these issues, some methods like electrode replacement, external electrodes are located at the end of the microfluidic device for direct ex-situ detection, and capacitive coupled contactless conductivity detection (C4D)[165] are employed. In the C4D approach, the high-frequency signal is foremost applied on the first electrodes and then signal to pass through the microfluidic channel, which is finally detected by the second electrodes and read electronically.

2.3.4 Other detection methods

The miniaturization of separation/detection methods aims to enable a single and straightforward experiment with higher resolution or higher output. Before the emergence of miniaturization, various other combinations have been proposed, where two or more complementary methods have been combined to fulfill the aim of higher resolution and high throughput result, Among them, fluorescence imaging with UV–Vis spectroscopy[95,170,171] and the integration of SDS-PAGE[97] with MS[172,173] are meaningful example of hybrid systems for protein detection. In another study, apart from MS, a light scattering detection system was also reported for quantification.[174] The light scattering detection system is very efficient in quantifying the sample with limited amounts.[175] This system is usually combined with HPLC and detects dry analysis when it interacts with a light beam. The results are analyzed as scattered light.

Furthermore, magneto-resistive sensors[176] are another appropriate method for detecting how functionalized magnetic particles are employed for protein separation that allows resolving protein as per their size and quantity. Dynamic light scattering (DLS) is another instrument that can be applied for protein detection. In a previous study, this method was applied to characterize hemoglobin when it was present in a mixture containing bovine serum albumin. In this method, silica-coated superparamagnetic nanoparticles[177] were first applied to the sample mixture, which can attract hemoglobin and form aggregates with NPs where proteins can be detected with the help of DLS. Finally, autoradiography has been proposed as a more specific method to detect radioactive phospho-peptides and phospho-amino-acids, which can be separated on thin-layer cellulose plates using electrophoresis and chromatographic method.[178]

3. Conclusion

The current advancement in microfluidic systems' development is achieving momentum and opening new avenues in developing innovative and hybrid methodologies/technologies. The ideal microfluidic system would extract and separate the highly purified proteins within a short time compared to the conventional methods. The integration of various techniques with the microfluidic system may help enhance the quality of results. Also, various materials comprising polymeric materials, pumps with electronic circuits, and sensors simplify microfluidic systems' preparation. Finally, a concept of microfluidic platforms is being developed in which various components can be easily assembled and enhance its versatility and utilization. Besides, standardization is essentially required to develop a cost-effective screening platform that can perform as a single-use (one-time use) technology and provide good quality results. To achieve this goal, the upcoming product will have to be standardized and enhance the level of automation. In developing standardized microfluidic systems, detection schemes such as electrochemical sensors and photonic lab-on-a-chip technologies can be integrated on-chip that will attract manufacturers to develop cost-effective upcoming products.

Inherent surface enhancement at the microfluidic platform is still a significant drawback for protein analysis on the chip-based system. It has also been highlighted in this chapter that the adsorption of biomolecules on the surface of the microfluidic channel can affect the protein transport on the microfluidic device/system. This undesired effect can benefit new and upcoming functionalized materials that can provide a stable and robust microfluidic platform with adjustable surface properties utilizing rapid and straightforward photo-grafting. When all these issues are sorted out, it may be expected that a miniaturized form of on-field detection device may provide results rapidly with a low-effective cost. These microfluid devices can be used for sensing applications and integrated with other powerful and potential technologies that can provide better results.

Soon, the microfluidic device may develop micro/nanofluidic electrophoretic techniques for single-cell analysis, be portable, operate on-site for point of care testing, and provide high throughput results with multiple analysis and versatile applications.

References

1. Vicente FA, Plazl I, Ventura SPM, Žnidaršič-Plazl P. Separation and purification of biomacromolecules based on microfluidics. *Green Chem*. 2020;22:4391–4410. https://doi.org/10.1039/C9GC04362D.
2. Hardt S, Hahn T. Microfluidics with aqueous two-phase systems. *Lab Chip*. 2012; 12:434–442. https://doi.org/10.1039/C1LC20569B.
3. Pohar A, Plazl I. Process intensification through microreactor application. *Chem Biochem Eng Q*. 2009;23:537–544.
4. Kaminski TS, Garstecki P. Controlled droplet microfluidic systems for multistep chemical and biological assays. *Chem Soc Rev*. 2017;46:6210–6226. https://doi.org/10.1039/C5CS00717H.
5. Shirai K, Mawatari K, Ohta R, Shimizu H, Kitamori T. A single-molecule ELISA device utilizing nanofluidics. *Analyst*. 2018;143:943–948.
6. Wang C, Shi Y, Wang J, Pang J, Xia X-H. Ultrasensitive protein concentration detection on a micro/nanofluidic enrichment chip using fluorescence quenching. *ACS Appl Mater Interfaces*. 2015;7:6835–6841.
7. Gornall AG, Bardawill CJ, David MM. Determination of serum proteins by means of the biuret reaction. *J Biol Chem*. 1949;177:751–766.
8. Lowry OH, Rosebrough NJ, Farr AL, Randall RJ. Protein measurement with the Folin phenol reagent. *J Biol Chem*. 1951;193:265–275.
9. Smith PK, Krohn RI, Hermanson GT, et al. Measurement of protein using Bicinchoninic acid. *Anal Biochem*. 1985;150:76–85.
10. Bradford MM. A rapid and sensitive method for the quantitation of microgram quantities of protein utilizing the principle of protein-dye binding. *Anal Biochem*. 1976;72:248–254.
11. You WW, Haugland RP, Ryan DK, Haugland RP. 3-(4-Carboxybenzoyl) quinoline-2-carboxaldehyde, a reagent with broad dynamic range for the assay of proteins and lipoproteins in solution. *Anal Biochem*. 1997;244:277–282.
12. Lorenzen A, Kennedy SW. A fluorescence-based protein assay for use with a microplate reader. *Anal Biochem*. 1993;214:346–348.
13. Jones LJ, Haugland RP, Singer VL. Development and characterization of the NanoOrange protein quantitation assay: a fluorescence-based assay of proteins in solution. *Biotechniques*. 2003;34:850–861.
14. Zhu K, Zhang Y, He S, et al. Quantification of proteins by functionalized gold nanoparticles using click chemistry. *Anal Chem*. 2012;84:4267–4270.
15. You C, Miranda OR, Gider B, et al. Detection and identification of proteins using nanoparticle-fluorescent polymer 'chemical nose' sensors. *Nat Nanotechnol*. 2007; 2:318–323.
16. Pihlasalo S, Kirjavainen J, Hanninen P, Harma H. Ultrasensitive protein concentration measurement based on particle adsorption and fluorescence quenching. *Anal Chem*. 2009;81:4995–5000.
17. Zhang CY, Johnson LW. Microfluidic control of fluorescence resonance energy transfer: breaking the FRET limit. *Angew Chem*. 2007;46:3482–3485.
18. Puleo CM, Yeh HC, Liu KJ, Wang TH. Coupling confocal fluorescence detection and recirculating microfluidic control for single particle analysis in discrete nanoliter volumes. *Lab Chip*. 2008;8:822–825.
19. Ji J, Nie L, Li YX, Yang PY, Liu BH. Simultaneous online enrichment and identification of trace species based on microfluidic droplets. *Anal Chem*. 2013;85:9617–9622.
20. Wang C, Ouyang J, Gao HL, et al. UV-ablation nanochannels in micro/nanofluidics devices for biochemical analysis. *Talanta*. 2011;85:298–303.

21. Chen W, Jin B, Hu YL, Lu Y, Xing HX. Entrapment of protein in nanotubes formed by a nanochannel and ion-channel hybrid structure of anodic alumina. *Small.* 2012;8:1001–1005. https://doi.org/10.1002/smll.201102117.
22. Wang C, Ouyang J, Wang YY, Ye DK, Xia XH. Sensitive assay of protease activity on a micro/nanofluidics preconcentrator fused with the fluorescence resonance energy transfer detection technique. *Anal Chem.* 2014;86:3216–3221.
23. Hu YL, Wang C, Wu ZQ, Xu JJ, Chen HY, Xia XH. Interconnected ordered nanoporous networks of colloidal crystals integrated on a microfluidic chip for highly efficient protein concentration. *Electrophoresis.* 2011;32:3424–3430.
24. Kim SM, Burns MA, Hasselbrink EF. Electrokinetic protein preconcentration using a simple glass/poly- (dimethylsiloxane) microfluidic chip. *Anal Chem.* 2006;78: 4779–4785.
25. Kim D, Raj A, Zhu L, Masel RI, Shannon MA. Non- equilibrium electrokinetic micro/nano fluidic mixer. *Lab Chip.* 2008;8:625–628.
26. Cannon Jr DM, Kuo TC, Bohn PW, Sweedler JV. Nanocapillary array interconnects for gated analyte injections and electrophoretic separations in multilayer microfluidic architectures. *Anal Chem.* 2003;75:2224–2230.
27. Kim SJ, Ko SH, Kang KH, Han J. Direct seawater desalination by ion concentration polarization. *Nat Nanotechnol.* 2010;5:297–301.
28. Wang YC, Stevens AL, Han J. Million-fold Preconcentration of proteins and peptides by nanofluidic filter. *Anal Chem.* 2005;77:4293–4299.
29. Wang C, Li SJ, Wu ZQ, Xu JJ, Chen HY, Xia XH. Study on the kinetics of homogeneous enzyme reactions in a micro/nanofluidics device. *Lab Chip.* 2010;10:639–646.
30. Wang YC, Han J. Pre-binding dynamic range and sensitivity enhancement for immuno-sensors using nanofluidic preconcentrator. *Lab Chip.* 2008;8:392–394.
31. Kwak R, Kim SJ, Han J. Continuous-flow biomolecule and cell concentrator by ion concentration polarization. *Anal Chem.* 2011;83:7348–7355.
32. Manz A, Graber N, Widmer HM. Miniaturized total chemical analysis systems: a novel concept for chemical sensing. *Sens Actuators B.* 1990;1:244–248.
33. Castro ER, Manz A. Present state of microchip electrophoresis: state of the art and routine applications. *J Chromatogr A.* 2015;1382:66–85.
34. Lockyear LLS, Colyer CL, Fan ZH, Roy KI, Harrison DJ. Effects of injector geometry and sample matrix on injection and sample loading in integrated capillary electrophoresis devices. *Electrophoresis.* 1999;20:529–538.
35. Harrison DJ, Fluri K, Seiler K, Fan ZH, Effenhauser CS, Manz A. Micromachining a miniaturized capillary electrophoresis-based chemical analysis system on a chip. *Science.* 1993;261:895–897.
36. Rocklin RD, Ramsey RS, Ramsey JM. A microfabricated fluidic device for performing two-dimensional liquid-phase separations. *Anal Chem.* 2000;72:5244–5249.
37. Chen P, Chen D, Li S, Ou X, Liu BF. Microfluidics towards single cell resolution protein analysis. *Trac-Trend Anal Chem.* 2019;117:2–12.
38. Mathies RA, Huang XC. Capillary array electrophoresis: an approach to high-speed, high-throughput DNA sequencing. *Nature.* 1992;359:167–169.
39. Wuethrich A, Quirino JP. A decade of microchip electrophoresis for clinical diagnostics—a review of 2008–2017. *Anal Chim Acta.* 2019;1045:42–66.
40. Rodríguez-Ruiz I, Babenko V, Martínez-Rodríguez S, Gavira JA. Protein separation under microfluidic regime. *Analyst.* 2018;143(3):606–619.
41. Kutter JP. Liquid phase chromatography on microchips. *J Chromatogr A.* 2012; 1221:72–82.
42. Ericson C, Holm J, Ericson T, Hjertén S. Electroosmosis- and pressure-driven chromatography in chips using continuous beds. *Anal Chem.* 2000;72:81–87.

43. Svehla D, Feldman S, Feldman J, et al. Nano-fabricated size exclusion chromatograph. In: *Micro Total Analysis Systems*. Dordrecht: Springer; 2002:545–547. https://doi.org/10.1007/978-94-010-0295-0_182.
44. Chirica G, Lachmann J, Chan J. Size exclusion chromatography of microliter volumes for online use in low-pressure microfluidic systems. *Anal Chem*. 2006;78:5362–5368. https://doi.org/10.1021/ac060258t.
45. Millet LJ, Lucheon JD, Standaert RF, Retterer ST, Doktycz MJ. Modular microfluidics for point-of-care protein purifications. *Lab Chip*. 2015;15:1799–1811. https://doi.org/10.1039/C5LC00094G.
46. Björkman H, Ericson C, Hjertén S, Hjort K. Diamond microchip capillary chromatography of proteins. In: van den Berg A, Olthuis W, Bergveld P, eds. *Micro Total Analysis Systems*. Dordrecht: Springer; 2000:187–190. https://doi.org/10.1007/978-94-017-2264-3_44.
47. Shapiro MS, Haswell SJ, Lye GJ, Bracewell DG. Design and characterization of a microfluidic packed bed system for protein breakthrough and dynamic binding capacity determination. *Biotechnol Prog*. 2009;25:277–285. https://doi.org/10.1002/btpr.99.
48. Chan AS, Danquah MK, Agyei D, Hartley PG, Zhu Y. A parametric study of a monolithic microfluidic system for on-chip biomolecular separation. *Sep Sci Technol*. 2014;49:854–860. https://doi.org/10.1080/01496395.2013.872144.
49. Yu C, Davey MH, Svec F, Fréchet JMJ. Monolithic porous polymer for on-chip solid-phase extraction and preconcentration prepared by photoinitiated in situ polymerization within a microfluidic device. *Anal Chem*. 2001;73:5088–5096. https://doi.org/10.1021/ac0106288.
50. Li Y, Tolley HD, Lee ML. Poly[hydroxyethyl acrylate-co-poly(ethylene glycol) diacrylate] monolithic column for efficient hydrophobic interaction chromatography of proteins. *Anal Chem*. 2009;81:9416–9424. https://doi.org/10.1021/ac9020038.
51. Gao C, Sun X, Wang H, Qiao W, Hu B. Affinity monolith-integrated microchips for protein purification and concentration. In: Tran N, Taverna M, eds. *Capillary Electrophoresis of Proteins and Peptides*. New York: Humana Press; 2016:85–92. https://doi.org/10.1007/978-1-4939-4014-1_7.
52. Peoples MC, Phillips TM, Karnes HT. A capillary-based microfluidic instrument suitable for immunoaffinity chromatography. *J Chromatogr B*. 2007;848:200–207.
53. Peoples MC, Karnes HT. Microfluidic capillary system for immunoaffinity separations of C-reactive protein in human serum and cerebrospinal fluid. *Anal Chem*. 2008;80:3853–3858.
54. Teste B, Malloggi F, Siaugue J-M, Varenne A, Kanoufi F, Descroix S. Microchip integrating magnetic nanoparticles for allergy diagnosis. *Lab Chip*. 2011;11:4207–4213.
55. Deshpande AG, Darton NJ, Yunus K, Fisher AC, Slater NKH. In situ fabrication of a microfluidic device for immobilized metal affinity sensing. *N Biotechnol*. 2012;29:494–501. https://doi.org/10.1016/j.nbt.2012.01.002.
56. Hollins C, Soper SA, Feng J. Enriching carbonylated proteins inside a microchip through the use of oxalyldihydrazide as a crosslinker. *Lab Chip*. 2012;12:2526–2532. https://doi.org/10.1039/C2LC40103G.
57. Zhong H, Rassi ZE. Monolithic silica capillary columns having immobilized lectins and surface bound polar functionalities for lectin affinity and normal phase nano-LC and CEC of glycoconjugates, respectively. *J Sep Sci*. 2009;32:1642–1653. https://doi.org/10.1002/jssc.200900127.
58. Sandison ME, Jensen KT, Gesellchen F, Cooper JM, Pitt AR. Magnetite-doped polydimethylsiloxane (PDMS) for phosphopeptide enrichment. *Analyst*. 2014;139:4974–4981. https://doi.org/10.1039/C4AN00750F.

59. Batalha IL, Roque ACA. Phosphopeptide enrichment using various magnetic nanocomposites: an overview. In: von Stechow L, ed. *Phospho-Proteomics*. New York: Springer; 2016:193–209. https://doi.org/10.1007/978-1-4939-3049-4_13.
60. Gao C, Sun X, Woolley AT. Fluorescent measurement of affinity binding between thrombin and its aptamers using on-chip affinity monoliths. *J Chromatogr A*. 2013;1291:92–96. https://doi.org/10.1016/j.chroma.2013.03.063.
61. Chen F-Y, Wang Z, Li P, Lian H-Z, Chen H-Y. Aptamer-based thrombin assay on microfluidic platform. *Electrophoresis*. 2013;34:3260–3266. https://doi.org/10.1002/elps.201300338.
62. Jen CP, Amstislavskaya TG, Chen K-F, Chen Y-H. Sample preconcentration utilizing nanofractures generated by junction gap breakdown assisted by self-assembled monolayer of gold nanoparticles. *PLoS One*. 2015;10(5), e0126641. https://doi.org/10.1371/journal.pone.0126641.
63. Chen YH, Wu HF, Amstislavskaya TG, Li CY, Jen CP. A simple electrokinetic protein preconcentrator utilizing nano-interstices. *Biomicrofluidics*. 2016;10(2):024121. https://doi.org/10.1063/1.4946768.
64. Alzahrani E, Welham K. Fabrication of an octadecylated silica monolith inside a glass microchip for protein enrichment. *Analyst*. 2012;137:4751–4759. https://doi.org/10.1039/C2AN16018H.
65. Alzahrani E, Welham K. Preconcentration of milk proteins using octadecylated monolithic silica microchip. *Anal Chim Acta*. 2013;798:40–47. https://doi.org/10.1016/j.aca.2013.08.035.
66. Gasilova N, Srzentić K, Qiao L, et al. On-chip mesoporous functionalized magnetic microspheres for protein sequencing by extended bottom-up mass spectrometry. *Anal Chem*. 2016;88:1775–1784. https://doi.org/10.1021/acs.analchem.5b04045.
67. Reichmuth DS, Shepodd TJ, Kirby BJ. Microchip HPLC of peptides and proteins. *Anal Chem*. 2005;77:2997–3000. https://doi.org/10.1021/ac048358r.
68. Jorgenson JW, Lukacs KD. Zone electrophoresis in open-tubular glass capillaries. *Anal Chem*. 1981;53:1298–1302.
69. Jorgenson JW, Lukacs KD. Capillary zone electrophoresis. *Science*. 1983;222:266–272.
70. Yang Z, Sweedler JV. Application of capillary electrophoresis for the early diagnosis of cancer. *Anal Bioanal Chem*. 2014;406:4013–4031.
71. Zhang M, Phung SC, Smejkal P, Guijt RM, Breadmore MC. Recent trends in capillary and microchip electrophoretic instrumentation for field-analysis. *Trends Environ Anal Chem*. 2018;18:1–10.
72. Kim J, Jensen EC, Stockton AM, Mathies RA. Universal microfluidic automaton for autonomous sample processing: application to the mars organic analyzer. *Anal Chem*. 2013;85:7682–7688.
73. Roman GT, Carroll S, McDaniel K, Culbertson CT. Micellar electrokinetic chromatography of fluorescently labeled proteins on poly(dimethylsiloxane)-based microchips. *Electrophoresis*. 2006;27:2933–2939.
74. Liang Y, Cong Y, Liang Z, Zhang L, Zhang Y. Microchip isoelectric focusing with monolithic immobilized pH gradient materials for proteins separation. *Electrophoresis*. 2009;30:4034–4039.
75. Wang R, Xie H, Xu Y, et al. Study on detection of mutation DNA fragment in gastric cancer by restriction endonuclease fingerprinting with capillary electrophoresis. *Biomed Chromatogr*. 2012;26:393–399.
76. Yu S, Xu J, Huang K, et al. A novel method to predict protein aggregations using two-dimensional native protein microfluidic chip electrophoresis. *Anal Methods*. 2016;8:8306–8313.

77. Shameli SM, Ren CL. Microfluidic two-dimensional separation of proteins combining temperature gradient focusing and sodium dodecyl sulfate-polyacrylamide gel electrophoresis. *Anal Chem*. 2015;87:3593–3597. https://doi.org/10.1021/acs.analchem.5b00380.
78. Li Y, Buch JS, Rosenberger F, DeVoe DL, Lee CS. Integration of isoelectric focusing with parallel sodium dodecyl sulfate gel electrophoresis for multidimensional protein separations in a plastic microfludic network. *Anal Chem*. 2004;76:742–748. https://doi.org/10.1021/ac034765b.
79. Hou N, Chen Y, Yu S, et al. Native protein separation by isoelectric focusing and blue gel electrophoresis-coupled two-dimensional microfluidic chip electrophoresis. *Chromatographia*. 2014;77:1339–1346. https://doi.org/10.1007/s10337-014-2728-3.
80. He M, Herr AE. Automated microfluidic protein immunoblotting. *Nat Protoc*. 2010;5:1844–1856. https://doi.org/10.1038/nprot.2010.142.
81. Hughes AJ, Herr AE. Microfluidic western blotting. *Proc Natl Acad Sci U S A*. 2012;109:21450–21455. https://doi.org/10.1073/pnas.1207754110.
82. Chan AS, Danquah MK, Agyei D, Hartley PG, Zhu Y. A simple microfluidic chip design for fundamental bioseparation. *J Anal Methods Chem*. 2014;2014:1–6. https://doi.org/10.1155/2014/175457.
83. Wang Z, Wang W, Chen G, Wang W, Fu F. Liquid chromatography on a monolithic column microfluidic chip coupled with "three-T" sample injection mode and amperometric detection. *J Sep Sci*. 2010;33:2568–2574. https://doi.org/10.1002/jssc.201000304.
84. Hara T, Futagami S, De Malsche W, et al. Chromatographic properties of minimal aspect ratio monolithic silica columns. *Anal Chem*. 2017;89:10948–10956. https://doi.org/10.1021/acs.analchem.7b02764.
85. Fuentes HV, Woolley AT. Electrically actuated, pressure-driven liquid chromatography separations in microfabricated devices†. *Lab Chip*. 2007;7:1524–1531. https://doi.org/10.1039/B708865E.
86. Jung YK, Kim J, Mathies RA. Microfluidic linear hydro-gel array for multiplexed single nucleotide polymorphism (SNP) detection. *Anal Chem*. 2015;87:3165–3170.
87. Duncombe TA, Herr AE. Use of polyacrylamide gel moving boundary electrophoresis to enable low-power protein analysis in a compact microdevice. *Anal Chem*. 2012;84:8740–8747.
88. Chao T, Bhattacharya S, Ros A. Microfluidic gel electrophoresis in the undergraduate laboratory applied to food analysis. *J Chem Educ*. 2011;89:125–129.
89. Tentori AM, Hughes AJ, Herr AE. Microchamber integration unifies distinct separation modes for two-dimensional electrophoresis. *Anal Chem*. 2013;85:4538–4545.
90. Vlassakis J, Herr AE. Joule heating-induced dispersion in open microfluidic electrophoretic cytometry. *Anal Chem*. 2017;89:12787–12796.
91. Pan Q, Yamauchi KA, Herr AE. Controlling dispersion during single-cell polyacrylamide-gel electrophoresis in open microfluidic devices. *Anal Chem*. 2018;90:13419–13426.
92. Karns K, Vogan JM, Qin Q, et al. Microfluidic screening of electrophoretic mobility shifts elucidates riboswitch binding function. *J Am Chem Soc*. 2013;135:3136–3143.
93. Duncombe TA, Herr AE. Photopatterned free-standing poly-acrylamide gels for microfluidic protein electrophoresis. *Lab Chip*. 2013;13:2115–2123.
94. Xia H, Murray K, Soper S, Feng J. Ultra-sensitive affinity chromatography on avidin-functionalized PMMA microchip for low abundant post-translational modified protein enrichment. *Biomed Microdevices*. 2012;14:67–81. https://doi.org/10.1007/s10544-011-9586-7.

95. Sarkar A, Hou HW, Mahan AE, Han J, Alter G. Multiplexed affinity-based separation of proteins and cells using inertial microfluidics. *Sci Rep*. 2016;6:23589. https://doi.org/10.1038/srep23589.
96. Peyman SA, Iles A, Pamme N. Rapid on-chip multi-step (bio)chemical procedures in continuous flow—manoeuvring particles through co-laminar reagent streams. *Chem Commun*. 2008;10:1220–1222.
97. Hu R, Feng X, Chen P, et al. Rapid, highly efficient extraction and purification of membrane proteins using a microfluidic continuous-flow based aqueous two-phase system. *J Chromatogr A*. 2010;1218:171–177. https://doi.org/10.1016/j.chroma.2010.10.090.
98. Meagher RJ, Light YK, Singh AK. Rapid, continuous purification of proteins in a microfluidic device using genetically-engineered partition tags. *Lab Chip*. 2008;8:527–532. https://doi.org/10.1039/B716462A.
99. Williams SKR, Lee D. Field-flow fractionation of proteins, polysaccharides, synthetic polymers, and supramolecular assemblies. *J Sep Sci*. 2006;29:1720–1732. https://doi.org/10.1002/jssc.200600151.
100. Whitesides GM. The origins and the future of microfluidics. *Nature*. 2006;442:368–373.
101. Napoli M, Eijkel JCT, Pennathur S. Nanofluidic technology for biomolecule applications: a critical review. *Lab Chip*. 2010;10:957–985.
102. Pennathur S, Santiago JG. Electrokinetic transport in nanochannels—1 theory. *Anal Chem*. 2005;77:6772–6781.
103. Gillespie D, et al. A physical mechanism for large-ion selectivity of ion channels. *Phys Chem Chem Phys*. 2002;4(19):4763–4769.
104. Hansen PL, Podgornik R, Parsegian VA. Osmotic properties of DNA: critical evaluation of counterion condensation theory. *Phys Rev E Stat Nonlin Soft Matter Phys*. 2001;64(2):021907.
105. Eisenburg RS. *Living Transistors: A Physicist's View of Ion Channels*; 2008:24. available on http://arxiv.org/ as q-bio/0506016v2.
106. Pricl S, et al. Multiscale modeling of protein transport in silicon membrane nanochannels. Part 1. Derivation of molecular parameters from computer simulations. *Biomed Microdevices*. 2006;8(4):277–290.
107. Amato F. Multiscale modeling of protein transport in silicon membrane nanochannels. Part 2. From molecular parameters to a predictive continuum diffusion model. *Biomed Microdevices*. 2006;8(4):291–298.
108. McGuffee SR, Elcock AH. Atomically detailed simulations of concentrated protein solutions: the effects of salt, pH, point mutations, and protein concentration in simulations of 1000-molecule systems. *J Am Chem Soc*. 2006;128(37):12098–12110.
109. Nelissen K, Misko VR, Peeters FM. Single-file diffusion of interacting particles in a one-dimensional channel. *Europhys Lett*. 2007;80(5):56004.
110. Martin F, et al. Tailoring width of microfabricated nanochannels to solute size can be used to control diffusion kinetics. *J Control Release*. 2005;102(1):123–133.
111. Durand NFY, Saveriades E, Renaud P. Detecting proteins complex formation using steady-state diffusion in a nanochannel. *Anal Bioanal Chem*. 2009;394(2):421–425.
112. Durand NFY, Renaud P. Label-free determination of protein-surface interaction kinetics by ionic conductance inside a nanochannel. *Lab Chip*. 2009;9(2):319–324.
113. Zeng H, Glawdel T, Ren CL. Microchip with an open tubular immobilized ph gradient for UV whole column imaging detection. *Electrophoresis*. 2015;36:2542–2545. https://doi.org/10.1002/elps.201500041.
114. Česla P, Fischer J, Jandera P. Separation of phenolic acids and flavone natural antioxidants by two-dimensional method combining liquid chromatography and micellar electrokinetic capillary chromatography. *Electrophoresis*. 2010;31:2200–2210. https://doi.org/10.1002/elps.200900689.

115. Česla P, Fischer J, Jandera P. Improvement of the sensitivity of 2D LC-MEKC separation of phenolic acids and flavonoids natural antioxidants using the online pre-concentration step. *Electrophoresis*. 2012;33:2464–2473. https://doi.org/10.1002/elps.201100594.
116. Porebski PA, Lynen F. Combining liquid chromatography with multiplexed capillary gel electrophoresis for offline comprehensive analysis of complex oligonucleotide samples. *J Chromatogr A*. 2014;1336:87–93. https://doi.org/10.1016/j.chroma.2014.02.007.
117. Wu Y, Wang Y, Gu X, Zhang L, Yan C. Two-dimensional strong cation-exchange liquid chromatography/reversed-phase pressurized capillary electrochromatography for separation of complex samples. *J Sep Sci*. 2011;34:1027–1034. https://doi.org/10.1002/jssc.201000815.
118. Hempe JM, McGehee AM, Chalew SA. Two-dimensional analysis of glycated hemoglobin heterogeneity in pediatric type 1 diabetes patients. *Anal Biochem*. 2013;442:205–212. https://doi.org/10.1016/j.ab.2013.07.016.
119. Newsome TE, Olesik SV. Planar electrochromatography using an electrospun polymer nanofiber layer. *Anal Chem*. 2014;86:10961–10969. https://doi.org/10.1021/ac503568a.
120. Niu XZ, Zhang B, Marszalek RT, et al. Droplet-based compartmentalization of chemically separated components in two-dimensional separations. *Chem Commun*. 2009;41:6159–6161. https://doi.org/10.1039/B918100H.
121. Ye L, Wang X, Han J, et al. Two dimensional separations of human urinary protein digest using a droplet-interfaced platform. *Anal Chim Acta*. 2015;863:86–94. https://doi.org/10.1016/j.aca.2015.01.006.
122. Mao Q, Pawliszyn J. Demonstration of isoelectric focusing on an etched quartz chip with UV absorption imaging detection. *Analyst*. 1999;124:637–641. https://doi.org/10.1039/A809756I.
123. Huang T, Ertl P, Wu X-Z, Mikkelsen S, Pawliszyn J. Microfabrication of microfluidic cartridge for isoelectric focusing by screen printing. *Sensor Mater*. 2002;14:141–150. https://myukk.org/SM2017/sm_pdf/SM477.pdf.
124. Vlčková M, Kalman F, Schwarz MA. Pharmaceutical applications of isoelectric focusing on microchip with imaged UV detection. *J Chromatogr A*. 2008;1181:145–152. https://doi.org/10.1016/j.chroma.2007.12.046.
125. Liu Z, Ou J, Samy R, et al. Side-by-side comparison of disposable microchips with commercial capillary cartridges for application in capillary isoelectric focusing with whole column imaging detection. *Lab Chip*. 2008;8:1738–1741. https://doi.org/10.1039/B807646D.
126. Ou J, Glawdel T, Samy R, et al. Integration of dialysis membranes into a poly (dimethylsiloxane) microfluidic chip for isoelectric focusing of proteins using whole-channel imaging detection. *Anal Chem*. 2008;80:7401–7407. https://doi.org/10.1021/ac8010928.
127. Huang T, Pawliszyn J. Microfabrication of a tapered channel for isoelectric focusing with thermally generated pH gradient. *Electrophoresis*. 2002;23:3504–3510. https://doi.org/10.1002/1522-2683(200210)23:20<3504::AID-ELPS3504>3.0.CO;2-X.
128. Björkman H, Ericson C, Hjertén S, Hjort K. Diamond microchips for fast chromatography of proteins. *Sens Actuators B*. 2001;79:71–77. https://doi.org/10.1016/S0925-4005(01)00837-1.
129. Pace CN, Vajdos F, Fee L, Grimsley G, Gray T. How to measure and predict the molar absorption coefficient of a protein. *Protein Sci*. 1995;4:2411–2423.
130. Tsai S-W, Loughran M, Hiratsuka A, Yano K, Karube I. Application of plasma-polymerized films for isoelectric focusing of proteins in a capillary electrophoresis chip. *Analyst*. 2003;128(3):237–244.

131. Dauriac V, Descroix S, Chen Y, Peltre G, Sénéchal H. Isoelectric focusing in an ordered micropillar array. *Electrophoresis*. 2008;29(14):2945–2952. https://doi.org/10.1002/elps.200800052.
132. Zilberstein GV, Baskin EM, Bukshpan S. Parallel processing in the isoelectric focusing chip. *Electrophoresis*. 2003;24(21):3735–3744. https://doi.org/10.1002/elps.200305663.
133. Zilberstein G, Korol L, Bukshpan S, Baskin E. Parallel isoelectric focusing chip. *Proteomics*. 2004;4:2533–2540. https://doi.org/10.1002/pmic.200300794.
134. Zilberstein GV, Baskin EM, Bukshpan S, Korol LE. Parallel isoelectric focusing II. *Electrophoresis*. 2004;25:3643–3651. https://doi.org/10.1002/elps.200406117.
135. Rodríguez-Ruiz I, Conejero-Muriel M, Ackermann TN, Gavira JA, Llobera A. A multiple path photonic lab on a chip for parallel protein concentration measurements. *Lab Chip*. 2015;15:1133–1139. https://doi.org/10.1039/C4LC01332H.
136. Araya-Farias M, Dziomba S, Carbonnier B, et al. A lab-on-a-chip for monolith-based pre-concentration and electrophoresis separation of phosphopeptides. *Analyst*. 2017;142:485–494. https://doi.org/10.1039/C6AN02324J.
137. Hofmann O, Che D, Cruickshank KA, Müller UR. Adaptation of capillary isoelectric focusing to microchannels on a glass chip. *Anal Chem*. 1999;71:678–686. https://doi.org/10.1021/ac9806660.
138. Tan W, Fan ZH, Qiu CX, Ricco AJ, Gibbons I. Miniaturized capillary isoelectric focusing in plastic microfluidic devices. *Electrophoresis*. 2002;23:3638–3645. https://doi.org/10.1002/1522-2683(200210)23:20<3638::AID-ELPS3638>3.0.CO;2-Y.
139. Guillo C, Karlinsey JM, Landers JP. On-chip pumping for pressure mobilization of the focused zones following microchip isoelectric focusing. *Lab Chip*. 2007;7:112–118. https://doi.org/10.1039/B609620D.
140. Cong Y, Zhang L, Tao D, Liang Y, Zhang W, Zhang Y. Miniaturized two-dimensional capillary electrophoresis on a microchip for analysis of the tryptic digest of proteins. *J Sep Sci*. 2008;31:588–594. https://doi.org/10.1002/jssc.200700444.
141. Zhang J, Das C, Fan ZH. Dynamic coating for protein separation in cyclic olefin copolymer microfluidic devices. *Microfluid Nanofluid*. 2008;5:327–335. https://doi.org/10.1007/s10404-007-0253-5.
142. Wang YC, Choi MH, Han J. Two-dimensional protein separation with advanced sample and buffer isolation using microfluidic valves. *Anal Chem*. 2004;76:4426–4431. https://doi.org/10.1021/ac0497499.
143. Xia L, Dutta D. High efficiency hydrodynamic chromatography in micro-and sub-micrometer deep channels using an on-chip pressure-generation unit. *Anal Chim Acta*. 2017;950:192–198. https://doi.org/10.1016/j.aca.2016.11.014.
144. MacConnell AB, Price AK, Paegel BM. An integrated microfluidic processor for DNA-encoded combinatorial library functional screening. *ACS Comb Sci*. 2017;19:181–192. https://doi.org/10.1021/acscombsci.6b00192.
145. Huang W, Chang CL, Brault ND, et al. Separation and dual detection of prostate cancer cells and protein biomarkers using a microchip device. *Lab Chip*. 2017;17:415–428. https://doi.org/10.1039/C6LC01279E.
146. Chabinyc ML, Chiu DT, McDonald JC, et al. An integrated fluorescence detection system in poly (dimethylsiloxane) for microfluidic applications. *Anal Chem*. 2001;73:4491–4498.
147. Uchiyama K, Xu W, Qiu J, Hobo T. Polyester microchannel chip for electrophoresis–incorporation of a blue LED as light source. *Fresenius J Anal Chem*. 2001;371:209–211. https://doi.org/10.1007/s002160100990.
148. Chin CD, Laksanasopin T, Cheung YK, et al. Microfluidics-based diagnostics of infectious diseases in the developing world. *Nat Med*. 2011;17:1015–1019. https://doi.org/10.1038/nm.2408.

149. Turgeon RT, Bowser MT. Improving sensitivity in micro-free flow electrophoresis using signal averaging. *Electrophoresis*. 2009;30:1342–1348. https://doi.org/10.1002/elps.200800497.
150. Huang X, Ren J. On-line chemiluminescence detection for isoelectric focusing of heme proteins on microchips. *Electrophoresis*. 2005;26:3595–3601. https://doi.org/10.1002/elps.200500076.
151. Liu YM, Cheng JK. Ultrasensitive chemiluminescence detection in capillary electrophoresis. *J Chromatogr A*. 2002;959:1–13. https://doi.org/10.1016/S0021-9673(02)00434-X.
152. Tsukagoshi K, Nakamura T, Nakajima R. Batch-type chemiluminescence detection cell for sensitization and simplification of capillary electrophoresis. *Anal Chem*. 2002;74:4109–4116. https://doi.org/10.1021/ac020171h.
153. Ren J, Huang X. Sensitive and universal indirect chemiluminescence detection for capillary electrophoresis of cations using cobalt (II) as a probe ion. *Anal Chem*. 2001;73:2663–2668. https://doi.org/10.1021/ac001414d.
154. Gao J, Xu J, Locascio LE, Lee CS. Integrated microfluidic system enabling protein digestion, peptide separation, and protein identification. *Anal Chem*. 2001;73:2648–2655. https://doi.org/10.1021/ac001126h.
155. Wen J, Lin Y, Xiang F, Matson DW, Udseth HR, Smith RD. Microfabricated isoelectric focusing device for direct electrospray ionization-mass spectrometry. *Electrophoresis*. 2000;21:191–197.
156. Mok ML-S, Hua L, Phua JB-C, Wee MK-T, Sze NS-K. Capillary isoelectric focusing in pseudo-closed channel coupled to matrix assisted laser desorption/ionization mass spectrometry for protein analysis. *Analyst*. 2004;129:109–110. https://doi.org/10.1039/B309610F.
157. Fujita M, Hattori W, Sano T, et al. High-throughput and high-resolution two dimensional mapping of pI and m/z using a microchip in a matrix-assisted laser desorption/ionization time-of-flight mass spectrometer. *J Chromatogr A*. 2006;1111:200–205. https://doi.org/10.1016/j.chroma.2005.11.117.
158. Svedberg M, Pettersson A, Nilsson S, et al. Sheathless electrospray from polymer microchips. *Anal Chem*. 2003;75:3934–3940. https://doi.org/10.1021/ac030045t.
159. Cheng H, Liu J, Xu Z, Wang Y, Ye M. Improving sensitivity for microchip electrophoresis interfaced with inductively coupled plasma mass spectrometry using parallel multichannel separation. *J Chromatogr A*. 2016;1461:198–204. https://doi.org/10.1016/j.chroma.2016.07.075.
160. Liu RH, Yang J, Lenigk R, Bonanno J, Grodzinski P. Self-contained, fully integrated biochip for sample preparation, polymerase chain reaction amplification, and DNA microarray detection. *Anal Chem*. 2004;76:1824–1831.
161. Martin RS, Gawron AJ, Lunte SM. Dual-electrode electrochemical detection for poly(dimethylsiloxane)-fabricated capillary electrophoresis microchips. *Anal Chem*. 2000;72:3196–3202. https://doi.org/10.1021/ac000160t.
162. Wang J, Pumera M, Chatrathi MP, et al. Towards disposable lab-on-a-chip: poly(methylmethacrylate) microchip electrophoresis device with electrochemical detection. *Electrophoresis*. 2002;23:596–601.
163. Zeng Y, Chen H, Pang D-W, Wang Z-L, Cheng J-K. Microchip capillary electrophoresis with electrochemical detection. *Anal Chem*. 2002;74:2441–2445. https://doi.org/10.1002/1522-2683(200202)23:4 < 596::AID-ELPS596 > 3.0.CO;2-C.
164. Fanguy JC, Henry CS. The analysis of uric acid in urine using microchip capillary electrophoresis with electrochemical detection. *Electrophoresis*. 2002;23:767–773. https://doi.org/10.1002/1522-2683(200203)23:5<767::AID-ELPS767>3.0.CO;2-8.

165. Ranjbar L, Gaudry AJ, Breadmore MC, Shellie RA. Online comprehensive two-dimensional ion chromatography × capillary electrophoresis. *Anal Chem*. 2015;87: 8673–8678. https://doi.org/10.1021/acs.analchem.5b01130.
166. Galloway M, Stryjewski W, Henry A, et al. Contact conductivity detection in poly(methyl methacylate)-based microfluidic devices for analysis of mono- and poly-anionic molecules. *Anal Chem*. 2002;74:2407–2415. https://doi.org/10.1021/ac011058e.
167. Žúborová M, Demianová Z, Kaniansky D, Masár M, Stanislawski B. Zone electrophoresis of proteins on a poly (methyl methacrylate) chip with conductivity detection. *J Chromatogr A*. 2003;990:179–188. https://doi.org/10.1016/S0021-9673(02)01805-8.
168. Edwards T, Gale BK, Frazier AB. Microscale purification systems for biological sample preparation. *Biomed Microdevices*. 2001;3:211–218. https://doi.org/10.1023/A:1011459213759.
169. Suni II. Impedance methods for electrochemical sensors using nanomaterials. *TrAC Trends Anal Chem*. 2008;27:604–611. https://doi.org/10.1016/j.trac.2008.03.012.
170. Huber DL, Manginell RP, Samara MA, Kim B-I, Bunker BC. Programmed adsorption and release of proteins in a microfluidic device. *Science*. 2003;301:352–354. https://doi.org/10.1126/science.1080759.
171. Stevenson PR, Dunlap BE, Powell PS, et al. Simultaneous chromatography and electrophoresis: two-dimensional planar separations. *Anal Bioanal Chem*. 2013;405: 3085–3089. https://doi.org/10.1007/s00216-013-6737-0.
172. Liu J, Chen C-F, Yang S, Chang C-C, DeVoe DL. Mixed-mode electrokinetic and chromatographic peptide separations in a microvalve-integrated polymer chip. *Lab Chip*. 2010;10:2122–2129. https://doi.org/10.1039/C003505J.
173. Fritzsche S, Hoffmann P, Belder D. Chip electrophoresis with mass spectrometric detection in record speed. *Lab Chip*. 2010;10:1227–1230. https://doi.org/10.1039/C000349B.
174. Andersen T, Skuland IL, Holm A, Trones R, Greibrokk T. Temperature-programmed packed capillary liquid chromatography coupled to evaporative light-scattering detection and electrospray ionization time-of-flight mass spectrometry for characterization of high-molecular-mass hindered amine light stabilizers. *J Chromatogr A*. 2004;1029:49–56. https://doi.org/10.1016/j.chroma.2003.12.043.
175. Hoffmann S, Norli H, Greibrokk T. Improved signal to noise ratios of the laser light-scattering detector with packed fused silica columns. *J Sep Sci*. 1989;12:260–264. https://doi.org/10.1002/jhrc.1240120417.
176. Weddemann A, Albon C, Auge A, et al. How to design magneto-based total analysis systems for biomedical applications. *Biosens Bioelectron*. 2010;26:1152–1163. https://doi.org/10.1016/j.bios.2010.06.031.
177. Lee SHS, Hatton TA, Khan SA. Microfluidic continuous magnetophoretic protein separation using nanoparticle aggregates. *Microfluid Nanofluid*. 2011;11:429–438. https://doi.org/10.1007/s10404-011-0808-3.
178. van der Geer P, Hunter T. Phosphopeptide mapping and phosphoamino acid analysis by electrophoresis and chromatography on thin-layer cellulose plates. *Electrophoresis*. 1994;15:544–554. https://doi.org/10.1002/elps.1150150173.

CHAPTER FOUR

Micro/nanofluidic devices for DNA/RNA detection and separation

Nidhi Verma[a], Sakshi Walia[b], and Alok Pandya[a,*]

[a]Department of Engineering and Physical Sciences, Institute of Advanced Research, Gandhinagar, Gujarat, India
[b]Department of Biological Sciences and Biotechnology, Institute of Advanced Research, Gandhinagar, India
*Corresponding author: e-mail address: alok.pandya@iar.ac.in

Contents

1. Introduction 86
2. Separation of nucleic acid using NF and MF 87
   2.1 Entropic based separation of nucleic acid 87
   2.2 Reputation based separation of nucleic acid (electrophoresis) 90
   2.3 Electro kinetic based separation of DNA and RNA 92
   2.4 Ratchet mechanism in separation of nucleic acid 93
3. Detection of nucleic acid using NF and MF using different fabrication techniques 94
   3.1 Polydimethylsiloxane (PDMS) based polymer material in detection of nucleic acid 96
   3.2 Glass based inorganic material in detection of nucleic acid 97
   3.3 Silica based inorganic substrate in detection of nucleic acid 98
   3.4 Paper based material in detection of nucleic acid 99
4. Future prospects 103
Acknowledgments 104
References 104

Abstract

The development and research have ramped up at a greater speed than ever in the field of diseases diagnosis. Still there is struggle in developing early detection techniques which uses complex biomolecules like RNA, DNA and proteins in order to detect diseases caused by bacteria, viruses or fungi. Until now separation techniques used before detection rely on traditional techniques like electrophoresis etc. which often require centralized services. Although efforts are made in developing devices that is capable enough on carrying out separation and detection based on microfluidic (MF) and nanofluidic (NF) or lab on chip. Hence, in this chapter, we have discussed about the advancement, limitations and future steps that needs to be taken to flourish the field of NF and MF for the detection and separation of nucleic acid.

Progress in Molecular Biology and Translational Science, Volume 186
ISSN 1877-1173
https://doi.org/10.1016/bs.pmbts.2021.07.009

1. Introduction

Microfluidics (MF) and Nanofluidic (NF) devices deals with transport of fluids and ionic species at micrometer and nanometer scale. It is used to extensively focus, trap, concentrate and separate particles of all type including DNA, RNA and proteins depending on sizes. Microfluidic and nanofluidic devices are even used constant water analysis to analysis the presence of a specific ions. MF and NF has come a long way starting from as nanofabrication of the device to making available point of care devices.

Disease diagnosis falls among one of the fields of focus pertaining to use of MF and NF. Traditionally, bioanalytical analysis related to diagnosis was done using high end instruments which are not only costly but also laborious and required trained personnel. In order to carry out biomolecular studies, it becomes imperative to separate the biomolecules with great efficiency and accuracy as biological samples are complex, low sample volume and tendency to degrade overtime before analysis could occur. Hence, it is desirable to develop devices that could provide quick, easy and accurate separation at the earliest. Another milestone to reach is developing low-cost separation techniques for biomolecules like DNA, RNA and proteins which are governed by multiple parameters like pH, temperature, size, concentration etc. Alongside affordable separation, miniaturization of the separation techniques is need of the hour as that would tandemly help in decrease of cost without comprising on the accuracy.

Taking advantage of inculcating MF and NF can help as they can easily overcome such problems as they are miniature devices with high efficiency and low cost as shown in Fig. 1. Although MF and NF having a large advantage, it does have few drawbacks like non-specific adsorption and precise injection of such small analyte amounts still remain unsolved for many applications.

This chapter provides information about the separation application under the influence of nano and micro environment based on the choice of analyte. The environment for the separation is greatly affected by the nanofluidic and microfluidics substrate chemistry. Here in also, we describe various substrate which have been used over the year for providing a more robust detection technique for biomolecule including of PDMS based material, glass as a substrate, silica, paper based microfluidic and nanofluidic devices.

Later, we conclude with a summary of advantages, limitations and future steps that needs to be taken to flourish the field of NF and MF.

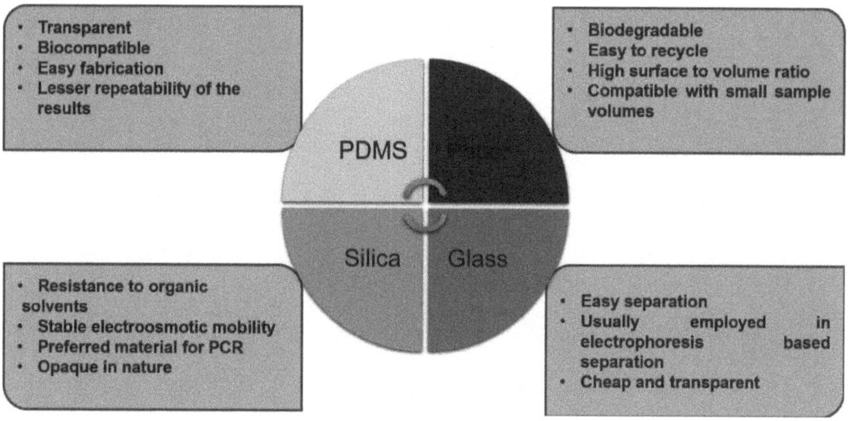

Fig. 1 Varied materials employed in nanofluidic and microfluidic devices of RNA/DNA.

2. Separation of nucleic acid using NF and MF

Nucleic acid act as an important biomolecule which is been routinely used in detection of diseases making it important for developing easy and point of care separation techniques so it can be easily used in routine testing without rigorous procedure. In order to achieve that integration of NF and MF alongside nucleic acid is one of the most pivotal steps which can be taken to achieve easy, point of care separation. Nucleic acids could be categorized as RNAs, plasmids and chromosomal. The procedure for separation of both DNA and RNA is usually overlapping even after structural variations with minor modification Fig. 2). NF and MF can help in achieving better separation efficiency of nucleic acid. Herein, we describe a few NF and MF based separation techniques for nucleic acid.

2.1 Entropic based separation of nucleic acid

The concept of entropic trapping can be dated back to 1999 when it was done using computer stimulation. Later, it was proposed that entropic traps could be integrated into an electrophoresis for better studying and understanding the entropic principle and help in separation of the DNA or RNA. It was than proposed that entropic traps could be studied extensively and more appropriately use of microfluidics. There have been many attempts ever since to make use of entropic traps in separation of nucleic acids by integrating them into nanofluidic and microfluidics.

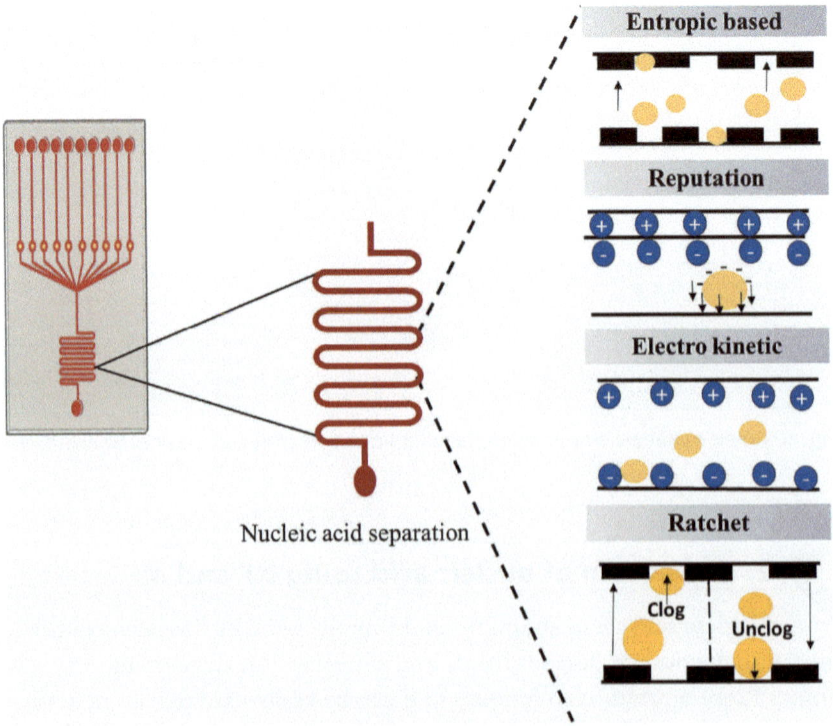

Fig. 2 Various techniques in microfluidic and nanofluidic used for separation of nucleic acid.

Silicon based lithography and etching techniques has now been included in entropic trapping of long DNA using an artificial channel. The channel consisted of thin and thick regions, having a thickness of 90 nm as a function of the applied electric field, the electrophoretic mobility of long DNA molecules was studied. The trapping of the DNA was attributed to the radius of gyration of the DNA used been larger on contrary to thin gap of the system on movement from thick to thin region leading to the trapping of the DNA molecules. It was elucidated from the study that the larger DNA moved faster as a result of larger contact area with the slit as compared to small DNA molecules enabling fast manipulation and separation of long polymers.[1] The behavior of DNA was detected in another nanofluidic device near the interface of two regions that produce entropies with different configuration. An electric field drives the molecules part way across the two regions. When the field was removed, it was observed that the molecules recoil to the higher-entropy region with a profile characteristic of a force localization at the interface and was length independent. This was in consistency with a confinement-mediated entropic force, distinct from entropic elasticity common to all polymers. This of great importance when phenomenon can be exploited to separation of long-strand polyelectrolytes in accordance to length in needed.[2] Another nanofluidic channel device, designed consisted of several entropic traps, for the separation of long DNA molecules. The proposed method resulted in efficient separation even in absence of gel matrix or pulsed electric fields since the channel enabled size-dependent confining of DNA at the onset of a constriction. The entropic trap caused an electrophoretic mobility differences among the DNA molecules resulting in effective separation of DNA molecule with size 5000–160,000 base pairs in 15-mm-long channels.[3] An alternative multistage fluidic device showed presence of dense array of nanopillars for separation of DNA molecules which acted as an entropically unfavorable region (an entropically unfavorable region). Here, it made use of a pulsed electric field molecules of unequal lengths in which both T2 coliphage DNA and T7 coliphage DNA were inserted in a way that shorter T7 molecules fully inserted into the region whereas longer T2 molecules to recoiled out of the unfavorable region. This method can be particularly important in case of unknown samples. The multistage device had separation time and extraction resolution as same as conventional separation techniques.[4] Array of microscopic sized entropic traps having both deep regions and narrow constrictions were used to study the mobility of DNA by using Brownian dynamics simulations. It was found that mobility was increased with increase in length of DNA. It was also observed that few chains diffuse out of the path and diffuse at the corner of the box and remains trapped. The trapping increases with the diffusion constant.[5] Fluorescent video microscopy was also used to measure the movement of group of DNA molecules as the mobility and electric field strength showed a nonlinear relationship as DNA molecule constrict in order to crosswise the boundary in the middle of the deep and shallow region fluctuations was observed in each case even though identical conditions were provided. Various variable was studied including fluorescent dye staining ratio DNA concentration, electroosmotic flow, ionic strength of the solvent and electric field strength to find the reason for fluctuation. It was later observed that mobility fluctuations were relatively improved in conditions of decreased electroosmotic flow and ionic strength.[6]

Development of a single staged chip-based system using entropic trapping in order to integrate it with long-read genomics technologies to concentrate and purify sample under applied pressure that will help in for sample injection. The electric field was applied in order to achieve better concentration and filtration of the sample. Weak entropic helped short DNA tends to pass through the filter. The single-stage prototype developed functions in a continuous pulsatile manner, achieving selectivity of 3.5 in case of λ-phage DNA (48.5 kilobase pairs) compared to a 2 kilobase pair standard based on data generated by the experiment for the fraction using pure samples of each species. The single stage device is fabricated in fused silica operating in a continuous pulsatile manner having a selectivity of about 3.5 for λ-phage DNA (48.5 kilobase pairs).[7] A gravity separation method was developed for entropic transport that resulted in enhanced geometric confinement and noise typically in case of narrow channel microfluidic device. Through tuning the geometry parameters or the amplitude and geometry parameters were tuned so that it can be modified confined in a nano structure to have an effective particle separation in accordance to their weights.[8] Theoretical models usually employed for explaining separation mechanisms are non-cross-linked polymer solutions, thermo responsive (viscosity-adjustable) polymer solutions and novel entropic trapping networks. In a typical micro-channels, a thermo responsive polymer solution is loaded at one temperature at lower viscosities and can be employed in separation at another. The separations based on entropic mechanism make use of arrays of obstacles which are placed at regular intervals acting as size-separations. These progresses have ability to integrate this into potentially into an microfluidic chip and automated capillary systems aiding in enhanced separation using microchannels, lower DNA separation time as well as higher matrix degradation.[9] Entropic trapping effect separates molecules based on size and have capability to even separate small sized molecules such as DNA or protein polymer using alternate thin and thick regions. The thick region is significantly larger than the molecule to be separated, while in case of thin region the than the size of the molecules is smaller to be separated. In addition, entropic traps are employed to accumulate, block and regulate molecules in the nanofluidic and microfluidic channel.[10]

2.2 Reputation based separation of nucleic acid (electrophoresis)

Electrophoresis is employed for separation of molecules for a long decade. It makes use of electric field in order to separate different molecules based on size. In solution, the migration of a molecule is highly influenced by the applied electric field and gives separation based upon its charge and size. Modification of the separation media can highly alter the separation achieved of the molecules.

Focus has now shifted on developing portable electrophoresis devices which are miniature in nature having lesser sample requirements and are portable. In this era, fluidics can act as a discrete device in order to develop microfluidic and nanofluidic based electrophoresis. In order to make use of MF and NF screening, different methods and materials can be employed.

Electrophoretic based separation is used over other methods of separation owing to its advantageous properties like low cost, low sample intake, rapid assessment, integration capability and high throughput. Microfluidic device make use of electrophoretic technologies like capillary based electrophoresis, gel based electrophoresis, dielectrophoresis or field (electric) flow fractionation.[11] A complete on chip detection including of preconcentration followed by separation, and detection for tracing of DNA. The proposed microchip consisted of in total three channels each having characteristic function. The first two channels used for the field-amplified sample stacking and subsequent for field-amplified sampled injection steps, while the third one is for the microchip gel electrophoresis (MGE) with electrochemical detection (ED) (MGE-ED). To advance the preconcentration and separation operations of the system, modification was done in both the stacking buffers and separation buffers using hydroxypropyl cellulose (HPC) matrix tailored with gold nanoparticles (AuNPs). The total sensitivity when compared with a traditional MGE-ED analysis was enhanced by ~25,000-fold.[12] Out of the various factor influencing the performance of microfluidics for separation and sample pre-concentration, electric field falls among on the most intriguing factor to be considered. Simulations were carried out to study the non-uniform collisions which are necessary to help in separation it stated that electric field gradient greatly influences the separation technique and DNA transport in microfluidic devices.[13] Another factor governing the separation and sample pre concentration is the design of the microfluidic device used for detection. It was studied by making use of electrophoretic based separation showing that both straight and convergent microchannels based pre concentration. It was observed that a preconcentration factor of 40 times was accomplished in the case of straight microchannel. Whereas the preconcentration factor improves to 50 times while use of convergent channel. The same was confirmed by carrying out studies using fluorescein isothiocyanate labeled bovine serum albumin (FITC-BSA) and tetramethyl rhodamine sample.[14] Achieving improved electrophoretic DNA separations was proposed by Chen and Dorfman in which titled electric filed was applied in regards to the lattice vectors of the array stating that a better separation was obtained at higher fields. A 20 kbp DNA baseline resolution was achieved and λ DNA (48.5 kbp) in a 4 mm channel, 5 min were required to reach a resolution of unity and had improvement over untitled post arrays.[15] An electrophoretic DNA focusing method employed resulted in 2700 folds' increase in nucleic acid enrichment concentration five individual electrodes system was developed and electrical waveform was applied such that DNAs moved sequentially to the collection electrodes at great speed. The device consists of two-stages, first stage was for extraction of DNAs with a capacity of 1 mL sample volume and second stage to enhance its concentration by 2790-fold. Efficiency of 99.7% was achieved for DNA capture. This novel fluidic platform can be used for DNA extraction, sample pre-concentration and integration for genetic analysis and bio sensing.[16] A unique and novel economical, compact and instant forensic human identification system was developed. The fabrication of a centrifugal microdevices capable of carrying out both reagent loading and DNA separation inspired by the print, cut and laminate (PCL) technique. A multiple polyester-toner fluidic layers, integrated gold leaf electrodes and a cyclic olefin copolymer separation domain the sample was separated with an effective separation length of only 4 cm in under 5 min. Whereas along with reagent loading the separation is completed in <37 min. The electrophoretic separation was confirmed with two base resolution for the STR fragments.[17]

2.3 Electro kinetic based separation of DNA and RNA

Electrokinetic is the study of liquid or particle motion under the action of an electric field. It is based on mechanisms such as electrokinetic instability, electrophoresis, electro kinesis, dielectrophoretic, electroosmosis, electrowetting and electrorotation in order to achieve separation. Electrokinetic separation does have drawback of instabilities caused due to can rapid mixing which can undesirable in case of critical sample separation hence having limits used in MF and NF.

The spatial confinement plays a major role in designing of microfluidic, nanofluidic device or biosensors on molecular behavior for understanding diverse biological phenomena. In Particular, the behavior of molecular binding under an applied electric field changes being a key in the development of electrokinetic biosensors for detection of DNA, RNA or any other biomolecule. In capillary electrophoresis confinement of DNA oligomers influences the binding kinetics is highly influenced and the same was studied using capillary electrophoresis of the DNA. It was concluded based on studies that binding increased with increase in confinement under the effect of the electric field.[18] The electric field gradients are produced by either active microelectrodes or inert micro-insulators, when electrokinetic particle-based separation techniques are employed. It is primarily dependent on dielectrophoresis. A bifurcating microchannel were used for continuous separation of particles based on electrokinetic method. Electrical lift necessary to focus particles towards the center line of the main-branch in the device which is sheath free based separation. Once in the main branch they are deflected based on the size in side-branch electric field-driven sheath less separation. This was in good agreement with the theoretical model making its way in application to achieve increased particle resolution.[19] The electrokinetic based techniques not only helps in detection of biomolecules it also helps in separation and pre concentration in many NF and MF platforms. One such application was seen in microRNAs (miRNAs) detection that act as a disease marker helping in progress of diagnosis and treatment of diseases. Triggerable Mutually Amplified Signal (TMAS) probe was fabricated and supplemented at the center of a microfluidic chip having a better quantitative detection when compared with PCR-based approaches or the hybridization chain reaction. The detection of microRNAs of the biomarker was achieved using surface enhanced Raman scattering (SERS) technology. An enzyme-free target-strand was used to produce double strands initially and displacement recycling reaction initiated by the target miRNA due to generation of superior SERS signal make effective mixing and concentration is achieved using electrokinetic flow technology using alternating current aiding in increased SERS signal intensity and DNA hybridization rate. Detection limit of 2.33 fM was achieved in 30 min for detection of miR-21 in human breast cancer cells.[20]

2.4 Ratchet mechanism in separation of nucleic acid

Ratchet is another mechanism to be explored in separation of biomolecules. Various parameters have been optimized in order to get the maximum output using ratchet mechanism. A triangular (ratchet) shaped obstacle was included in a rachet device for separation of mixture of five different species. The externally applied voltage and time-dependent voltage were tailored exclusively using algorithm. Opposite migration was enforced for selected particles to all the remaining species in a way that species particle simultaneously moves in one direction to other. Simply adapting the voltage protocol can be applied to separate the selected species at a desired time.[21] A non-thermal ratchet where the trajectory of particles in a certain size range is irreversible when the sign of the pressure gradient is reversed at a low Reynolds number. A triangular post is employed in the array rather than circular post array to displaces particles in order to transport them through the array. There is no net displacement of the fluid when the ratchet irreversibly moves particles of a certain size range in a direction orthogonal to an oscillating flow. The asymmetric fluid velocity distribution through the gap between the triangular posts and irreversible particle-post interactions are employed here as diffusion is not employed at all for separation making it available to use in high rates of flow.[22] An oscillatory flow to generate a ratcheting effect to distinguish the path flows of different cell types like circulating tumor cell (CTC), erythrocytes and leukocytes. A successful separation of deformed was achieved to separate CTC from whole blood based on size although the cells are almost similar in size in a label-free manner. The proposed microfluidic device is able to captured more than 90% of cancer cells and achieved 104-fold enrichment of target cells relative to leukocytes. When compared to Cell Search system 25 times higher yield was reported. Oscillatory flow exploiting ratchet mechanism to separate cells was put forward to overcome problems associated with the filtration-based separation. The proposed mechanism had the potential to separate the cells based on deformities and size. The results produced were reproducible although not reversible in nature.[23] A Brownian motor was used contact along with a ratcheted channel which overcome the problems of the existing ratchet which includes of the limit of strong thermal noise. The Brownian motor achieved directed transport against an applied load. The proposed ratcheted electrophoresis can be included in nanoscale machines capable of transport, electrochemically powered and actuation of nanoscale components.[24]

3. Detection of nucleic acid using NF and MF using different fabrication techniques

The traditional detection techniques of nucleic acid (NA) made use of bench-top, macroscale techniques which had drawbacks such as requirement of equipment, large sample sizes, trained personnel, laborious and costly, thus restraining them to be used only in centralized laboratory facilities. Consequently, there is a need for an effective strategy to develop compact point-of-need devices that can be used in limited resource-settings. Thus, the concept of a micro and nanofluidic was introduced to provides substantial development in performance. The key benefits of miniaturizing

NA detection assays that would need minute sample volumes, utilize quantities of reagents, and would be quicker and more sensitive. Additionally, the detection can be completed within a single device and the chance of contamination would be reduced and the system can be pre-set.

An important aspect to consider is the material used, which serve as the main constituent of various microfluidic/nanofluidic devices. It is also referred as *substrates* used to fabricate the devices. Until now, inorganic, polymer and cellulose based substrate are widely used in fabricating microfluidic devices bearing diverse functionalities. Thus, specific substrate for development of microfluidic device must be adequate and present the appropriate properties necessary to fulfill the desired application. A rigorous interplay of engineering and materials science governs microfluidic technologies and must be considered for detection of biomolecules since different substrate brings about difference in detection of the biomolecule as shown in Fig. 3. Here, different substrates are compared in terms of benefits and drawbacks as well as their performance in particular detection.

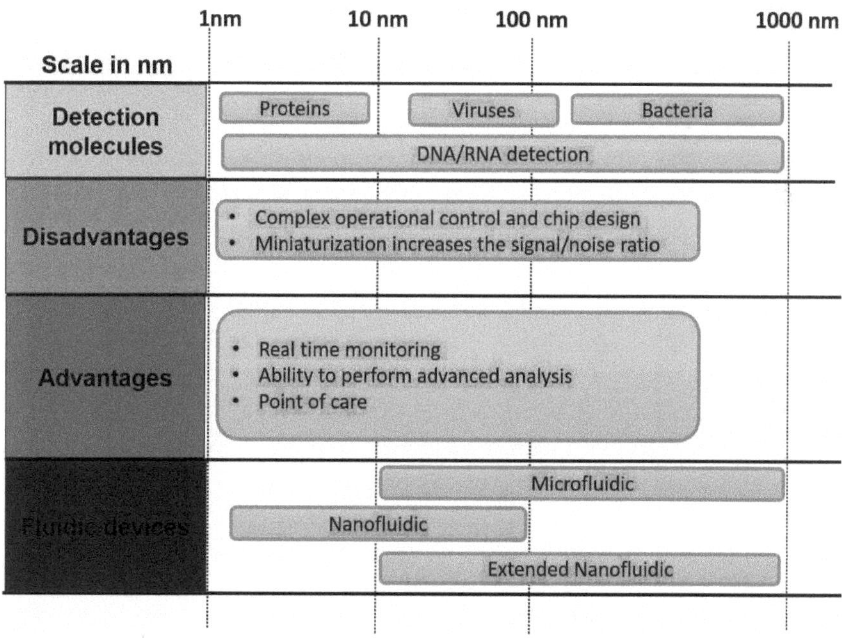

Fig. 3 Representative materials separated based on size in nm on various fluidic devices.

3.1 Polydimethylsiloxane (PDMS) based polymer material in detection of nucleic acid

Dimethicone or polydimethylsiloxane a polymer generally refereed as PDMS is usually a material of choice for varied application in fabrication of microfluidic and nanofluidic devices, as it fulfills one of the discrete requirements of creating inexpensive microfluidic and nanofluidic devices. PDMS have another advantage of developing reproducible devices with easy handling. PDMS is used along with a cross linking agent providing a transparent background making it a material of choice in application where electrode are to be integrated into wearable device, solar cells etc.[25] A cost-effective PDMS chips with controllable channel size and layout method was proposed which could be integrated both single or multiple nanochannels. Artificial defects is used to position the single nanocrack arrays generated initially using solvent-induced method, on a polystyrene surface nanocrack arrays and replicated onto the polymer layers to form the nanochannel molds. The simple soft lithography was employed on the nanochannel molds to create the bi-layer PDMS microchannel–nanochannel to develop an alignment system under an optical microscope. The proposed method ensured development of excellent PDMS nanofluidic chips.[26] PDMS as a substrate is extensively used in developing chips which help in detection of DNA, RNA and even nanoparticles in the field of medicine, chemical and biological applications. Peng and Li reported developing a PDMS based chip which could be used in detection of DNA or nanoparticles consisting of two-layer PDMS slabs in which the microchannels were fabricated using photolithography method. The label free detection was done using Resistive Pulse Sensing technique (RPS). The RPS chip detected nanoparticles of size 20 nm giving a high signal to noise ratio as the nanoparticle passed through the sensing gate of the chip using a single stage differential amplifier. The proposed device was capable enough to distinguish between two different sized nanoparticles with small difference such as 60 and 83 nm. The chip had results comparable to that of dynamic light scattering. The RPS chip was able to detect translocation of dsDNA molecules.[27] Another PDMS based slabs with micro/nanochannels were achieved by transfer process to study the electrical characteristics of KCl solution (concentration of 1 mM) indicating efficient and economical fabrication using PDMS. The PDMS based nanofluidic devices had nanochannels made using nano milling technique comprising of an AFM system along with piezoelectric actuator to control the size.[28] A PDMS microfluidic devices were developed for separation of pre term birth biomarkers. Device consisted of injection, separation channels plus the control layer. An externally actuated on-chip peristaltic pump is affixed on the control layer consisted of whereas sample injection was done using the four pneumatic valves around the T-intersection the fluidic and control layers had PDMS membrane layer sandwiched in between as the actuated component in pumps and valves. Another fabrication was done using soft lithographic techniques the devices had the same set up of the peristaltic pump design but difference in the valve spacings ranging from 100 to 400 μM. Characterization of the developed devices were done utilizing fluorescent imaging of captured plugs which detected using fluorescein-labeled mixture of amino acid and via microchip electrophoresis separations. A peristaltic pump actuation rate (50 ms), an injection time of 30 s and valve spacing of 200-μM was found to be optimum after analysis based on height of the peak, separation efficacy and time of analysis. A pressure-driven and electrokinetic injection were compared and it was seen that pressure driven separation resulted in increase in peak height by 3.9-fold with comparable separation efficiency. The pre-term birth were separated using the developed microchip systems.[29] PDMS solely have been used extensively although it have a few drawbacks like high absorption rate of small hydrophobic molecules, leaching of oligomers, poor chemical compatibility and loss of reaction mixture due to evaporation in case of open reactors.[30] In order to overcome such problem surface modified PDMS is used for microfluidic devices. Gitlin *et al.* reported using concentrated sulfuric acid to surface modify the bulk PDMS. The surface modification drastically reduced the uptake of small organic molecule at low electroosmotic properties. The separation efficiency was indicated to increase on contrary to unmodified PDMS devices. Similarly, in order to achieve effective nucleic acid amplification, Ramalingam and co-workers developed a high-throughput microchip devices. The developed device had ability to overcome loss of the amplification mixture due to evaporation and cross-contamination. The MF device consisted of unsealed reactors along with a single-step capillary-based flow scheme for sequential distribution of amplification mixture into an array of reactors along with heating scheme for the reactors.[31] Although surface modification helped in improved separation but still reported efficiency lower than commercial glass microfluidic devices.[32]

3.2 Glass based inorganic material in detection of nucleic acid

In order to overcome problems faced by using PDMS such as low chemical compatibility, absorption of small hydrophobic molecules and leaching of oligomers using a material which was could possibly overcome the problem using glass as a substrate. Glass has excellent properties like transparency, reusability, easy surface modification and resistance to wide range of solvent. Glass along with being inert is hard in nature making it ideal for real application. Fluorescent labeled DNA oligonucleotides. Glass made nanofluidic device was developed to monitor real time conformational at equilibrium of a DNA hairpin along with observation DNA synthesis constantly. A T-shaped junction of nanochannels using pressure driven flow through array of parallel nanochannels were used. Fluorescently labeled DNA oligonucleotides were analyzed with high throughput using glass-based substrate. The glass based nanofluidic device could act as a robust system in studying the kinetics at molecular level.[33,34] In order to increase the efficiency of using glass a substrate using it in conjugation with other inert material can enhance its property and make it the substrate of choice. Integration of glass substrate with silicon was proposed using low temperature annealing and oxygen plasma activation processes. The proposed method to develop silicon and glass based devices resulted in developing detachable and reusable devices which can be used in varied applications.[35] Another insight in employing glass nanochips were in calculating transport proteins and water molecules using the molecular dynamics. Glass nanochip was developed using ultraviolet lithography and wet chemical etching. Etching time was optimized in order to adjust the channel depth. The system developed could possibly be used and integrated into various fields like DNA, RNA based application using nanofluidic.[36] Glass based nanochip was used in analysis of DNA in terms of translocation, transient intra event fluctuations and the duration of the translocation. The glass chip made use of solid state nanopore platform for detection and showed that characteristic translocation rates are at least 10 times faster than previously recorded.[37] Another device made use of silicon nitride membrane along with glass chip and was used to demonstrate DNA translocation measurements. The device overcome the problems of low bandwidth stemming and high current noise due to extreme capacitance of the membrane chip. The developed glass chip had capacitance below 1 pF.[38] Andresen and co-workers developed an On Chip amplification (On Chip-HDA) for single and multiple amplifications for detection of pathogens like *N. gonorrhoeae* and *S. aureus* using specific primers capable to be used in point of care diagnostics.[39]

3.3 Silica based inorganic substrate in detection of nucleic acid

Silicon based chip detection was developed and integrated into a point of need nucleic acid-based amplification tests (NAT) playing role in identifying infections caused by viruses. As there are limitations of high tech and inadequate clinical settings for detection. The silicon chip had the capability of detecting both DNA and RNA based viruses on the single silicon microchip platform. One-step RT-qPCR and qPCR assays for detection of HPV 16, HIV, HCV, Zika virus and HPV 18 on a benchtop real-time PCR instrument was initially carried out. A silicon microchip was designed consisting of meandering microreactor, integrated aluminum heaters, thermal insulation trenches and microfluidic channels each having a precise function to perform. The overall reaction time was of 25 min with high sensitivity and reproducibility. It was later concluded that the silicon chip detection of DNA and RNA viruses could be integrated into a full-fledged technology consisting of nucleic acid extraction solutions to be used in Point of need setting (Fig. 4).[40] Another application of silicon as a substrate was seen in using it in developing Biofilter devices. One such miniaturized device was developed for extraction of HBV (Hepatitis B Virus) DNA from biological sample. The result was dependent on the pillars geometrical dimensions and increases proportionally with the surface/volume ratio. The genetic non-laboratory-based devices could be developed and combined with real time PCR amplification chip. Similarly, Lee *et al.* developed silicon-based microfluidic device in centrifugation-free extraction of cell-free DNA a biomarker in cancer management as shown in Fig. 5. Under vacuum pressure extraction is completed with 15 min using an immiscible solvent. The developed microfluidic chip have performance comparable to that of QIAamp kit.[41] Use of silicon is not limited to be used as a substrate in microfluidics as it has extensively been used to develop silica-based beads for selective trapping of amplicons for nucleic acid amplification in combination of rolling circle amplification coupled with a thin-film photodiode over an area of 200×200 μm having a fluorescence readout. Optimization was done considering the amplicon solution flow rate and time of trapping. The limit of detection was below 0.5 fM. The device had application in detecting RNA from Ebola viruses as well as influenza virus.[42]

3.4 Paper based material in detection of nucleic acid

Paper-based microfluidic diagnostics originally came to light as low-cost medical diagnostics tool which is accessible to all in across the world. Although cellulose and cellulose based materials were already in use it faced problems like poor detection limit.[43] Microfluidic paper-based analytical tools are referred to as μPADs and have developed tremendously over the

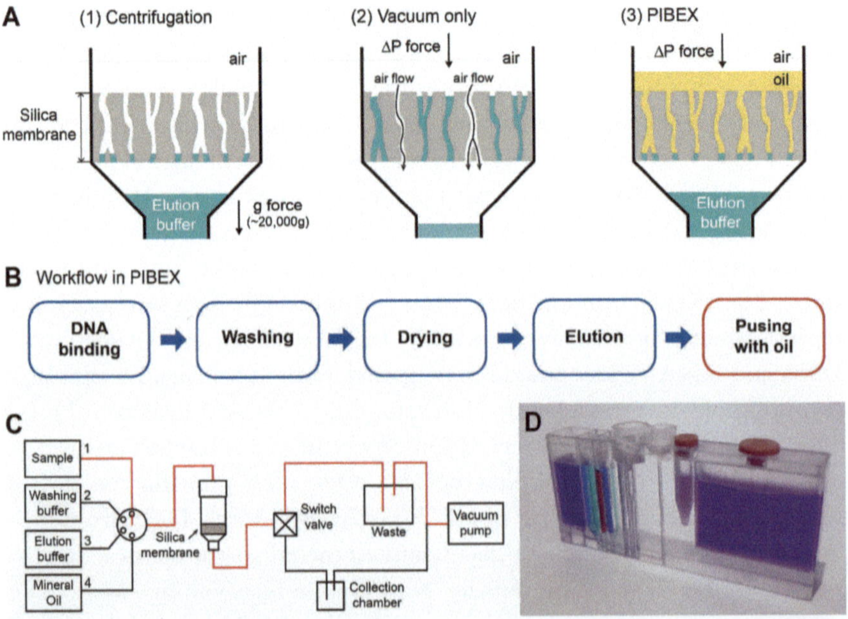

Fig. 4 Integrated Microfluidics (PIBEX) for cfDNA extraction. (A) Comparison of cfDNA extraction methods. The schematic illustrates mechanisms of extracting residual eluents in silica membrane using centrifugation, vacuum, and immiscible fluid and vacuum (PIBEX), respectively. This work is reproduced under CC BY 4.0/partial use and word modifications from the original. (B) Workflow of the proposed microfluidic process of cfDNA extraction. The microfluidic PIBEX method is identical to the conventional spin column method, except for the last additional phase of "pushing with oil," which collects eluents that have remained inside the silica membrane using an immiscible solvent. (C) Schematic of the flow channel at the DNA binding step. The flow pathway is selectively connected from sample chamber to collection chamber, as depicted by a red line. (D) Photograph of an integrated microfluidics for cfDNA extraction into PIBEX chip. *Reprinted from Lee H, Park C, Na W, Park KH, Shin S. Precision cell-free DNA extraction for liquid biopsy by integrated microfluidics.* NPJ Precis Oncol *2020;4(1):1–10.*

last few years. Cellulose and modified cellulose has been used for a long time but faces drawbacks like poor sensitivity and inability to provide quantifiable measurements. μPADs have capability to overcome problems by using paper in order to establish microfluidic channel networks that directs the flow in specific regions of the device without having requirement for use of external force. Furthermore, the devices made from cellulose or modified cellulose are cost effective when compared to traditional materials like glass substrate and PDMS substrate. Other benefit of using paper based material is that they are disposable in nature and does not require any external pumps to carry out

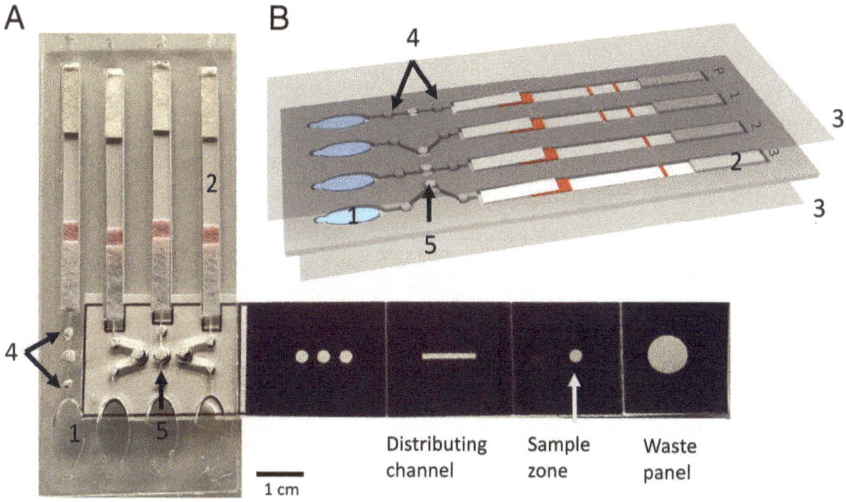

Fig. 5 Paper-based microfluidic device that enables multiplex LAMP-based detection of malaria in blood. (A) An assembled paper–plastic lateral flow diagnostic device, used for amplification and detection. (B) A schematic of the plastic microfluidic device. *Reprinted from Reboud J, Xu G, Garrett A, et al. Paper-based microfluidics for DNA diagnostics of malaria in low resource underserved rural communities.* Proc Natl Acad Sci U S A *2019;116:4834–4842.*

the analysis.[44] One such paper-based microchip was proposed for detection of kidney-based biomarkers like creatinine. The system was established with great ease using hand drawn pencil electrodes in order to perform capacitively coupled conductivity detection. The system had benefits of easy of channel separation, better injection-to-injection repeatability and low cost. In long terms the system has potential to be used in detection of more complex biomolecules like DNA or RNA.[45,46] A more complex biomolecule detection like nucleic acid was carried out by utilizing wax chamber on a cellulose paper. The cellulose microchip had potential to detect and quantify dsDNA by giving a color alter from colorless to purple when nucleic acid is present by using crystal violet dye. The dye has affinity towards DNA and detected *Sus scrofa* (porcine) DNA at a concentration as low as 1 pg/μL (3.43×10^{-1} copies/μL) whereas, Bacillus subtilis (bacteria) DNA at a concentration as low as and 10 pg/μL (2.2×10^{3} copies/μL) respectively. The extracted DNA was amplified employing loop-mediated isothermal amplification (LAMP).[47]

Utilizing paper-based substrate for a range of detection ranging from colorimetric detection to a more advanced approach like integration into an electrode system for detection via electrical sensing. Such approach was utilized in order to detect HIV virus where lab are under developed and detection is of paramount importance. In this work, cellulose paper-based microchip with graphene-modified electrodes was developed for HIV detection with real sample analysis through electrical sensing.[48] Another interesting application of paper in microfluidics in detection of DNA was done by Reboud *et al.* for detection of malaria. The proposed microfluidic technology combines vertical flow along with an isothermal amplification utilizing lateral flow detection mechanism and easy visualization as shown in Fig. 5. The developed DNA-based sensors is of importance in developed and under developed countries where testing facilities are limited.[49]

Alongside various patterning techniques have also been optimized in order to develop a more robust and effective microfluidic system for detection of various biomarker. Various commercial permanent markers were employed for detection and assessed for their water-resistant abilities in order to create hydrophobic barriers to develop paper-based microfluidics via using a desktop pen plotter integrated with a custom-made, low-cost paper feeder.[50,51] With greater advancement in the field more attention is now given to multiplexed detection of multiple RNA or DNA components. One such approach was used in detection of nuclei acid initially by performing Recombinase Polymerase Amplification (RT-RPA) and isothermal reverse transcription of Ebola virus. Later it was utilized in detection of Ebola virus in human sample. The device is able to perform techniques allowing early detection of infectious diseases based on nucleic acid detection RT-RPA technique integrated into paper-based microfluidics devices to identify the occurrence of Ebola virus in human sample. RNA extracts with using freeze-dried reagents on paper having 90.0% sensitivity. Furthermore, the devices have ability to detect three distinct RNA sequences.[52] The paper based microfluidics has come a long way in detection of nuclei acid and simultaneously extracting and amplifying the nucleic acid component. Such a setting was developed and used in practical application of paper based microfluidic devices. In this device, a filter disc is incorporated into a plastic molded microfluidic device which acted as a DNA capture for PCR and DNA extraction. The components of microdevice consisted of Fusion 5 filter as well as a PDMS (polydimethylsiloxane) membrane sandwiched in between two PMMA (poly(methyl methacrylate)) layers. The device had ability to alone extract DNA extraction in 7 min by successively aspirating HCl, NaOH and water through the filter. The DNA was higher on contrary to QIAamp® DNA Micro kits. Samples like dried blood stains, whole blood, on Whatman® 903 paper, dried blood stains on FTA™ cards, saliva, buccal swabs, and cigarette butts, were processed within 8 min. Such studies provides the grounds for developing a more robust and compact characterizes a considerable step towards the functional application of on-chip DNA extraction methods, as well as the advancement of fully unified genetic analytical systems.[53]

4. Future prospects

The development and research have progressed at a higher pace than ever in the field of diseases diagnosis although there are constant efforts, there is need to develop more robust in order to decrease mortality rate due to deadly diseases. Still there is struggle in developing early detection techniques which uses complex biomolecules like RNA, DNA and proteins order to detect diseases caused by bacteria, viruses or fungi. Until now separation techniques used before detection rely on traditional techniques like electrophoresis etc. which often require centralized services. Although efforts are made in developing devices that is capable enough on carrying out separation and detection based on lab on chip in need of the hour. Even recently RT-PCR based systems are used which have multiple steps and require well developed facility to carry out detection of the NF and MF based new diagnostic techniques are in the early stage of development. Developing point of care-based NF and MF as it would have advantages over traditional techniques in developing nations were the resources are limited. Although the process of technology development is progressing, integration in necessary for better output. In our view, following advancements can make an improved nanofluidic and microfluidic tool for detection (Fig. 6). (1) Making isothermal amplification-based NF and MF. (2) Integration of paper for real-time and multiplexed detection of DNA/RNA targets for developing highly sensitive and specific detection. (3) Incorporation of nanoparticles for effective separation of nucleic acids can help in reaching lower limit of detection. (4) Development of advanced models for early detection and point of care separation of the biomolecule. (5) Integration of smartphone which are omnipresent, for establishing a more technology forward approach. (6) Combination of CRISPR-Chip along with electronics for detection can help us to detect the analyte even at lesser concentration. (7) Using electrochemical based approach can help in detection of even narrow antigenic load. (8) Proposing new separation techniques having advanced principles to fabricate active devices is a essential to

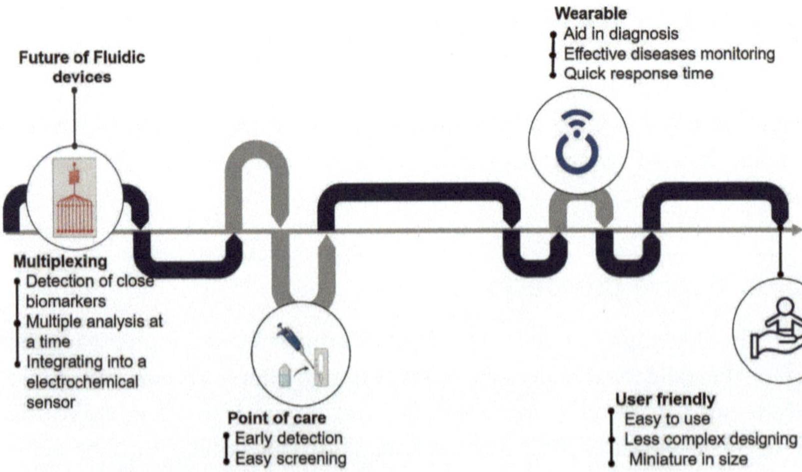

Fig. 6 Future prospects of fluidic devices.

achieve such advanced functions. (9) Minimization of contaminants adsorbed on the walls of the fluidic channels by passivation of the device surfaces can greatly affecting the performance of the device. Technology integration will make it easy not only to separate but also to detect diseases and take measure for better diseases management.

Acknowledgments

We gratefully acknowledge Gujarat State Biotechnology Mission (GSBTM Project ID: AJJV48) and SHODH: Scheme of Developing High Quality Research Scholarship (Ref No: 201901440009) for financial support.

References

1. Han J, Craighead HG. Entropic trapping and sieving of long DNA molecules in a nanofluidic channel. *J Vac Sci Technol A*. 1999;17(4):2142–2147.
2. Turner S, Cabodi M, Craighead HG. Confinement-induced entropic recoil of single DNA molecules in a nanofluidic structure. *Phys Rev Lett*. 2002;88(12):128103.
3. Han J, Craighead HG. Separation of long DNA molecules in a microfabricated entropic trap array. *Science*. 2000;288(5468):1026–1029.
4. Cabodi M, Turner SW, Craighead HG. Entropic recoil separation of long DNA molecules. *Anal Chem*. 2002;74(20):5169–5174.
5. Streek M, Schmid F, Duong TT, Ros A. Mechanisms of DNA separation in entropic trap arrays: a Brownian dynamics simulation. *J Biotechnol*. 2004;112(1–2):79–89.
6. Wu L, Levy S. Fluctuations of DNA mobility in nanofluidic entropic traps. *Biomicrofluidics*. 2014;8(4):044103.
7. Agrawal P, Bognár Z, Dorfman KD. Entropic trap purification of long DNA. *Lab Chip*. 2018;18(6):955–964.
8. Zheng C, Mei D. Gravity separation by entropic transport. *EPL (Europhysics Letters)*. 2015;109(1):10006.

9. Xu F, Baba Y. Polymer solutions and entropic-based systems for double-stranded DNA capillary electrophoresis and microchip electrophoresis. *Electrophoresis*. 2004;25(14): 2332–2345.
10. Han J, Craighead HG. *Entropic Trapping and Sieving of Molecules*. Google Patents; 2003.
11. Ou X, Chen P, Huang X, Li S, Liu BF. Microfluidic chip electrophoresis for biochemical analysis. *J Sep Sci*. 2020;43(1):258–270.
12. Shiddiky MJ, Shim Y-B. Trace analysis of DNA: preconcentration, separation, and electrochemical detection in microchip electrophoresis using Au nanoparticles. *Anal Chem*. 2007;79(10):3724–3733.
13. Ou J, Cho J, Olson DW, Dorfman KD. DNA electrophoresis in a sparse ordered post array. *Phys Rev E*. 2009;79(6):061904.
14. Chen Y-Y, Chiu P-H, Weng C-H, Yang R-J. Preconcentration of diluted mixed-species samples following separation and collection in a micro–nanofluidic device. *Biomicrofluidics*. 2016;10(1):014119.
15. Thomas JD, Dorfman KD. Tilted post arrays for separating long DNA. *Biomicrofluidics*. 2014;8(3):034115.
16. Qiao W, Wang C, Ding Z, Song J, Wei X, Lo Y-H. A two-stage electrophoretic microfluidic device for nucleic acid collection and enrichment. *Microfluid Nanofluidics*. 2016;20(5):77.
17. Thompson BL, Birch C, Nelson DA, et al. A centrifugal microfluidic device with integrated gold leaf electrodes for the electrophoretic separation of DNA. *Lab Chip*. 2016;16 (23):4569–4580.
18. Downs AM, McCallum C, Pennathur S. Confinement effects on DNA hybridization in electrokinetic micro-and nanofluidic systems. *Electrophoresis*. 2019;40(5):792–798.
19. Li D, Lu X, Song Y, Wang J, Li D, Xuan X. Sheathless electrokinetic particle separation in a bifurcating microchannel. *Biomicrofluidics*. 2016;10(5):054104.
20. Wang Z, Ye S, Zhang N, Liu X, Wang M. Triggerable mutually amplified signal probe based SERS-microfluidics platform for the efficient enrichment and quantitative detection of miRNA. *Anal Chem*. 2019;91(8):5043–5050.
21. Bogunovic L, Eichhorn R, Regtmeier J, Anselmetti D, Reimann P. Particle sorting by a structured microfluidic ratchet device with tunable selectivity: theory and experiment. *Soft Matter*. 2012;8(14):3900–3907.
22. Loutherback K, Puchalla J, Austin RH, Sturm JC. Deterministic microfluidic ratchet. *Phys Rev Lett*. 2009;102(4):045301.
23. McFaul SM, Lin BK, Ma H. Cell separation based on size and deformability using microfluidic funnel ratchets. *Lab Chip*. 2012;12(13):2369–2376.
24. Kowalik M, Bishop KJ. Ratcheted electrophoresis of Brownian particles. *Appl Phys Lett*. 2016;108(20):203103.
25. Moraes C, Labuz JM, Shao Y, Fu J, Takayama S. Supersoft lithography: candy-based fabrication of soft silicone microstructures. *Lab Chip*. 2015;15(18):3760–3765.
26. Peng R, Li D. Fabrication of polydimethylsiloxane (PDMS) nanofluidic chips with controllable channel size and spacing. *Lab Chip*. 2016;16(19):3767–3776.
27. Peng R, Li D. Detection and sizing of nanoparticles and DNA on PDMS nanofluidic chips based on differential resistive pulse sensing. *Nanoscale*. 2017;9(18):5964–5974.
28. Wang J, Yan Y, Geng Y, Gan Y, Fang Z. Fabrication of polydimethylsiloxane nanofluidic chips under AFM tip-based nanomilling process. *Nanoscale Res Lett*. 2019; 14(1):1–14.
29. Sahore V, Kumar S, Rogers C, Jensen J, Sonker M, Woolley A. Pressure-actuated microfluidic devices for electrophoretic separation of pre-term birth biomarkers. *Anal Bioanal Chem*. 2016;408(2):599–607.
30. Wang Y, Chen S, Sun H, Li W, Hu C, Ren K. Recent progresses in microfabricating perfluorinated polymers (Teflons) and the associated new applications in microfluidics. *Microphysiol Syst*. 2018;2:1–11.

31. Ramalingam N, San TC, Kai TJ, Mak MYM, Gong H-Q. Microfluidic devices harboring unsealed reactors for real-time isothermal helicase-dependent amplification. *Microfluid Nanofluidics*. 2009;7(3):325.
32. Gitlin L, Schulze P, Ohla S, Bongard HJ, Belder D. Surface modification of PDMS microfluidic devices by controlled sulfuric acid treatment and the application in chip electrophoresis. *Electrophoresis*. 2015;36(3):449–456.
33. Fontana M, Fijen C, Lemay SG, Mathwig K, Hohlbein J. High-throughput, non-equilibrium studies of single biomolecules using glass-made nanofluidic devices. *Lab Chip*. 2019;19(1):79–86.
34. Ohta R, Mawatari K, Takeuchi T, Morikawa K, Kitamori T. Detachable glass micro/nanofluidic device. *Biomicrofluidics*. 2019;13(2):024104.
35. Wang C, Fang H, Zhou S, et al. Recycled low-temperature direct bonding of Si/glass and glass/glass chips for detachable micro/nanofluidic devices. *J Mater Sci Technol*. 2020;46:156–167.
36. Liu K, Hao M, Meng D, et al. Investigation on the electrochemical manipulation of bio-molecules in glass nanofluidic channels. *J New Mat Electr Syst*. 2015;18(4):243–249.
37. Chien C-C, Shekar S, Niedzwiecki DJ, Shepard KL, Drndić M. Single-stranded DNA translocation recordings through solid-state nanopores on glass chips at 10 MHz measurement bandwidth. *ACS Nano*. 2019;13(9):10545–10554.
38. Balan A, Chien C-C, Engelke R, Drndić M. Suspended solid-state membranes on glass chips with sub 1-Pf capacitance for biomolecule sensing applications. *Sci Rep*. 2015; 5:17775.
39. Andresen D, von Nickisch-Rosenegk M, Bier FF. Helicase dependent OnChip-amplification and its use in multiplex pathogen detection. *Clin Chim Acta*. 2009;403 (1–2):244–248.
40. Powell L, Wiederkehr RS, Damascus P, et al. Rapid and sensitive detection of viral nucleic acids using silicon microchips. *Analyst*. 2018;143(11):2596–2603.
41. Lee H, Park C, Na W, Park KH, Shin S. Precision cell-free DNA extraction for liquid biopsy by integrated microfluidics. *NPJ Precis Oncol*. 2020;4(1):1–10.
42. Soares RR, Neumann F, Caneira CR, et al. Silica bead-based microfluidic device with integrated photodiodes for the rapid capture and detection of rolling circle amplification products in the femtomolar range. *Biosens Bioelectron*. 2019;128:68–75.
43. Gong MM, Sinton D. Turning the page: advancing paper-based microfluidics for broad diagnostic application. *Chem Rev*. 2017;117(12):8447–8480.
44. Carrell C, Kava A, Nguyen M, et al. Beyond the lateral flow assay: a review of paper-based microfluidics. *Microelectron Eng*. 2019;206:45–54.
45. Chagas CL, de Souza FR, Cardoso TM, et al. A fully disposable paper-based electrophoresis microchip with integrated pencil-drawn electrodes for contactless conductivity detection. *Anal Methods*. 2016;8(37):6682–6686.
46. Nilghaz A, Guan L, Tan W, Shen W. Advances of paper-based microfluidics for diagnostics—the original motivation and current status. *ACS Sensors*. 2016;1(12): 1382–1393.
47. Roy S, Mohd-Naim NF, Safavieh M, Ahmed MU. Colorimetric nucleic acid detection on paper microchip using loop mediated isothermal amplification and crystal violet dye. *Acs Sensors*. 2017;2(11):1713–1720.
48. Safavieh M, Khetani S, Kaul V, Kuritzkes DR, Shafiee H. A graphene-modified cellulose paper microchip for HIV detection. In: *Paper Presented at Advances in Global Health Through Sensing Technologies*; 2015.
49. Reboud J, Xu G, Garrett A, et al. Paper-based microfluidics for DNA diagnostics of malaria in low resource underserved rural communities. *Proc Natl Acad Sci U S A*. 2019;116:4834–4842.

50. Ghaderinezhad F, Amin R, Temirel M, Yenilmez B, Wentworth A, Tasoglu S. High-throughput rapid-prototyping of low-cost paper-based microfluidics. *Sci Rep*. 2017;7(1):1–9.
51. Magro L, Escadafal C, Garneret P, et al. Paper microfluidics for nucleic acid amplification testing (NAAT) of infectious diseases. *Lab Chip*. 2017;17(14):2347–2371.
52. Magro L, Jacquelin B, Escadafal C, et al. Based RNA detection and multiplexed analysis for Ebola virus diagnostics. *Sci Rep*. 2017;7(1):1–9.
53. Gan W, Zhuang B, Zhang P, Han J, Li C-X, Liu P. A filter paper-based microdevice for low-cost, rapid, and automated DNA extraction and amplification from diverse sample types. *Lab Chip*. 2014;14(19):3719–3728.

CHAPTER FIVE

Paper based microfluidics: A forecast toward the most affordable and rapid point-of-care devices

Anirban Sinha*, Mainak Basu, and Prerna Chandna
MeFy Care Private Ltd., Pune, India
*Corresponding author: e-mail addresses: anirban.sinha@mefy.care; ani@gapp.nthu.edu.tw

Contents

1. Introduction and historical perspective of paper-based microfluidic devices 110
 1.1 Principle and prospect for paper-based microfluidic devices 112
 1.2 The chemistry behind paper 112
 1.3 The physics behind paper 113
2. Fabrication methods 117
3. Fluid flow control and liquid handling 125
 3.1 Geometrical approaches 126
 3.2 Chemical processes 130
 3.3 Mechanical methods 132
4. Applications of paper-based microfluidic devices 133
5. Current limitations and future works 142
 5.1 Market analysis 142
 5.2 Competitive advantages 145
 5.3 The path to becoming a key technological contributor 145
 5.4 Regulatory challenges 147
6. Conclusion and final thoughts 148
Acknowledgment 149
References 149

Abstract

The microfluidic industry has evolved through years with acquired scientific knowledge from different, and already developed industries. Consequently, a wide range of materials like silicon from the electronic industry to all the way, silicone, from the chemical engineering industry, has been spotted to solve similar challenges. Although a typical microfluidic chip, fabricated from glass or polymer substrates offers definite benefits, however, paper-based microfluidic analytical devices (μPADs) possess numerous special benefits for practical implementation at a lower price. Owing to these features, in recent

Progress in Molecular Biology and Translational Science, Volume 186
ISSN 1877-1173
https://doi.org/10.1016/bs.pmbts.2021.07.010

years, paper microfluidics has drawn immense interest from researchers in industry and academia alike. These devices have wider applications with advantages like lower cost, speedy detection, user-easiness, biocompatibility, sensitivity, and specificity etc. when compared to other microfluidic devices. Therefore, these sensitive but affordable devices fit themselves into point-of-care (POC) testing with features in demand like natural disposability, situational flexibility, and the capability to store and analyze the target at the point of requirement. Gradually, advancements in fabrication technologies, assay development techniques, and improved packaging capabilities, have contributed significantly to the real-time identification and health investigation through paper microfluidics; however, the growth has not been limited to the biomedical field; industries like electronics, energy storage and many more have expanded substantially. Here, we represent an overall state of the paper-based microfluidic technology by covering the fundamentals, working principles, different fabrication procedures, applications for various needs and then to make things more practical, the real-life scenario and practical challenges involved in launching a device into the market have been revealed. To conclude, recent contribution of μPADs in the 2020 pandemic and potential future possibilities have been reviewed.

1. Introduction and historical perspective of paper-based microfluidic devices

Paper was invented around 105 CE[1] and was initially used for writing, later it evolved to usage in printing and painting.[1] It has traversed a long journey into this century and applications have changed with various developments in the chemical and biochemical industry. Gay Lussac developed litmus paper in the early 19th century.[2] From 1850 to 1950s, paper-based devices were fabricated to be used for various applications such as urine tests, recognition of dyes and pigments,[3] paper chromatography,[4] etc. The Nobel Prize in Chemistry (1952) was awarded to Martin and Synge[5] for the invention of paper chromatography, and this breakthrough led to a major drive in the recognition and development of paper-based sensors. A few of the earlier technologies,[3] used a paraffin patterned filter paper, however, this approach changed in 2007 when the Whiteside's group (Harvard University) re-invented the approach of the paper-based microfluidic devices.[6] They fabricated a paper-based microfluidic device called 2D-μPAD[7] in which three different outlets with hydrophobic walls were fabricated using photolithography on a chromatography paper and was utilized for the detection of albumin and glucose. This offered a new direction of thought in the paper microfluidic cartridge development and afterwards, precise fluid control and handling were utilized in various fields including environmental monitoring, diagnostics, food analysis, etc.[6]

Conventional approaches of disease diagnoses initially suffered from various loopholes like the requirement of skilled labor, the necessity of sophisticated instruments, large turn-around time, and ultimately a major financial obligation on the overall health care system.[8] The World Health Organization (WHO), 2015 reported that around 1.8% of the population spent about one-thirds of their income on healthcare. In different circumstances, a contagious disease that originates in the technologically primitive developing or underdeveloped countries, claims thousands of lives due to lack of advanced medical facilities and proper diagnostic tools.[9] Bill and Melinda Gates Foundation, in alliance with the NIH (National Institute of Health, USA) in 2003 stated that the major challenges faced by the diagnostic industry are the pricey laboratory apparatus and shortage of skilled labor. In remote areas where resources are limited, the fabrication, packaging, and distribution of simple, reliable, cost-effective, and portable devices is an extreme necessity to meet healthcare needs. Point-of-care diagnostic (POC) devices use the criteria called "ASSURED" which stands for "**a**ffordable, **s**ensitive, **s**pecific, **u**ser-friendly, **r**apid and **r**obust, **e**quipment-free, and **d**eliverable," as its benchmark.[10] Therefore, POC is a cost-effective and miniaturized system with various functions for the timely and accurate diagnosis and/or clinical management of diseases in developed or underdeveloped countries.[11] This ultimately reduces the requirement for a well-equipped central laboratory. POC devices are also very useful in preventing the outbreak of new diseases and early diagnosis of old ones leading to timely treatment. In 2016, the global POC marketplace was evaluated at 1764.6 million USD which is anticipated to expand by about 1.5% each year throughout 2017–2025.[12]

Since paper is biodegradable, portable, and produced in mass quantities, it satisfies all the conditions mentioned for an "ASSURED" device stated by the WHO.[6,13] Generally, active microfluidic devices require an external power supply, either in the form of an active pump[14–16] or other fluid manipulating techniques[6] for operation. The main advantage of paper-based passive microfluidic devices is its' power-free operation and low sample input, which makes them simple and user-friendly. Therefore, the ideal POC diagnostic device represents itself by being inexpensive, affordable, and the ability to generate fast results. The concept for such types of devices are not new and not even developed very recently. If we look into history, we could trace a lot of earlier developments. The first POC glucose detection kit for human blood was developed by Ames in 1957.[17] Unipath launched its one-step lateral flow assay-based POC pregnancy test[18] in

1988. Major developments have transpired in paper-based devices for the identification of diabetes,[19] infectious diseases,[9] detection of pathogens[20] cholesterol,[21] and many more.[22–25] Recently, a review of the research progress and application for paper-based microfluidic chips provide an overall current state of the art for the μPADs projecting toward various applications in different fields like chemical, biochemical, POC, etc.[26] As previously discussed, paper-based microfluidic devices encompass various capabilities and benefits as compared to conventional channel-based microfluidic devices like fixed assay time, storage of reagents in various forms, low sample intake, and low cost with no power requirement and highly specific analytical outcome.[27]

To realize the current scenario of paper-based microfluidic technology developments along with applications in present and or in the future, this section has been divided into three parts. In the first section, we have explained the principles and their various prospects. The second part deals with the chemical composition of paper and in the third section, we delve into the basic physical laws so that it will be easy for everyone to comprehend the parameters governing device development.

1.1 Principle and prospect for paper-based microfluidic devices

The basic principle behind paper-based microfluidic devices is to create a hydrophobic boundary that acts as a physical barrier in the hydrophilic zone for sample dispersion. Each detection zone fabricated on the μPAD is coated with the respective organic receptor to detect the targeted molecules. For performing multiple analyses on the filter paper, the reagents travel through the patterned path by passive capillary action. As the analytes flow through the detection area containing organic receptors, a rapid color change is observed as an indication of the reaction, which makes μPADs very useful for POC application. The color change could be observed by naked eyes or captured electronically by photodetectors or digital cameras. The fabrication of μPADs is a much simpler process than that of lab-on-chip devices. Filter papers are used to develop μPADs, thus it makes the device extremely commercially viable.[28] The key features for a paper-based microfluidic device as μPADs are highlighted in Fig. 1.

1.2 The chemistry behind paper

Paper is produced by pressing cellulose fibers into a thin sheet. Various sources of cellulose like jute, wood, grass, flax, bamboo, hemp, ramie, straw, or cotton

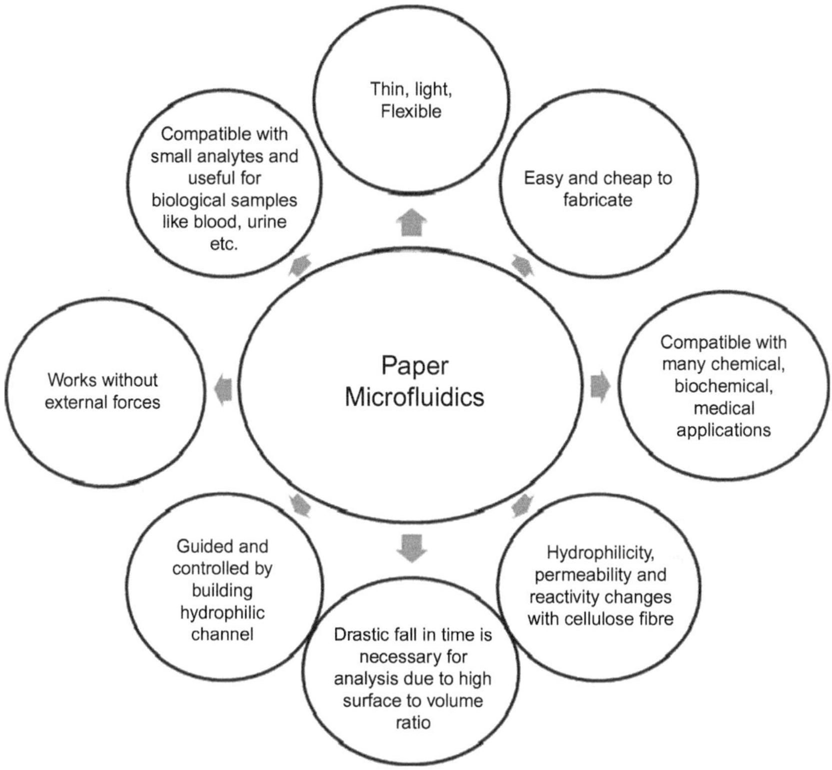

Fig. 1 Key points behind choosing paper for developing μPADs.

(chromatography paper and filter), and many more have been used.[29] Nitric acid and sulfuric acid are used in this process where nitric acid is used for the esterification of cellulose, i.e., the conversion of cellulose to nitrocellulose. Nitronium ions are produced using sulfuric acid as a catalyst and esterification vary with the quantity of water in the reaction. The intensity of nitration defines the flammability and solubility of the paper device and nearly all nitrocellulose compounds dinitrate to nitrogen content of 10.5–13.5%.[29] Protran membranes alternatively range between 1.9 and 2.4 after nitration.[30] In the current scenario, non-wood plant source is expensive, therefore 90% of raw material consists of wood.[30] Due to the high concentration of lignin, the paper eventually discolors and degrades.[30] Paper made from 100% cotton is primarily used for diagnostic purposes.[30]

1.3 The physics behind paper

This section reveals the physics behind the wicking of paper, and it provides an understanding of how it could be used to enhance efficiency while designing a paper-based device for a specific application. Fluidic flow in a paper device is governed by several factors, for instance, the size, thickness, permeability (porosity), and capillary flow rate. The grade of paper is also a determining factor for any application as there is no "one size fits all" rule. Therefore, when modeling a device for commercialization, scientists investigate the porosity, flow rate, and particle retention.

A high-quality cotton liner that has an α-cellulose content of a minimum of 98% is used to manufacture chromatography and filter paper. Most of the filter papers are intended for general separation techniques and thus the level of retention is as low as 2.5 mm.[31] Paper is composed of an alignment of multiple fibers,[32] these bundles of fibers create microchannels through which the liquid passes. Capillary flow through the horizontal bundle of the parallel tubes which is used as a template for a porous medium like paper[33] could be explained by the Lucas-Washburn equation. This is represented in Eq. (1) and described in Fig. 2A.

$$l(t) = \sqrt{\frac{\gamma r \cos \theta t}{2\mu}} \tag{1}$$

where $l(t)$ is the rise of the height of the liquid due to capillary pressure into the channel, t is the time of rising, γ is the surface tension, r is the average radius of the channel or effective pore radius of the paper, θ is the contact angle and μ is the dynamic viscosity of the fluid. In Eq. (1), the effect of gravity during capillary filling is considered negligible.[34]

A closer look at Eq. (1) tells us that various parameters like viscosity, surface tension, and contact angle are constant for a specific liquid. Therefore, by increasing the effective pore radius, we can increase the wicking of a

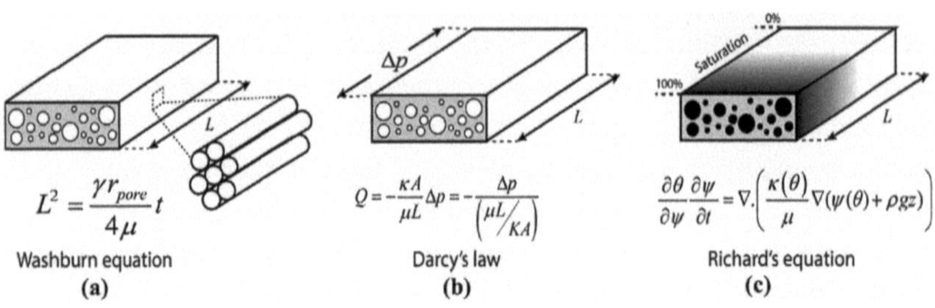

Fig. 2 Demonstration of capillary filling inside paper fibers using the (A) Lucas-Washburn model; (B) Darcy's Law; and (C) the equation presented by Richard. *Reprinted/adapted by permission Toley BJ, Das D, Ganar KA, et al. Multidimensional paper networks: a new generation of low-cost pump-free microfluidic devices.* J Indian Inst Sci. *2018:98:103–136. https://doi.org/10.1007/s41745-018-0077-1.*

paper channel. It has been observed that the pore radius of a particular paper is fixed, and to increase the effective pore radius, stacking of two paper-based channels on top of each other could be done. From the literature survey, it is observed that as two paper-based channels get stacked, it generates a small gap between the layers which acts as a capillary of larger radius than the effective pore radius of the individual ones.[34]

However, a new approach (Fig. 2B), Darcy's law, has been successfully used in modeling sintering polymer wick or bank of fiber[35] based upon the assumption that pores in the rigid porous medium and the size of constituent particles are constant. Darcy's Law is represented as,

$$\left\langle \vec{V} \right\rangle = -\frac{K}{\mu}\vec{\nabla}\langle P\rangle \tag{2}$$

Here, $\left\langle \vec{V} \right\rangle$ and $\langle P\rangle$ are the average liquid velocity and average pore modified pressure, respectively, K is the permeability of the porous paper and μ is the viscosity of the liquid. Saturated flow in the porous medium is explained by Darcy's law, however, Richard's equation (as shown in Fig. 2C) explains the fluid movement through an unsaturated porous medium. Richard's equation is represented as,

$$\frac{\partial \theta}{\partial t} = \frac{\partial}{\partial z}\left[K(\theta)\left(\frac{\partial h}{\partial z} + 1\right)\right] \tag{3}$$

Here, K is the hydraulic conduction, h is the metric head induced by capillary action, z is the rise over a vertical limit, t is the time and θ is the volumetric water. This modern approach is used to understand the rise of liquid level inside the paper when a paper is dipped in the liquid.

Most of the findings involved in paper-based microfluidic devices are from chromatography or filter paper but various research labs have used unconventional paper for different POC applications. The material consideration of printing paper for fabrication of hydrophilic channels is not suitable for liquid transport because of its low porosity and high surface tension.[32] Moreover, high porosity leads to uncontrolled hydrophilization and material diffusion. For example, cellulose fiber filter paper with a large pore size may swell which might constrain the capillary flow. The surface area of the paper strip is also crucial due to its impact on sensitivity, reproducibility, specificity, and reagent deposition. If other factors are constant, then the amount of surface area drops with increasing pore size in a nonlinear manner and approaches linearity with the increases in thickness. Another important parameter for paper-based microfluidic devices is capillary flow rate which is specified as the speed of front movement of the sample moving along the membrane strip's span. The key factor in performance of the assay is the capillary flow rate as the effective concentration of the analytes which is present in the sample is in reverse proportionality to the square of the flow rate shift.[36] For achieving consistent sensitivity, the capillary flow rate is important which varies with the position of the detection region on the test strip. The transportation of analytes is slower if the inlet is placed far away from the detection region because it needs to pass through the capture lines. Capillary flow measurement is not an appropriate method in the manufacturing industry, because when the front liquid travels along the membrane, its speed reduces exponentially during the flow time. This parameter is needed to record the time needed for the sample to move through such that the membrane gets saturated within the defined length and is measured as s/cm.

To fabricate a microfluidic device, selecting the pore size and pore distribution of the porous membrane is very important. The retention of a particle is dependent on the pore size. Membranes are characterized by nominal pore size. The largest pore dimension in the filtration category is called nominal pore size. Pore size is determined experimentally by forcing hard particles through the membrane. Owing to the complexities of the above method, alternatively, a bubble point has been chosen to determine the pore size. Determination of bubble point is performed by forcing air (or isopropanol) through a wetted substrate and measuring the pressure requirement in doing so. If the pore size is between 0.1 and 3 mm, the relationship is linear but as the pore size becomes greater than 3 mm the relation becomes more complex. For membranes that have a larger pore size, it is difficult to handle membrane consistent with this correlation. The pore size distribution is the range of pores inside the membrane. Different range of pore size is also responsible for regulating the capillary flow rate which is a function of cumulative pore size. Thus, for membranes with identical pores, pore size distribution is a useful factor to differentiate the membrane type.[37]

Porosity defined as the amount of air present in three dimensional areas of the membrane is also considered to be an important parameter.[30] Initially, it was described as the void membrane volume which could be estimated by measuring the total volume of sample required to soak it. It was observed that when factors such as distribution of pore size and thickness are kept constant the porosity of the membrane increases, and therefore the capillary flow rate increases linearly. The paper thickness also influences various parameters such as bed volume, tensile strength, and signal visibility. The total volume available for the sample in the membrane is called bed volume. Bed volume is a function of the porosity and dimension of the membrane.[37] The bed volume increases as the thickness of the paper increases. When the absorbent pad is absent then bed volume is important because it can be used to calculate the actual amount of sample wicked up. The membrane thickness partly depends upon the control line and width of the testing chamber.[37]

For colorimetric assays, during the process of image analysis, one of the most important parameters is the color of the paper. Ideally, the paper should have no shade and must be white but when it gets exposed to humidity, high temperature, and sunlight, the color changes to yellowish and brownish. During the analysis phase, the identification and compensation of the acceptable limit of color should be taken into consideration.[38]

2. Fabrication methods

The fabrication techniques of paper-based microfluidic devices utilized the already existing technologies from other industries and could be categorized according to their morphology. The fabricated paper-based chips could be of two types, e.g., 2D or 3D. Basically, in 2D, the liquid transports horizontally, whereas in 3D the fluidic flow is not restricted to the horizontal plane and it travels vertically in different layers packed on top of each other. Various fabrication techniques have been used in fabric and paper-based microfluidic devices such as (1) photolithography,[39] (2) plotting with an analogue plotter,[40] (3) inkjet etching,[41] (4) plasma treatment[42], (5) wax printing,[43] (6) paper cutting,[44] (7) inkjet printing,[46] (8) flexography printing,[47] (9) screen printing[47] and (10) laser treatment,[48] etc. The fundamental principle is to create hydrophilic-hydrophobic contrast to fabricate micron-scale capillary channels on paper. Table 1 illustrates the different fabrication processes and Table 2 describes a comparative study of different fabrication methods, substrate, equipment, consumables, and their limitations.

Based upon the channel fabrication methods, all paper devices could be divided into two general categories: (1) physical blocking of pores,[56] where pores are blocked systematically with non-permeable material leaving the channel to create a predefined fluidic path and (2) two-dimensional shaping/cutting[49] where the permeable papers are cut into predefined small structures that works as a liquid channel. In the cutting technique, paper cutting is done by using a plotter of the CO_2 laser cutter for fabricating

Table 1 Fabrication category.

Planner device fabrication method (2D)	*Wax screen printing*[49]: In this method patterns are imprinted onto the printing surface by forcing ink across a mesh. Wax works as a boundary of the liquid channel.
	Embossing[50]: Omniphobic paper is modified by treating with highly fluorinated alkylsilane (Salinization) which forms a channel for the flow of liquid.
	Flexographic Printing[47]: Thin sheet of paper is printed with a thin roll with a different pattern.
	Inkjet Printing[46]: Inkjet printing is used to make hydrophobic barriers on paper.
	Photolithography[39]: This is a standard procedure similar to silicon microchip micro-fabrication on paper.
	PDMS screen printing[51]: PDMS is used to create a fluidic flow channel.
	Printed Circuit technology[52]: An electronic circuit design software is utilized to design a fluidic flow pattern and copied to paper from a copper sheet.
	Two-dimensional shaping[44]: A computer-controlled knife plotter is used for manipulating various porous media.
	Plasma Treatment[42]: Alkyl ketene dimer (AKD) dispersed in n-heptane is employed for modifying the filter paper.
	Laser treatment[48]: A paper coated with hydrophobic material is treated with laser to form hydrophilic channels.
	Fast Lithographic Activation of Sheets[7]: A photoresist is soaked into the paper and UV is exposed and a pattern is created to control the fluidic flow direction.
Multidimensional device fabrication method (3D)	*Stacking of patterned paper*[39]: Multiple patterned pads are stacked with double-sided tape to control three-dimension fluidic flow.
	3D wax printing[53]: A traditional wax printing is extended to multiple layers to form three-dimensional fluidic flow channels.

Table 2 Fabrication process, consumables, equipment, advantages, and limitations of developing paper-based microfluidic devices.

Fabrication technique	Equipment	Consumables	Advantages	Drawbacks	Ref.
Cutting	Computer, X–Y plotter, or CO_2 laser	None	Sharp defined features	Requires expensive equipment; waste of raw material; yield low mechanical stability	54
Photolithography	Lithography equipment, UV lamp filter, hot plate	SU-82010 photoresist, Ultraviolet resin	High resolution of microfluidic channels	Requires expensive equipment, complex steps, expensive reagents	55
PDMS printing	Desktop plotter	PDMS	Cheap patterning agents	Inconsistent control over a hydrophobic barrier	40
Inkjet etching	Modified inkjet printer	Polystyrene	Low cost; requires only a single printing apparatus to create microfluidic channels	Requires multiple printing steps for creation of microfluidic channels; involves a complex number of steps	41
Wax dipping	Customized mask, heating unit	Solid wax	Very cheap and easy method	Inconsistent with batches due to the variation in dipping	56
Flexographic printing	Customized printing equipment	Polystyrene	Avoid heat treatment of the printed patterns	High cost; complex reagents and templates	47
Screen printing	Customized masks	Solid wax, polystyrene	Low-cost simple fabrication steps	Low resolution of microfluidic channels, rough barriers	49
Wax printing	Wax printer, hot plate	Solid wax	Manufacture devices with much simple and fast fabrication process	Poor availability in resources; low resolution during detection	43

Continued

Table 2 Fabrication process, consumables, equipment, advantages, and limitations of developing paper-based microfluidic devices.—cont'd

Fabrication technique	Equipment	Consumables	Advantages	Drawbacks	Ref.
Stamping	PDMS or stainless-steel stamp	Paraffin or commercial ink	Low cost and simple	Inconsistent process; low resolution while detection; oxidation of paper is necessary	57
Plasma treatment	Customized fabricated mask; hot plate or oven; vacuum chamber	Alkenyl ketene Dimer (AKD); Poly(hydroxybutyrate) (PHB)	Uses very cheap patterning agent AKD or fluorocarbon	High cost; requires fabricating different mask template for developing different microfluidic channels	42
Inkjet printing	Modified ink jet printer	AKD; UV purified acrylate	High resolution; requires only a desktop printer for printing the patterns	Requires modified ink-jet printer and different ink compositions	58
Vapor phase deposition	Deposition chamber	Poly(chloro-p-xylene)	Simple steps; complex pattern can be established	Involves high-cost equipment	59
Wet etching	Paper mask; glass slides	Trimethoxyoctadeculsilane (TMOS)	Simple and fast	High cost need a paper mask with a specific design	60

the paper channels. In most cases, higher strength and structure are provided to the device by blocking the channels using sticky tapes. Fabrication costs increase during the cutting method as special equipment is needed for mass production. Various methods of fabrication processes collected from literatures have been illustrated in Fig. 3. Fig. 3A represents the schematic representation of the wax screen printing method.[49] Fig. 3B[50] and Fig. 3C[47] describes the process of embossing and flexographically printed technique respectively. Inkjet printing and flash techniques are also used for the fabrication of paper-based microfluidic device which is shown in Fig. 3D[41] and Fig. 3E.[61] Mohammadi *et al.* described the process of fabrication of paper-based devices using PDMS as represented in Fig. 3F.[51] Fig. 3G[52] and Fig. 3H[44] represent printed circuit technology and two-dimensional shaping paper-based microfluidic devices. Plasma treatment and laser treatment are also used among various techniques which have been shown in Fig. 3I[62] and Fig. 3J,[48] respectively. Fig. 3K[39] and Fig. 3L[53] describe the process of layered paper-tape and three-dimensional wax printing.

As for general understanding, while fabricating a paper-based microfluidic device, the paper must be patterned into hydrophilic and hydrophobic zones. This could be achieved in many ways and could be also categorized into three types (1) blocking pores; (2) deposition of hydrophobization agent onto the fiber surface; (3) chemically alteration of the paper fibers. For the first two types, the reaction between the agent and paper is not a chemical reaction. It involves processes like physical blocking with photoresist or PDMS, physical deposition of wax, polystyrene, or commercial ink.[50] Due to the hydrophobicity of the agent, wetting properties of the paper are altered as it not only fills up the pores of the paper but also adsorbs to the fibers of the paper.[63] The pattern made by photolithography is based upon traditional electronic industry technologies, i.e., light-induced hydrophobization followed by de-hydrophobization of selected areas on the paper. This method of patterning by using SU-8 over the chromatographic paper was first introduced by the Whiteside's group at Harvard University.[7] This technique leads to the formation of a solid chemically resistant barrier at small dimensions in high resolution. However, it is not suitable for mass production at this moment as the steps involved are tedious and the process is expensive. Later on, various materials like epoxy or cyclized polyisoprene derivatives were introduced as negative photoresist to simplify the photolithography process and UV lamps and hotplate were introduced to get rid of the expensive and complicated lithography machine.[61] Furthermore, a laser directing method was introduced to avoid the use of a photomask, therefore

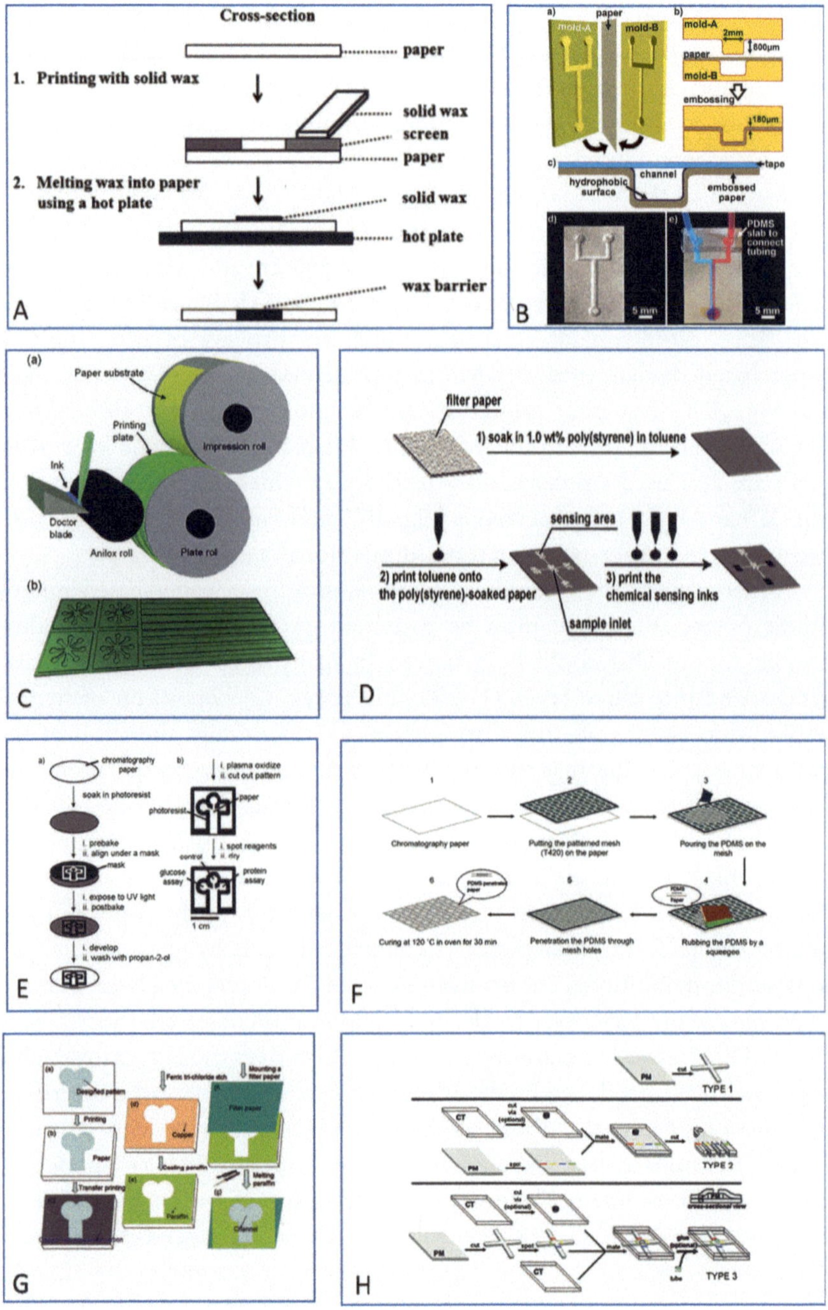

Fig. 3 Different types of fabrication techniques for developing μPADs. (A) Wax screen printing. The schematic diagram describes the steps involved in the procedure of the wax screen printing method. (B) Embossing or cut-and-stack process,[50] describes the fabrication of low-cost open channel paper-based microfluidic device that has been constructed by embossing channel on Whatman filter. (C) Flexographically printed[47]

(Continued)

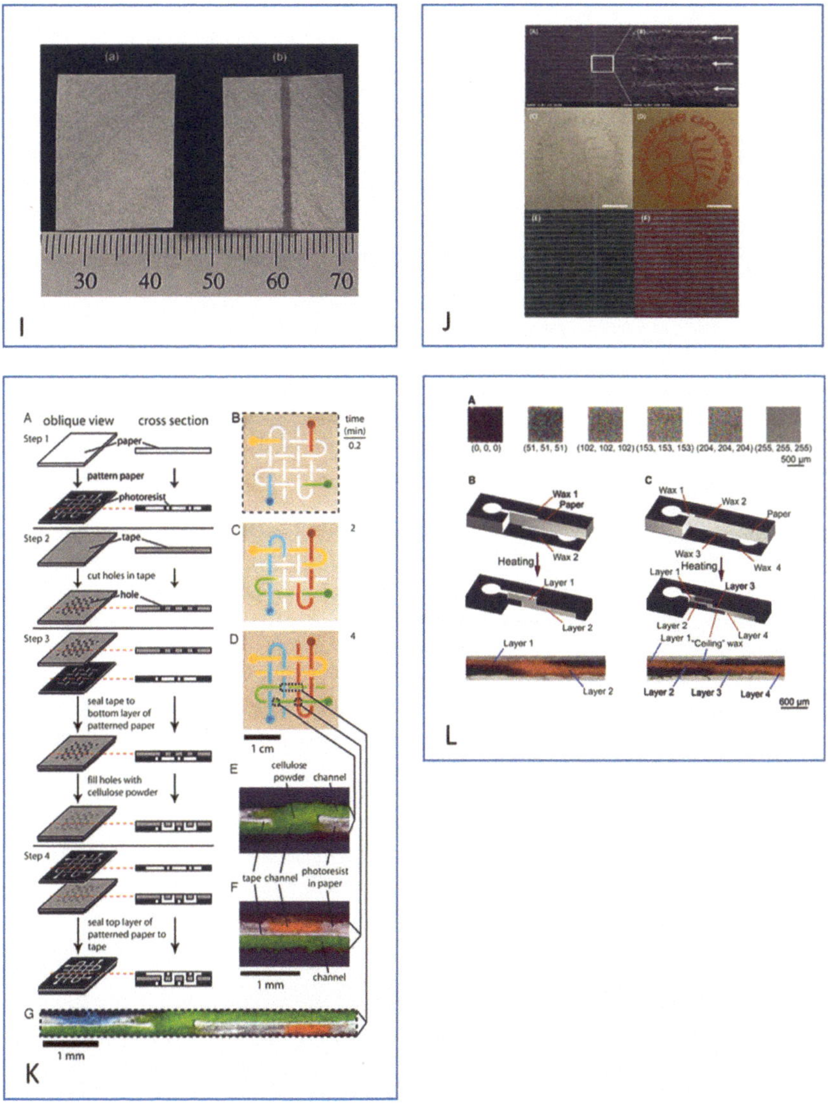

Fig. 3—Cont'd liquid guiding hydrophobic barrier structures are printed onto porous paper with the help of flexography units. (D) Inkjet printing[41] describes the process of fabrication of a microfluidic optical sensing device made from paper that requires single printing apparatus. (E) FLASH technique,[61] where paper-based microfluidic device is prepared from photolithography technique but the most important significance of this technique is that it requires only UV lamp and hot plate. (F) PDMS process,[51] which involves the fabrication of μPADs onto chromatography paper by patterning hydrophilic channels with PDMS through screen printing. (G) Printed circuit technology[52] is the technique which involves the fabrication of paper-based microfluidic device using printed circuit technology that successfully generates hydrophilic patterns onto the filter paper with a straight hydrophobic wall. (H) Two-dimensional shaping[44] is the procedure for the fabrication of lateral flow device formed from paper and nitrocellulose which is based on

(Continued)

Fig. 3—Cont'd two-dimensional shaping by computer-controlled knife. (I) Plasma treatment[62] is the method in which the paper-based microfluidic devices are formed with hydrophilic channels using plasma treatment. This process is effective for fabricating various components such as switches, microreactors, and microfilters as it does not affect the topography and flexibility of the paper surface. (J) Laser treatment[48] is the process of fabrication of hydrophilic patterns on paper-based microfluidic device on a hydrophobic surface using laser treatment. (K) Layered paper and tape[39] is the technique where a 3D microfluidic device is fabricated by creating multiple layers with adhesive tape and patterned paper. (L) Three-dimensional wax printing is the process of fabricating 3D microchannels in single layer of cellulose paper by using wax printing. Multiple channels are patterned at different layers along the thickness of paper which are interlinked with each. *Panel (A): Reprinted/adapted by permission Dungchai W, Chailapakul O, Henry CS. A low-cost, simple, and rapid fabrication method for paper-based microfluidics using wax screen-printing.* Analyst *2011;136:77–82. https://doi.org/10.1039/c0an00406e. Panel (B): Reprinted/adapted by permission from Thuo MM, Martinez RV, Lan WJ, et al. Fabrication of low-cost paper-based microfluidic devices by embossing or cut-and-stack methods.* Chem Mater. *2014;26(14):4230–4237. https://doi.org/10.1021/cm501596s. Copyright © 2014 American Chemical Society. Panel (C): Reprinted/adapted by permission from Olkkonen J, Lehtinen K, Erho T. Flexographically printed fluidic structures in paper.* Anal Chem. *2010;82(24):10246–10250. https://doi.org/10.1021/ac1027066. Copyright © 2010 American Chemical Society. Panel (D): Reprinted/adapted by permission from Abe K, Suzuki K, Citterio D. Inkjet-printed microfluidic multianalyte chemical sensing paper.* Anal Chem. *2008;80(18):6928–6934. https://doi.org/10.1021/ac800604v, Copyright © 2008 American Chemical Society. Panel (E): Reprinted/adapted by permission from Martinez AW, Phillips ST, Wiley BJ, Gupta M, Whitesides GM. FLASH: a rapid method for prototyping paper-based microfluidic devices.* Lab Chip *2008;8:2146–2150. https://doi.org/10.1039/b811135a, Copyright © Royal Society of Chemistry (2008). Panel (F): Reprinted/adapted by permission (CC BY 3.0) from Mohammadi S, Maeki M, Mohamadi RM, Ishida A, Tani H, Tokeshi M. An instrument-free, screen-printed paper microfluidic device that enables bio and chemical sensing.* Analyst. *2015;140:6493–6499. https://doi.org/10.1039/c5an00909j. Panel (G): Reprinted/adapted by permission from Zhang AL, Zha Y. Fabrication of paper-based microfluidic device using printed circuit technology.* AIP Adv. *2012;2:022171, https://doi.org/10.1063/1.4733346; licensed under a Creative Commons Attribution (CC BY) license. Panel (H): Reprinted/adapted by permission from Fenton EM, Mascarenas MR, López GP, Sibbett, SS. Multiplex lateral-flow test strips fabricated by two-dimensional shaping.* ACS Appl Mater Interfaces. *2009;1(1):124–129. https://doi.org/10.1021/am800043z, Copyright © 2008 American Chemical Society. Panel (I): Reprinted/adapted by permission from Li X, Tian J, Nguyen T, Shen W. Paper-based microfluidic devices by plasma treatment.* Anal Chem. *2008;80(23):9131–9134. https://doi.org/10.1021/ac801729t, Copyright © 2008 American Chemical Society. Panel (J): Reprinted/adapted by permission from Chitnis G, Ding Z, Chang CL, Savran CA, Ziaie B. Laser-treated hydrophobic paper: an inexpensive microfluidic platform.* Lab Chip *2011;11(11):1161–1165. https://doi.org/10.1039/c0lc00512f, Copyright © Royal Society of Chemistry (2011). Panel (K): Reprinted/adapted by permission from Martinez AW, Phillips ST, Whitesides GM. Three-dimensional microfluidic devices fabricated in layered paper and tape.* Proc Natl Acad Sci. U S A. *2008;105(50):19606–19611. https://doi.org/10.1073/pnas.0810903105, Copyright (2008) National Academy of Sciences, USA. Panel (L): Reprinted/adapted by permission from Li X, Liu X. Fabrication of three-dimensional microfluidic channels in a single layer of cellulose paper.* Microfluid Nanofluid. *2014;16:819–827. https://doi.org/10.1007/s10404-014-1340-z, Springer Nature, Copyright © 2014, Springer-Verlag Berlin Heidelberg.*

making the system simple, robust and thereby lowering the cost of the entire fabrication process.[64,65] Wax-based fabrication techniques use nontoxic patterning agents that are low-cost. "Wax printing" is a recent impregnation technology first introduced by Lu *et al.*,[43] which is highly compatible with mass production, although it involves a two-step process that includes printing and heating.[65] The drawback of this process is the high cost of the wax printer and the requirement of some extra time to melt the wax.

An alternate to wax printing is the ink stamping method, which was introduced by Curto *et al.*[57] Besides, another approach is the double-sided sticking method used by Akyazi *et al.*[66] These techniques are simple but require manual operation. As a result, the reproducibility of the devices is low. Despite the limitations, many authors have suggested that these processes could be scaled for mass production.[66]

For the third type of channel formation on paper-based devices, the paper surface is chemically modified with the addition of cellulose reactive agents (generally AKD) that forms hydrophobic regions when it reacts with the "–OH" group of cellulose.[63] Plasma treatment[42] and inkjet printing[58] methods are used for the formation of hydrophobic AKD channels. Ink-jet printing has evolved as an alternative for AKD channel formation.[67] Other methods of patterning hydrophilic/hydrophobic channels have also been studied such as PDMS patterning,[68] flexographic printing,[47] ink-jet etching,[41] screen printing,[69] printed circuit technology,[52] subtractive laser treatment.[48] The above-mentioned techniques are not considered as ideal fabrication methods for mass scale-up due to multiple fabrication steps like, the requirement of expensive equipment, and multistep post-processing.

3. Fluid flow control and liquid handling

Fluids can be contained in μPADs by using hydrophobic material for patterning but their efficiency to control fluidic flow in the device has not been demonstrated. Flow management is needed at various usages such as time-dependent addition of reagent/sample, sequential additions or dilutions[70]; thus new structures such as valves, switches, and fluidic timers are established to manage flow within the μPADs. Improvements have been made to control the fluid in commercially available filter paper-based devices, but a lot of progress is required for the accurate control of fluid transport kinetics. A modified lab-engineered paper sheet was developed by Böhm and his co-workers to modulate and control the fluid flow in a microfluidic channel.[71] These lab-engineered papers tuned the fiber density by controlling the paper density and porosity such that it prevents the side effects caused by chemical paper additives. Different fibers such as cotton linter pulp and eucalyptus sulfate, possessing different channel widths and porosities were used to investigate the capillary-driven transport of aqueous solutions.

The most challenging part is to control the fluid flow behavior.[71] Jeong *et al.* reported that the main challenge for flow management in μPADs is the lack of sufficient fluid on the paper.[72] Different parameters are responsible for the smooth functioning of a paper-based device but the guiding force for liquid wicking in cellulose fiber is the capillary motion on the paper. It has been well established that due to capillary action, no external pump[15] is required for fluid transportation on the paper. The only drawback is that the transportation of fluid is non-directional due to the paper's isotropic wicking behavior. Commercially available paper varies in properties like selective porosity, presence of chemical additives, and the types of fiber used. The device capability can be improved by sealing and enclosing the μPADs. It isolates the channel and reagents from the external environment to protect it from contamination which makes the device simple to handle and cost-effective by eliminating the need for packaging. It is one of the most effective control flow methods as it enhances the wicking rate of fluids in the channel and decreases solvent evaporation which makes the device accurate and efficient. Schilling *et al.*[73] used the low-cost method to seal the top and bottom sections of the fluidic channel by using laser printer toner. Wax printing is used to generate fully enclosed channels without using sealant.[74] Both processes result in improved capabilities of successful design. These methods accelerate or delay fluid transport as they do not change the hydrophilic channel shape or paper's porous structure. In paper-based microfluidic devices, the following methods could be used to control the fluid flow: (i) geometrical approach, (ii) chemical processes, and (iii) mechanical method. The process details are referenced in Figs. 4–6.

3.1 Geometrical approaches

This strategy helps to control the flow rate by optimizing the dimensions and structure of the microchannels. It can be achieved by varying the length and width of the channel. It can also be used to fabricate different volume metered pads so that different volume of reagents is released into the common channel. Lutz *et al.*[84] fabricated a 2D microfluidic network with pre-programmable disconnectors to attain a paper-based valve. These 2D

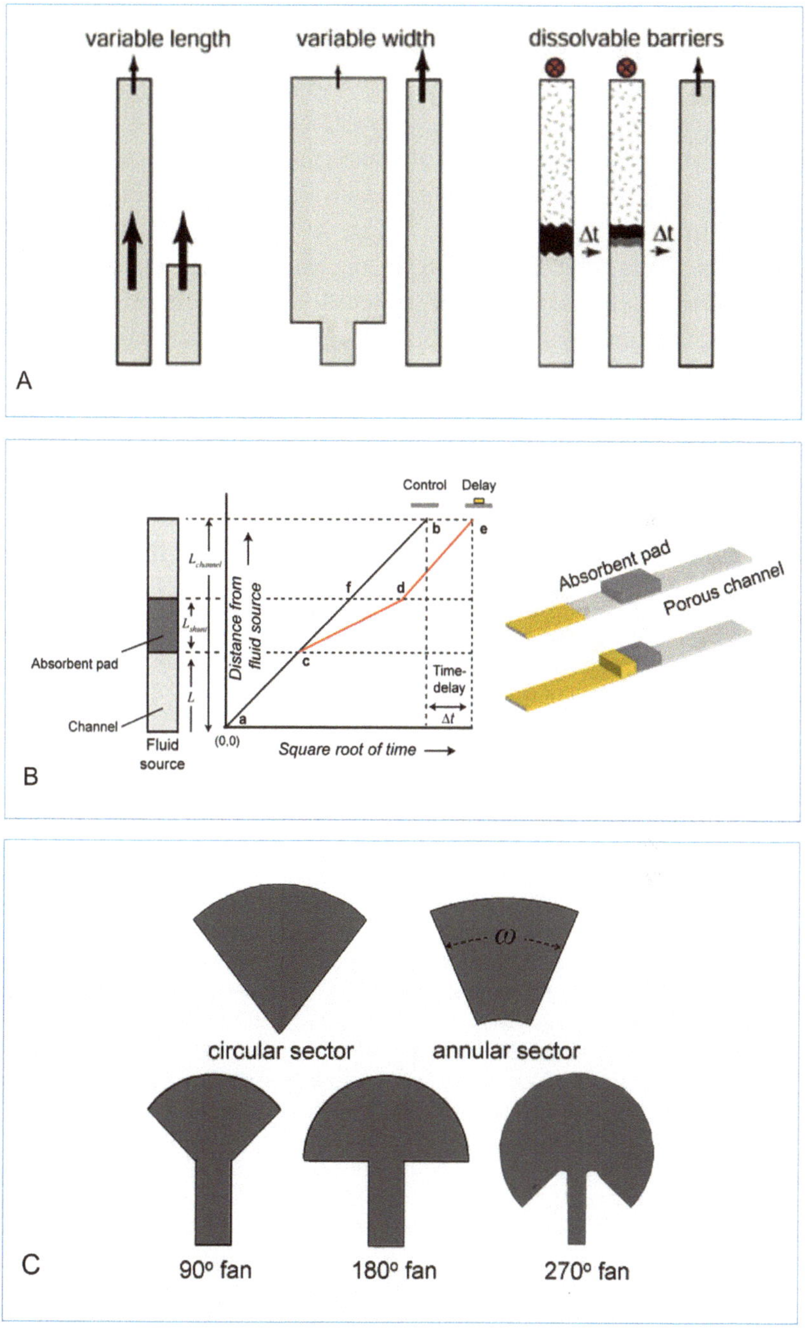

Fig. 4 See figure legend on next page.

networks consisted of legs of various lengths that get dipped in the fluid reservoir. Once the fluid level drops, the legs get disconnected in a predefined manner. When more complex fluidics is considered, this approach might not be similarly effective. Fu *et al.*[75] demonstrated the arrival time as well as the transportation time of multiple fluids in the sensing zone, it can be monitored by modifying the length or width of the microchannel as described in Fig. 4A, Toley *et al.*[76] (Fig. 4B) used a tunable absorbent pad-based shunt located on top of the main channel which adjusts the liquid flow process. Additionally, such regulated flow delay with novel "shunt" valving mechanism having delays between 3 and 20 min could be further modified by varying the parameters such as shunt material's dimensions. The significant advantage is that the same absorbent shunt can deliver a large volume of fluid. Moreover, to check their POC compatibility, they have been used to perform sequential delivery in μPAD device to detect a malaria antigen, but their demonstration has not been done under different environmental conditions such as high humidity or temperature. In case of lateral flow devices, quasi stationary flow has been generated with the use of 2 dimensional thin, porous membranes as shown in Fig. 4C).

Recently, 3D paper-based microfluidic devices were fabricated to control the fluid flow by mechanically activated valves to link or divide channels.[72] Toley *et al.*[81] fabricated a toolkit that contains automated paper microfluidic regulators that had used movable paper-based channels and fluid-stimulated swelling modules. The actuation time of the valve was

Fig. 4 Geometry control: (A) Controlled reagent transportation,[75] Describes the phenomena of controlling fluid transport by using variable length, width, and also dissolvable barriers in a 2D paper-based microfluidic device. (B) Tunable-Delay Shunts,[76] where fluid flow in paper-based microfluidic devices can be controlled through a porous channel by placing absorbent pad-based shunt in contact with the channel. (C) Complex shape dependent[77] represents different types of 2 dimensional thin porous membranes that help in generating quasi-stationary flow in lateral flow devices. *Panel (A): Reprinted/adapted by permission from Fu E, Lutz B, Kauffman P, Yager P. Controlled reagent transport in disposable 2D paper networks.* Lab Chip. *2010;10:918–920. https://doi.org/10.1039/b919614e, Copyright © Royal Society of Chemistry (2010). Panel (B): Reprinted/adapted by permission from Toley BJ, Mckenzie B, Liang T, Buser JR, Yager P, Fu E. Tunable-delay shunts for paper micro fluidic devices.* Anal Chem. *2013;85(23):11545–11552, Copyright © 2013 American Chemical Society. Panel (C): Reprinted/adapted by permission from Mendez S, Fenton EM, Gallegos GR, et al. Imbibition in porous membranes of complex shape: quasi-stationary flow in thin rectangular segments.* Langmuir. *2010;26(2):1380–1385. https://doi.org/10.1021/la902470b, Copyright © 2009 American Chemical Society.*

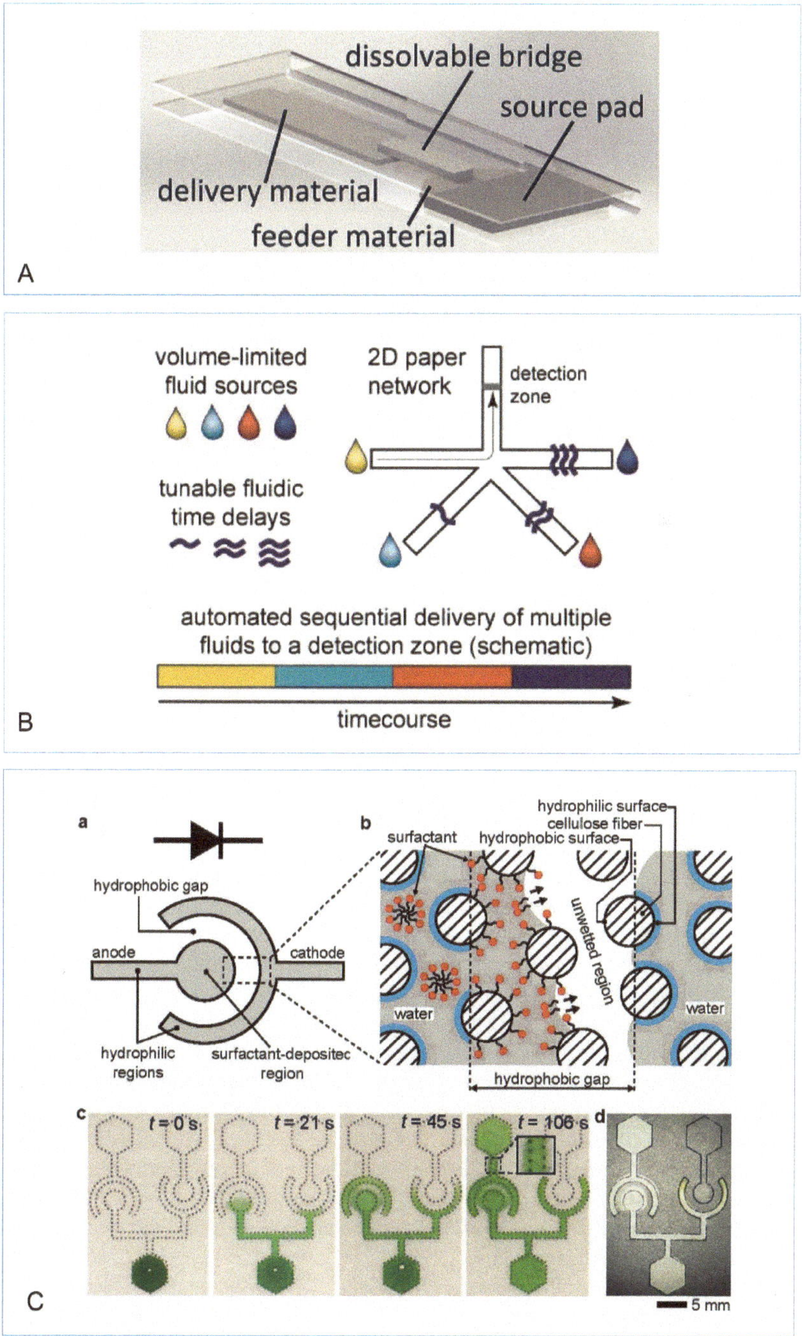

Fig. 5 See figure legend on next page.

controlled by shifting the length of the wicks. It was also demonstrated that these valves could execute a multi-step test for malaria pathogen detection. This opened the doors for multistep, serial analysis on μPADs; however, the mass production of these tools might be labor-intensive, time-consuming, and complicated.

3.2 Chemical processes

This method uses chemicals to create temporary flow channels or barriers such that it allows users to create time delays and control the flow rate. Lutz and his team fabricated the sugar fluidic resistor for creating fluid delay in multistep programming assays for paper diagnostics[79] as shown in Fig. 5B. For the diagnosis of the biomarker for malaria, a signal-amplified sandwich POC immunoassay was also designed.[85] Herein, solid sugar blocked the flows, and thereafter the flow resumed once the sugar was dissolved. Jahanshahi-Anbuhi *et al.*[86] fabricated the valve with a rapidly dissolving polymer named pullulan, to create a regulated time flow shut off valve system. Successful integration of this valve system was done in a complex pesticide assay where the timely sequential release of reagents in a limited resource setting was achieved to develop a complex assay. Similarly, Houghtaling *et al.* fabricated dissolvable bridges to control flow as described in Fig. 5A.[78] It was also reported that fluid control in a 2D paper-based device was done by a fluidic diode which consists of two terminal modules

Fig. 5 Chemical controlled. (A) Dissolvable bridges[78] is a method of fabricating dissolvable bridges for controlling fluid volume in different locations of paper-based microfluidic devices. (B) Dissolvable fluidic time delays[79] is a method to create fluidic time delays by applying sucrose to paper strips at different concentrations(10–70% saturation) and when it combines with paper network topology it creates time delays which help to build multi-step fluidic protocols. (C) Fluidic diode through hydrophobic gap[80] is fabricated onto the single layer of paper, since a diode is useful to stop or promote fluid wicking along the channel. Along with the diode, a trigger and a delay valve are fabricated to control two types of fluids sequentially. *Panel (A): Reprinted/adapted by permission from Houghtaling J, Liang T, Thiessen G, Fu E. Dissolvable bridges for manipulating fluid volumes in paper networks.* Anal Chem. *2013;85(23):11201–11204. https://doi.org/10.1021/ac4022677, Copyright © 2013 American Chemical Society. Panel (B): Reprinted/adapted by permission from Lutz B, Liang T, Fu E, Ramachandran S, Kauffman P, Yager P. Dissolvable fluidic time delays for programming multi-step assays in instrument-free paper diagnostics.* Lab Chip. *2013;13:2840–2847. https://doi.org/10.1039/c3lc50178g, Copyright © Royal Society of Chemistry (2013). Panel (C): Reprinted/adapted by permission from Chen H, Cogswell J, Anagnostopoulos C, Faghri M. A fluidic diode, valves, and a sequential-loading circuit fabricated on layered paper.* Lab Chip. *2012;12:2909–2913. https://doi.org/10.1039/c2lc20970e, Copyright © Royal Society of Chemistry (2012).*

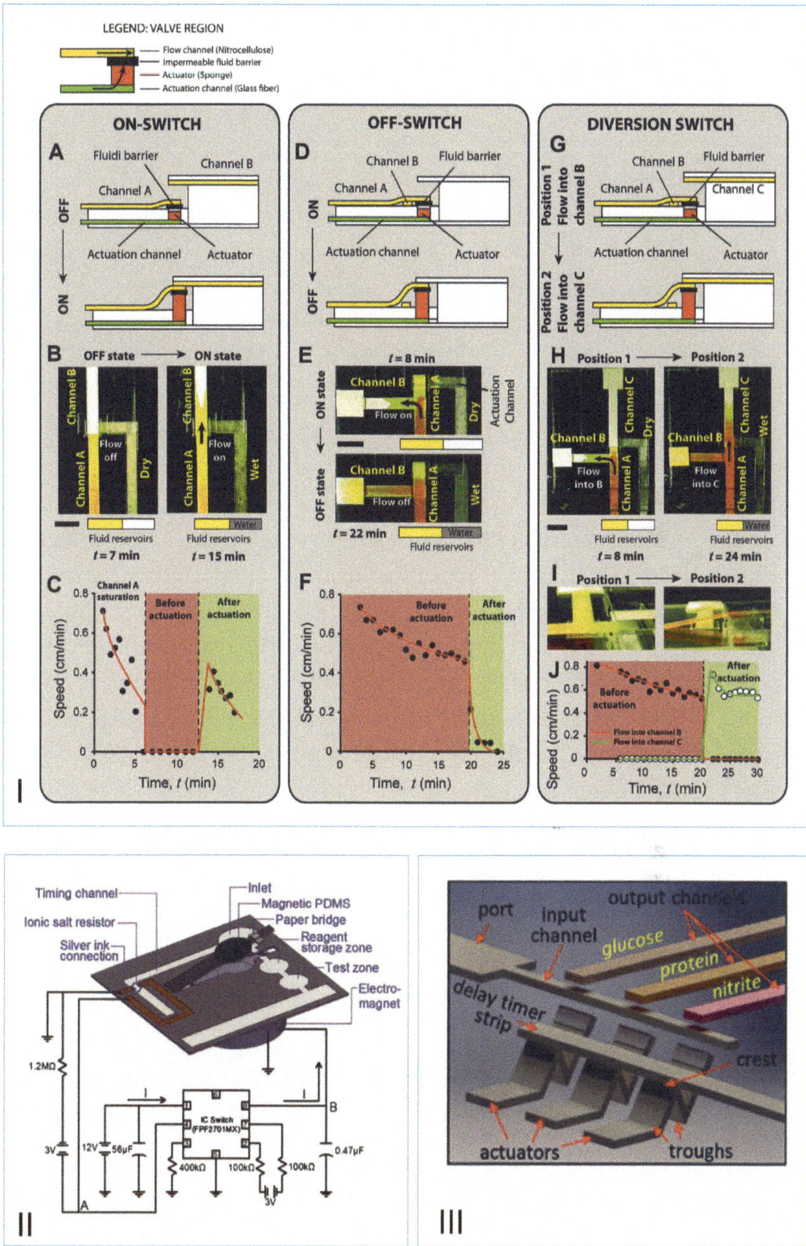

Fig. 6 Mechanical controlled: (I) Through valving toolkit[81] device consisting of a paper-based microfluidic valve which is automatically actuated by using movable paper strips and fully triggered expanding elements. (II) magnetic timing valve[82] is fabricated using paper-based magnetized cantilever valve which is driven by an electromagnet.

(Continued)

that prevent wicking alongside the path[80] as described in Fig. 5C. Noh *et al.*[87] developed a method to control the flow rate by controlling the amount of paraffin wax deposited in the channel.

3.3 Mechanical methods

This method uses mechanical contraptions in various forms to link or separate channels for fluid flow. Toley *et al.*[81] fabricated a tool kit with automatic valve actuation capabilities by using movable paper strips and fluid triggered expanding elements as shown in (Fig. 6A). Li *et al.*[82] fabricated a magnetic timing valve by the use of electromagnetic and ferromagnetic particles embedded in PDMS. It consists of an ionic resistor with a paper timing channel, that can detect the flow of solution and triggers the electromagnet which opens or closes the bridge valve as shown in (Fig. 6B). Similarly, Kong *et al.* fabricated a control valve using multiple chromatography papers to form crest and troughs as shown in (Fig. 6C). Valving system can be activated by the movement of folds in the paper-based channel by pressurizing a specific region for programmed sample delivery. Thus, the liquid delay was created in the pressurized area by increasing and decreasing the permeability and fluid resistance, respectively. By changing the applied pressure, it was demonstrated that the flow level was reduced in the pressurized area. This method developed the flow rate monitoring in a polypropylene sheet-based device for multiple fluid samples that perform a multistep calorimetric assay. Houghtaling *et al.*[78] developed a dissolvable structure of bridge with a shutoff valve that delivers various volumes from constant supply to different paths in a paper microfluidic network. Various parameters

Fig. 6—Cont'd It controls the timely flow of fluid by triggering the electromagnet to open or close the paper-based cantilever valve. (III) Folded actuator strip[83] is a paper-based actuator which is fabricated to control the fluid flow between input and output channels. Chromatography paper is used with multiple folds that function as a programmable switch of different fluidic connections. *(I) Reprinted/adapted by permission from Toley BJ, Wang JA, Gupta M, et al. A versatile valving toolkit for automating fluidic operations in paper microfluidic devices.* Lab Chip. *2015;15:1432–1444. https://doi.org/10.1039/c4lc01155d, Copyright © Royal Society of Chemistry (2015). (II) Reprinted/adapted by permission from Li X, Zwanenburg P, Liu X. Magnetic timing valves for fluid control in paper-based microfluidics.* Lab Chip. *2013;13:2609–2614. https://doi.org/10.1039/c3lc00006k, Copyright © Royal Society of Chemistry (2013). (III) Reprinted/adapted by permission from Kong T, Flanigan S, Weinstein M, Kalwa U, Legner C, Pandey S. Lab on a Chip A fast, reconfigurable flow switch for paper microfluidics based on selective wetting of folded paper actuator strips. 2017;17:3621–3633. https://doi.org/10.1039/c7lc00620a, Copyright © Royal Society of Chemistry (2017).*

could be adjusted in the bridge structure like geometry and composition of the material that carries volume in the range from 10 to 80 mL. It helps generate more efficient and sophisticated assays as it enables simple user loading. However, many mechanical fluid control systems in the microfluidic domain fail due to complex fabrication and operation protocols. Moreover, the cost is high for the fabrication of sophisticated fluidic control systems. Due to comparatively lower costs and ease of operation, μPADs could have a better reach to the end-users especially in resource limited countries as compared to conventional microfluidic devices.

4. Applications of paper-based microfluidic devices

μPADs have countless applications including point of care (POC), in healthcare diagnostics, solid, and food analysis, etc. In epidemic or pandemic situations, μPAD test strips have had a huge contribution toward rapid detection of specific antigens and these detections were majorly used in recent viral infections like COVID, SARS, etc. Although these devices have a very fast response time, various constraints exist in designing such μPADs since the design for each case is different and specific to the applications. The area of the application is categorized into three major groups, e.g., (1) Health care diagnostics, (2) Environmental assessment and (3) Food and beverage as shown in Table 3. The application might also be categorized (refer to Table 4) based upon its working mechanism and process and these are

Table 3 Areas of application for paper-based microfluidic devices.

Applications	Methods	References
Healthcare diagnostics	Analysis of biomolecules like proteins, hormones, neurotransmitters, etc. Analysis of small molecules like glucose, nucleic acid, etc. Pregnancy tests Blood grouping and testing Drug sensing	88,89,90,91,92,93,94,23,95,96,97,98
Environmental assessment	Soil, water, air analysis, and metal detection, etc.	99
Food and beverages control	Foodborne, pesticides, water, and wine quality analysis, etc.	100,101,102

Table 4 Applications of paper-based microfluidic devices based upon their mechanism.

Category	Sub-category	Applications	References
Multiplexed color change chemistries	Clinical diagnostics	Body fluids analysis, e.g., urine, saliva, sputum, blood component analysis, etc.	103,104,63,105,106, 107,108,109
	Pharmaceutical analysis	4-Aminophenol, beta-lactam antibiotics, tetracycline, paracetamol and ibuprofen, ascorbic acid (AA), and dopamine (DA), oxytetracycline, etc.	110,111,112,113,114,115,116,117,118,29
	Environmental analysis	Metal ion detection, e.g., Ni, Fe, Cu, Cr, Pb, Cr, and Cd in different sources	38,119,120,121,100,122,123,124
Signal-enhanced immunoassays	Two-dimensional Paper Network	2D network for multiple reagent manipulation: hCG (human chorionic gonadotropin) and PfHRP2 (malarial antigen), influenza A and B	125,126,127,128,129
	Paper ELISA	A pen-based high-performance device, SERS immunoassay, eosin-labeled detection, enzymatic-based, nanoparticle-based, and polymerization-based color reactions, magnetic ELISA-based detection of dengue	130,131,132,93,133,134,135
	Sliding devices	Reduce pipetting steps: Demonstrated for Rabbit IgG, neonatal sepsis, also demonstrated sequential delivery of multiple reagents	136,137,138

Nucleic acid amplification tests	Bacteria and viruses and genetic codes identification through LAMP, RPA, HDA, RCA, iSDA in a Whatman, chromatography paper, glass fiber, PES, FTA membrane.	139,140,141,142,143,144,145,146,147,148
Blood plasma separation	Isolating plasma from whole blood to avoid blood cell interference during analyzing analytes like glucose.	149,150,151,152
Tissue engineering and cancer diagnostics	As an ideal biocompatible, biodegradable, mechanically strong base to grow cells on it and helps to detect cancer. Human adipose-derived stem cells (hADSCs)	153,154,155,156,157,158,159,160,25,161
Energy generation and storage	Energy storage device (battery) using methanol and potassium hydroxide. RTIL electrochemical supercapacitor etc.	162,163,164,165

(1) Multiplexed color-changing chemistry, (2) Signal-enhanced immunoassays, (3) Nucleic acid amplification tests, (4) Blood Plasma Separation, (5) Tissue engineering and cancer diagnostics, (6) Energy generation and storage. There are also other ways to differentiate the applications. Additionally, the type of biological sample used for detection should also be considered while designing the μPADs for projected applications. Finally, the sensing modalities and readout methods are important; whether detection is based upon colorimetric, electrochemical, or other analytical methods. Cost of fabrication, portability, ease of use should be the main criteria for the application of μPADs in diagnostics. Ideally, an affordable diagnostic device could be mass-produced and readily available to consumers. In resource-limited settings, low-cost, simple, and portable diagnostics is the need of the hour to monitor and control diseases, and μPADs emerge as the most viable option (Fig. 7).

In short, the application of paper-based devices in diagnostic technologies has seen a rapid growth, specifically for rapid detection kits.[6] Most of the paper-based devices have been used for colorimetric detection and it has been well documented in various literatures.[45,104,112,137] Herein, the study is more focused on other unique applications which are expected to grow and gain popularity soon.

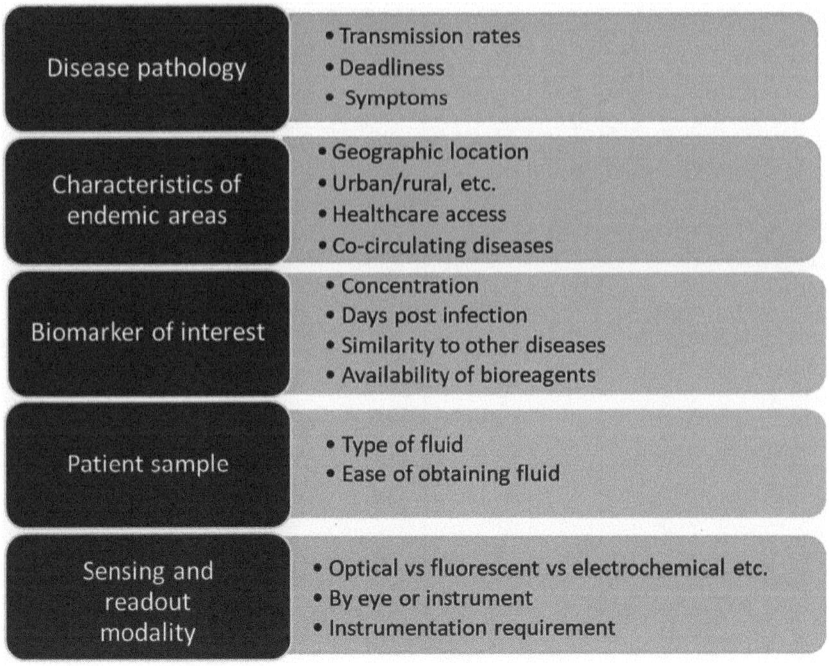

Fig. 7 Design consideration for point of care (POC) devices for disease diagnostics.

It is already known that the paper-based rapid tests are quick but sometimes lack proof of itself as a gold standard like a genetic test (RT-PCR). In that scenario, the requirement of nucleic acid amplification tests (Fig. 8) comes into the picture. Recently, with advancements in other supporting technologies, μPADs are being used to decipher genetic codes of foreign species through nucleic acid amplification (LAMP, RPA, HDA, RCA, iSDA). Various materials have been developed (Whatman, chromatography paper, glass fiber, PES, FTA membrane) to perform those amplification reactions and a few of them have been referenced in Fig. 8 such as the nucleic acid amplification test developed by Rohrman and Richards-Kortum (Fig. 8-(I)) on the microfluidic device fabricated using a combination of paper, glass, and plastic.[139] Another device was fabricated to incorporate LAMP technology as shown in Fig. 8-(II),[140] which is integrated with various steps like sample preparation, amplification, and molecular diagnostics on a single device. PDMS –paper-based device was fabricated for LAMP to detect infectious disease (Fig. 8-(III)).[166] Fig. 8-(IV)[142] and (V)[143] describe the process of Ebola RNA detection and DNA amplification for *Mycobacterium tuberculosis*. Paper-based device was used for the detection of Influenza A (H1N1) as shown in Fig. 8-(VI).[146] Fig. 8-(VII) represents the integrated molecular diagnostic chip for nucleic acid detection[144] and Fig. 8-(VIII) shows an integrated lateral paper based device along with biosensor for DNA amplification and detection in POCT.[147]

Likewise, the applicability of μPADs is not restricted to medical diagnostics; they have been used for monitoring environmental safety, i.e., for the detection of water, soil, or air contaminants. In resource-limited territories, the ability to determine the quality of the drinking water with a low-cost paper-based microfluidic device would be very advantageous. Food quality monitoring in the supermarket or within the household refrigerator is akin to quality control near the production or packaging units. Furthermore, veterinary diagnostics is another application area for μPADs as it involves human health via human-animal interaction, food, animal transportation, and clothing[27] and the regulatory compliance is less strict for animals compared to human trials. The application does not restrict to detection and diagnostics, moreover, there is huge scope to develop sample preparation, energy generation, and cell/tissue culture systems. Such capabilities are just the highlights of the future POC devices; a few works such as blood separation (Fig. 9), cancer diagnostics (Fig. 10), tissue engineering (Fig. 10), and energy harvesting (Fig. 11) has been shown graphically from different literatures. (See Fig. 12.)

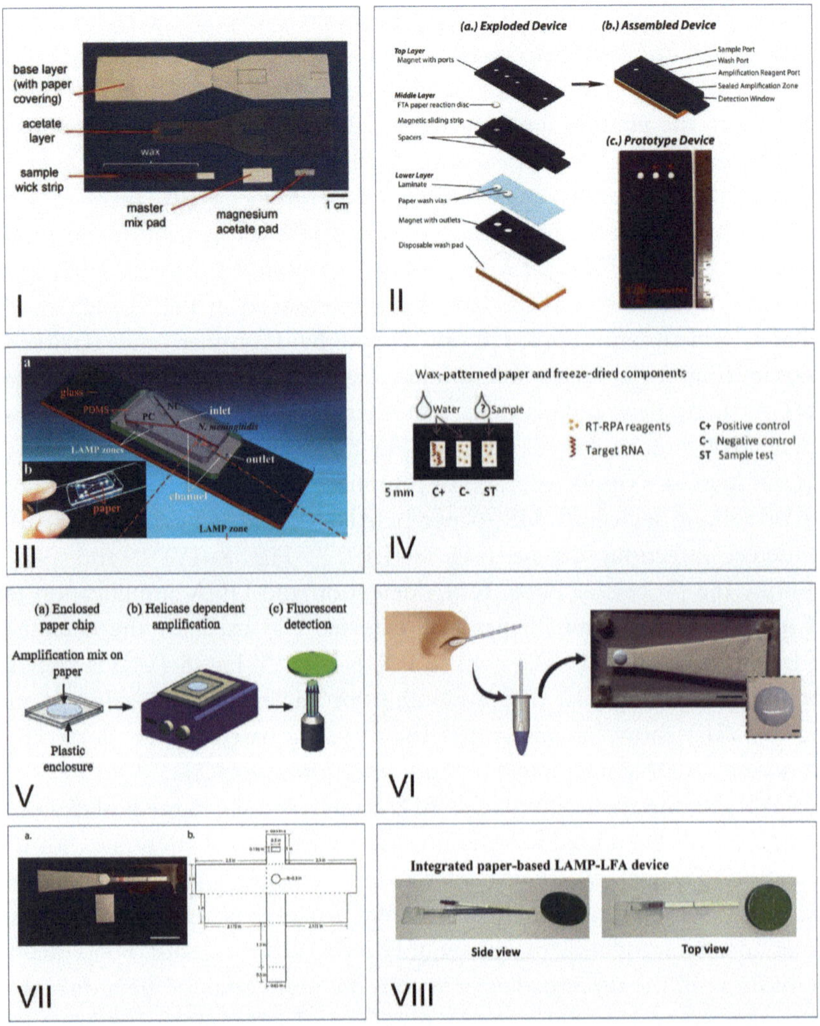

Fig. 8 Nucleic acid amplification test (I) Recombinase polymerase amplification of HIV DNA.[139] The device was fabricated by the combination of paper, glass fiber, and plastic and can perform isothermal, enzymatic amplification of HIV DNA. (II) Full integration of sample preparation with LAMP and limit of detection of 5 cells.[140] This paper-based microfluidic device integrates with sample preparation, amplification, and molecular diagnostics into a single device. (III) PDMS-paper hybrid with LAMP to detect infectious disease. (IV) Detection of RNA and multiplexed analysis for Ebola virus.[142] Multilayer geometry was fabricated in paper using wax printing and different types of dyes were used for detection. (V) DNA amplification of *Mycobacterium tuberculosis* on paper.[143] (VI) Paper-based microfluidic device was used for the extraction of RNA with in-situ isothermal amplification for the detection of Influenza A (H1N1) from clinical specimens.[146] (VII) Integrated paper-fluidic molecular diagnostic chip for extraction, amplification, and

(Continued)

In most of the developing or underdeveloped countries where 1/3 of deaths occur from curable diseases such as malaria, tuberculosis, lower respiratory infections, and diarrhea, etc., μPAD's could be of great importance. The characteristics of the place with infections also need to be considered, such as whether the area is urban or rural, the availability of healthcare, and if similar diseases coexist in the area. During such time, the technical aspects of detection should be considered, such as the specificity of detection, the concentration range, and the affinity of biomarkers. Sometimes, it does not reach the masses due to a lack of infrastructure, financial support, and lack of trained personnel. The paper-based diagnostic assay is simple, low cost, easy to operate, robust, affordable, and portable; the combination of all these features could be lifesaving in resource-limited countries.

Fig. 8—Cont'd detection of nucleic acids from clinical samples.[144] (VIII) Integrated lateral paper-based flow device with biosensor was fabricated with FTA card, glass fiber, and LAMP for DNA amplification and detection in the POCT.[147] *(I) Reprinted/adapted by permission from Rohrman BA, Richards-Kortum RR. A paper and plastic device for performing recombinase polymerase amplification of HIV DNA.* Lab Chip. *2012;12:3082–3088. https://doi.org/10.1039/c2lc40423k, Copyright © Royal Society of Chemistry (2012). (II) Reprinted/adapted by permission from Dou M, Dominguez DC, Li X, Sanchez J, Scott G. A versatile PDMS/paper hybrid microfluidic platform for sensitive infectious disease diagnosis.* Anal Chem. *2014;86(15):7978–7986. https://doi.org/10.1021/ac5021694, Copyright © 2014 American Chemical Society. (III) Reprinted/adapted by permission from Connelly JT, Rolland JP, Whitesides GM. "Paper machine" for molecular diagnostics.* Anal Chem. *2015;87(15):7595–7601. https://doi.org/10.1021/acs.analchem.5b00411, Copyright © 2015 American Chemical Society (further permissions related to the material excerpted should be directed to the ACS). (IV) Reprinted/adapted by permission (CC BY) from Magro L, Jacquelin B, Escadafal C, et al. Paper-based RNA detection and multiplexed analysis for Ebola virus diagnostics.* Sci Rep. *2017;7:1347. https://doi.org/10.1038/s41598-017-00758-9. (V) Reprinted/adapted by permission from Shetty P, Ghosh D, Singh M, Tripathi A, Paul D. Rapid amplification of Mycobacterium tuberculosis DNA on a paper substrate.* RSC Adv. *2016;6:56205–56212. https://doi.org/10.1039/c6ra07529k, Copyright © Royal Society of Chemistry (2016). (VI) Reprinted/adapted by permission from Rodriguez NM, Linnes JC, Fan A, Ellenson CK, Pollock NR, Klapperich CM. Paper-based RNA extraction, in situ isothermal amplification, and lateral flow detection for low-cost, rapid diagnosis of influenza A (H1N1) from clinical specimens.* Anal Chem. *2015;87(15):7872–7879. https://doi.org/10.1021/acs.analchem.5b01594, Copyright © 2015 American Chemical Society. (VII) Reprinted/adapted by permission from Rodriguez NM, Wong WS, Liu L, Dewar R, Klapperich CM. A fully integrated paperfluidic molecular diagnostic chip for the extraction, amplification, and detection of nucleic acids from clinical samples.* Lab Chip *2016;16:753–763. https://doi.org/10.1039/c5lc01392e, Copyright © Royal Society of Chemistry (2016). (VIII) Reprinted/adapted by permission from Choi JR, Hu J, Gong Y, et al. An integrated lateral flow assay for effective DNA amplification and detection at the point of care.* Analyst. *2016;141:2930–2939, Copyright © Royal Society of Chemistry (2016).*

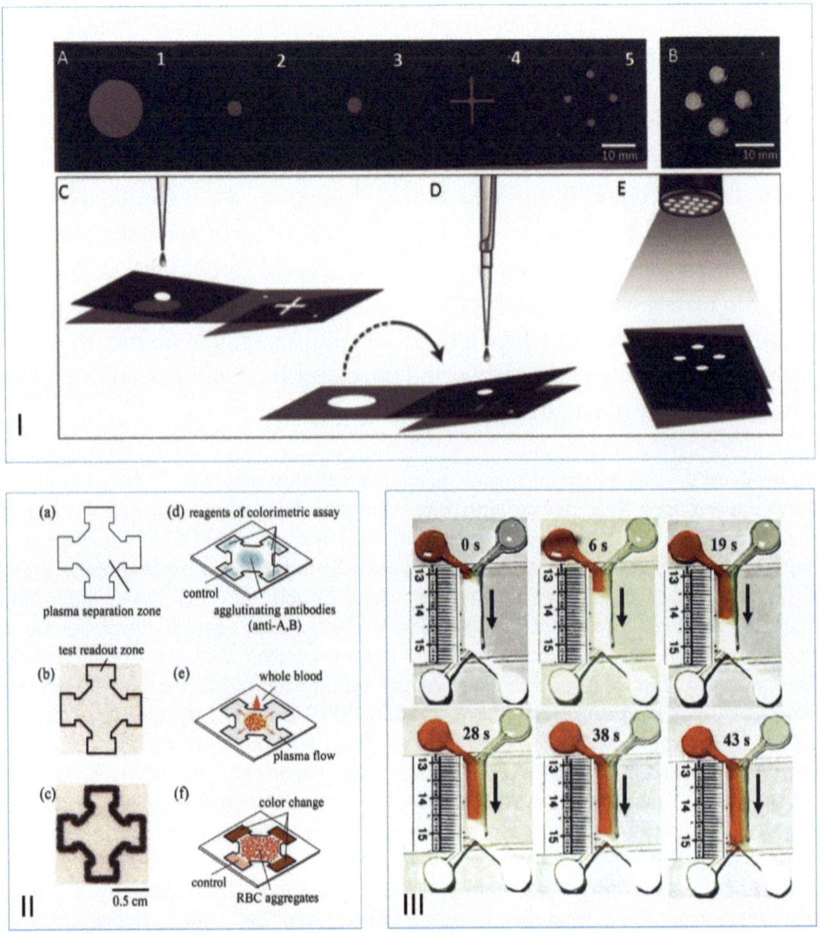

Fig. 9 Blood plasma separation (I) Paper-Origami-based multiplexed malaria diagnostics from whole blood[149] where paper-based microfluidic device is fabricated by the combination of paper folding origami technique, capillary wicking, LAMP, and array-based fluorescence detection. (II) Integrated separation of blood plasma from whole blood on microfluidic paper-based analytical devices.[150] (III) Capillary-driven blood plasma separation on paper-based devices with H channels.[151] *(I) Reprinted/adapted by permission from Xu G, Nolder D, Reboud J, et al. Paper-origami-based multiplexed malaria diagnostics from whole blood.* Angew Chem Int Ed. *2016;55:15250–15253. https://doi.org/10.1002/anie.201606060 (CC BY 4.0) © 2016. Published by Wiley-VCH Verlag GmbH & Co. KGaA. (II) Reprinted/adapted by permission from Yang X, Forouzan O, Brown TP, Shevkoplyas SS. Integrated separation of blood plasma from whole blood for microfluidic paper-based analytical devices.* Lab Chip. *2012;12:274–280. https://doi.org/10.1039/c1lc20803a, Copyright © Royal Society of Chemistry (2012). (III) Reprinted/adapted by permission from Kar S, Maiti TK, Chakraborty S. Capillarity-driven blood plasma separation on paper-based devices.* Analyst. *2015;140:6473–6476. https://doi.org/10.1039/c5an00849b, Copyright © Royal Society of Chemistry (2015).*

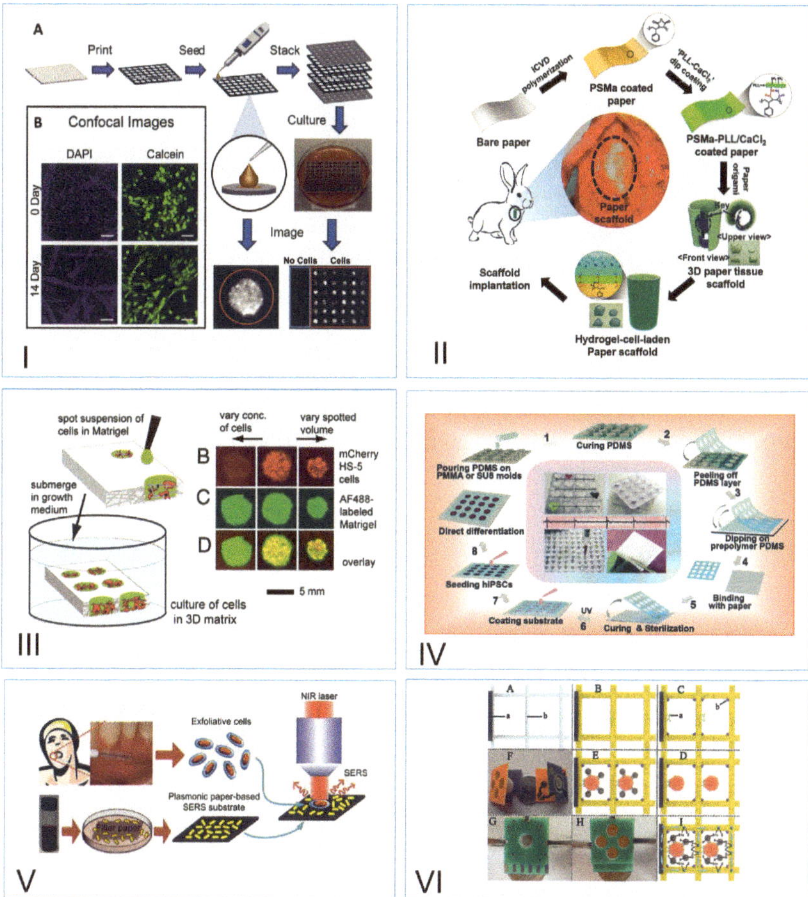

Fig. 10 Tissue engineering: (I) A filter paper-based microfluidic device was fabricated and enhanced to culture VIC for disease model of the aortic valve.[154] (II) Kim *et al.*[156] described the process of tissue regeneration by multifold scaffold system using the origami-based approach. (III) Derda *et al.*[155] fabricated a 3D microfluidic device that consists of stacked layers of paper or porous material which supports hydrogel-containing cells that help in the growth of cells. (IV) Schematic diagram of the fabrication steps of a paper-based microfluidic device, where the paper array is created by the combination of PDMS mold and paper substrate for the culture, proliferation, and direct differentiation of human iPSC cells.[157] (V) Liu *et al.*[160] describes the phenomena of combination of exfoliative cytology and SERS technology for cancer screening. (VI) Wu *et al.*[25] lab on a paper device was fabricated using ECL methodology for cancer cell detection. *(I) Reprinted/adapted by permission from Sapp MC, Fares, HJ, Estrada, AC, Grande-Allen KJ. Multilayer three-dimensional filter paper constructs for the culture and analysis of aortic valvular interstitial cells.* Acta Biomater. *2015;13:199–206. https://doi.org/10.1016/j.actbio.2014.11.039, Copyright © Elsevier (2014). (II) Reprinted/adapted by permission from Kim SH, Lee HR, Yu SJ, et al. Hydrogel-laden paper scaffold system for origami-based tissue engineering.* Proc Natl Acad Sci U S A. *2015;112:15426–15431. https://doi.org/10.1073/pnas.1504745112, Copyright (2015) National Academy of Sciences, USA. (III) Reprinted/adapted*

(Continued)

5. Current limitations and future works

Large-scale production of μPADs could utilize printing methods such as flexography, screen printing, inkjet printing, etc. However, to perform successful mass production of these devices, various other fabrication techniques are also required at the current stage. In other words, for fabricating μPADs with required multilayer complex structures, large-scale printing methods should be investigated such that μPADs could be produced at low cost, high resolution, using easy process steps. μPADs have expanded rapidly, however, there are still few challenges that need to be addressed. One of the key challenges is to find a suitable bed, a new easy to use cheaper porous material with good biological compatibility is highly required, so that μPADs could be used for a wide range of applications in the future. The challenges and opportunities are not only limited to technicality but other aspects also as discussed in the next few sections.

5.1 Market analysis

The worldwide IVD market size is expected to rise by about 11.5 billion USD in the next 5 years to reach ~96 billion by 2025, at a CAGR of 2.6%.[12] The IVD workspace is a very interdisciplinary field of research and different streams of science and technological innovations are involved. Recent developments in the field of physics, biotechnology, and electronics have made Point-of-Care (POC) IVD devices more robust, accurate, and as powerful as a bench-top system in a well-established pathology laboratory. These research achievements have led to the growth of the overall IVD

Fig. 10—Cont'd by permission from Derda R, Laromaine A, Mammoto A, et al. Paper-supported 3D cell culture for tissue-based bioassays. *Proc Natl Acad Sci USA*. 2009;106(44):18457–18462. https://doi.org/10.1073/pnas.0910666106, Copyright (2009) National Academy of Sciences, USA. (IV) Reprinted/adapted by permission from Wang L, Xu C, Zhu Y, et al. Human induced pluripotent stem cell-derived beating cardiac tissues on paper. *Lab Chip*. 2015;15:4283–4290. https://doi.org/10.1039/c5lc00919g, Copyright © Royal Society of Chemistry (2015). (V) Reprinted/adapted by permission from Liu Q, Wang J, Wang B, et al. Paper-based plasmonic platform for sensitive, non-invasive, and rapid cancer screening. *Biosens Bioelectron*. 2014;54:128–134. https://doi.org/10.1016/j.bios.2013.10.067, Copyright © Elsevier (2013). (VI) Reprinted/adapted by permission from Wu L, Ma C, Ge L, et al. Paper-based electrochemiluminescence origami cyto-device for multiple cancer cells detection using porous AuPd alloy as catalytically promoted nanolabels. *Biosens Bioelectron*. 2015;63:450–457. https://doi.org/10.1016/j.bios.2014.07.077, Copyright © Elsevier (2015).

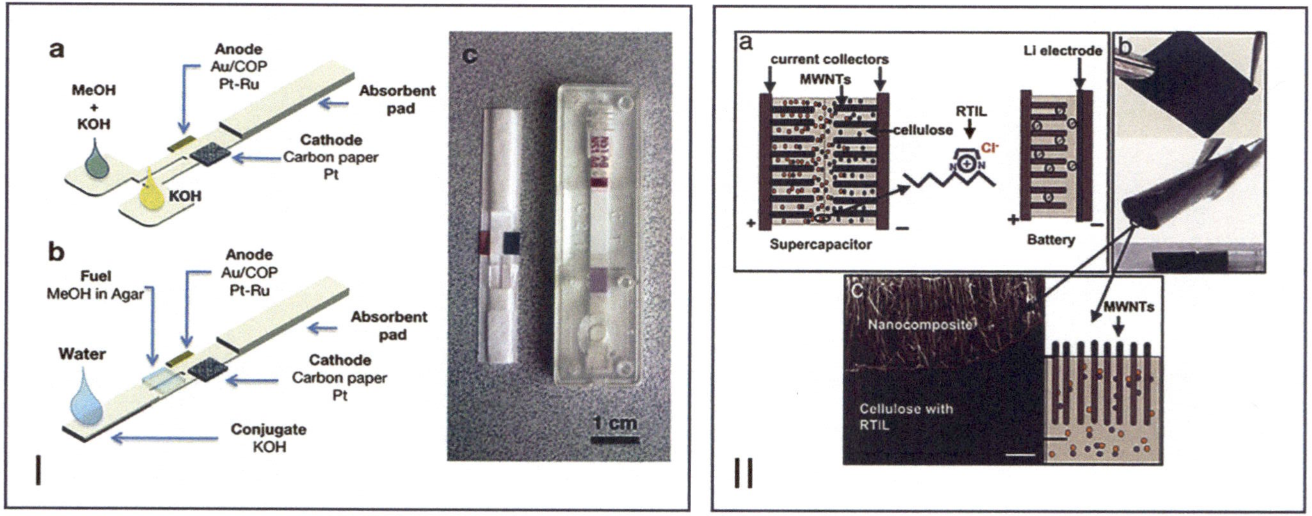

Fig. 11 Energy harvesting using paper-based microfluidic devices. (I) Microfluidic fuel cells on paper-based upon lateral flow device.[163] The device is a capillary-based autonomous sensing system that delivers power density in the range of 1–5 mW/cm^2. (II) Nano composite paper-based flexible energy storage devices.[164] The device can function over a wide range of temperature, deformation, and electrolytes. *(I) Reprinted/adapted by permission from (CC BY 3.0) Esquivel JP, Del Campo FJ, Gómez De La Fuente JL, Rojas S, Sabaté N. Microfluidic fuel cells on paper: meeting the power needs of next generation lateral flow devices.* Energy Environ Sci. *2014;7:1744–1749. https://doi.org/10.1039/c3ee44044c, Copyright © Royal Society of Chemistry (2014). (II) Reprinted/adapted by permission from Pushparaj VL, Shaijumon MM, Kumar A. Flexible energy storage devices based on nanocomposite paper.* Proc Natl Acad Sci U S A. *2007;104(34):13574–13577. https://doi.org/10.1073/pnas.0706508104, Copyright (2007) National Academy of Sciences, USA.*

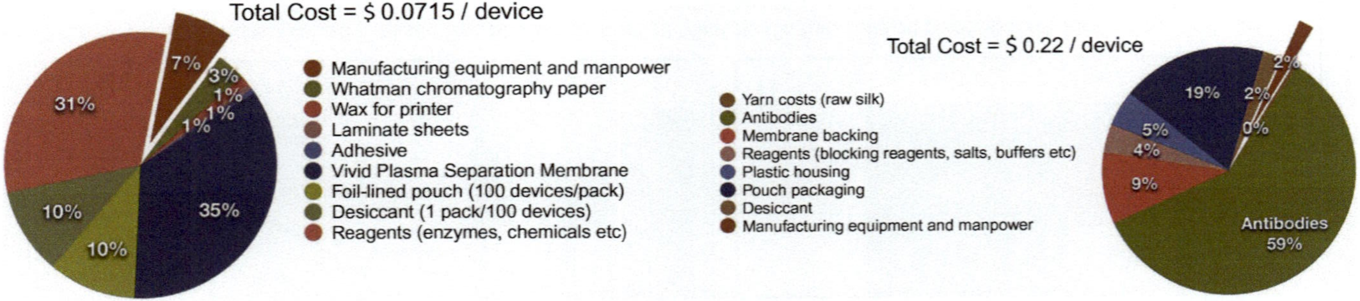

Fig. 12 Competitive study on paper-based microfluidic devices in different industries. *Reprinted/adapted by permission from Yetisen AK, Akram MS, Lowe CR. Paper-based microfluidic point-of-care diagnostic devices.* Lab Chip. *2013;13:2210–2251. https://doi.org/10.1039/c3lc50169h, Copyright © Royal Society of Chemistry (2013).*

market. Considering the recent market-dominating diagnostic products, like glucose monitoring kits or pregnancy detection kits which are commonly accessible, are principally chosen by affected individuals or specific population groups to lessen the requirement of hospital visits. Undoubtedly, the significance of such POC diagnostic inventions has intensified during the COVID-19 pandemic, as the need for accessible diagnostic test kits that could be utilized at home or remote care locations has sky-rocketed. Other factors like the growing popularity and knowledge of other diseases and disease detection factors have led to growing consciousness among the population, of diseases like CVDs, cancer, and diabetes. Thus, the demand and the rising disposable income levels are likely to boost the development of this marketplace. The current trend has also impacted microfluidic research and development as it is one of the key technologies involved in advanced POC devices.

5.2 Competitive advantages

One of the fastest expanding areas is the incorporation of molecular diagnostics with POC, and therefore there is an urgent demand for affordable and robust microfluidic devices. Surprisingly, paper devices offer competitive advantages to parallelly developed microfluidic devices. Even at the innovation stages, with the shortage of infrastructure, and well-equipped laboratory services, the ease of device fabrication attracts a lot of researchers. Several market analysis reports[27] predicted and analyzed the cost of microfluidic devices which is no more than 0.0715 USD for transaminase and 0.22 USD for immunoassay.[27]

5.3 The path to becoming a key technological contributor

It is indeed a challenge for a biotechnology or pharma company, an academic research group, or an investor to understand the value of these types of paper-based microfluidics devices when there is already an existing market for other microfluidic devices. Although with obvious potential benefits in terms of reduction in costs, and use of cheaper manufacturing units, commercialization is dependent on various other factors. Unique approaches should be considered based on the geographical placement (developmental nature of the country) of the merchandise and on the developmental stage of that kind of product in the beneficial area. If the market is already saturated with targeted therapeutic products, then anticipating a transfer from current diagnostic approaches to paper-based microfluidics is not realistic.[27] The organization, willing to make the shift would have to breathe hefty expenditure of shifting present assembly lines (set-up lines) and creating new trademarked expertise across paper or in certifying standing patents, which may also involve alterations to already patented counterparts (reagents or a process). These risks are viable only when there are major technological benefits and improvements involved.[27] This type of shift has not been witnessed yet. If the market is new and no competition exists in the market, then the company could follow any traditional method. This obvious effect is observed in the current market, where the top 15 firms account for 90% of the worldwide IVD market reach, in which only 6 are predominantly identified as diagnostic companies, the remaining are pharmaceutical companies with a diagnostic unit.[27] Historical trends show that majority of the larger diagnostic businesses have been purchased by other big companies e.g., Dade Behring, Diagnostic Product Company, Theresa, i-STAT, and MediSense. Additionally, it can also be witnessed from the acquisition trend that most of the big companies have funded small or mid-sized diagnostic businesses and have generated healthy returns from their trade. Considering these incidents, the diagnostic market is a promising sector. The availability of funds in the research and development sector in the developed world has offered sustainable business opportunities in developing economies.[27] The shift in global economics is also responsible for investment by big corporations in a blooming market as their next venture to grow their asset. It was forecasted that 7 developing countries; major players being China, India, and Russia will triple their GDP by 2050; Asian countries will be responsible for 51% of the whole world's GDP.[27] Since the healthcare system in these countries has not yet developed, there is a lot of room to set up new systems, clinical laboratories, diagnostic kits, etc. In these places government-funded healthcare system through health insurance is nearly non-existent. In such countries, the divide between rich and poor is increasingly significant, and the already existing expensive diagnostic approaches further increase the inequality gap. This strongly suggests that diagnostic solutions should be modern, affordable, and designed with the buyers' financial position in mind. Paper-based devices made for biomedical applications are extremely cheap to fabricate and develop. This product category should be brought into focus to solve some of the major health challenges in developing economies around the world.

5.4 Regulatory challenges

A good understanding of regulatory norms while developing low-cost diagnostic kits for the market is crucial as, without it, any device could be turned off from the market even if it is an outcome of one of the best researches in the world.[167] The overall influence of the diagnostic assessment or impact of the device on the wellbeing of the patients, particularly in consequence of false positives or false negatives are regulated through different essential validation norms. Certified merchandise also confirms whether the claims made by the company are true or not. Most countries have their own regulatory body to control and check upcoming products in that region and these regulations are direct or slightly modified versions of an international body like the International Organization for Standardization (ISO) which is an international standard-setting body instituted in 1947, with members from several national standard administrations. The organization promotes worldwide proprietary, industrial, and commercial standards. Diagnostic devices are part of medical products and are generally classified into three categories[168] considering their risk factors. As per the associated risk factors, devices are classified from Class I to III, where Class I device is prone to lesser and Class III are associated with increased risk to patients and must undergo strict regulatory processes. Thus, regulatory affairs involve an extra layer of intricacy to the development. In the United States,[169] the consent to market availability of medical device goes through either the 510(k) pathways, appealing that the new technology is analogous to the current FDA-approved technology, or premarket authorization, which necessitates more detailed investigative backing that the technology is innocuous and operative with enough benefit to risk. As a result, obtaining regulatory authorization itself takes substantial time and significant investment; not only the diagnostic characteristics of the device but features like risks, usability, the process of manufacturing, packaging, labeling,[168] etc. need to be addressed. The development phase of such a brand-new diagnostic test or gadget requires at least 7 years.[11,27] The turnaround period also differs from the method undertaken for the authorization procedure. In the European region, the product also requires "Conformity Assessment Procedure" to obtain a CE marking that allows companies for experimentation or sales of their device in the European region. Paper-based devices are simpler than other active or passive type microfluidic devices with fewer associated risk factors which create more scope for marketing paper-based chips.

6. Conclusion and final thoughts

Paper-based microfluidics is still in its initial expansion phase and no multidimensional paper-based device has been commercialized yet. Like any new technology, this also appeared with already existing technological trends like dipsticks and lateral flow with recognized assay structures for POC analysis. The very first paper-based diabetes dipstick test quantified glucose from urine, which was suggested in the 1950s and was commercialized in the 1960s. The introduction of a micro total analysis system was revolutionary in bringing microfluidics to light in the early 1990s. Thereafter, the advancement of paper-based microfluidics seems to be the next "mini-revolution." Numerous aspects have been involved[27] in this revolution namely (i) genuine worldwide attention in conducting investigations in global health, (ii) overall growing attention in "frugal engineering," (iii) a drive from funding bodies to conduct studies that "gives back to society," and (iv) participation of some key research groups that started the microfluidics revolution in the 1990s.

A recent example of the applicability of paper-based microfluidic devices during the 2020 pandemic caused by SARS-CoV2 must be mentioned here which infected and killed millions around the globe. Due to the robust design and ease of mass-production in limited-resource settings, this technology was quickly adapted in an emergency. At times where lockdown and isolation were the only options to tackle the situation, without any available vaccines, these paper devices based on rapid lateral flow microfluidic cartridges played a key role in POC disease diagnostics. Most of the diagnostic companies had the opportunity to develop their own rapid viral test kits using lateral flow paper microfluidics to support the health care system and helped millions of suspected patients before they could get a more accurate RT-PCR test. These paper-based microfluidic devices were the first to appear and could be found in people's homes more widely compared to other categories of microfluidic cartridges. However, there are roadblocks to clear, and it is an interdisciplinary science and technological field, where assay development, device development, and at the same time their packaging is equally important. Each component is so intertwined with the other since the development of one without the other is almost impossible and must incorporate knowledge and expertise from various backgrounds.

It must be mentioned that POC diagnostics is still in its infancy and arranging funds for research and development alongside marketing will be difficult due to volatile future goal statements but if the recent pandemic has taught us anything, the healthcare infrastructure of even the most developed countries on earth was brought down to its knees. This might be a good opportunity to take responsibility and raise awareness, innovate and apply to the overall growth of POC technologies. Research grant providers and diagnostic companies should also focus and invest more in developing various state-of-the-art technologies for the improvement of society.

Acknowledgment

The author(s) is grateful to MeFy Care (www.mefy.care) and the entire R&D team for their support in several aspects. A special thanks to Mr. Sanjay Sinha (Founder and CEO, MeFy Care) and Mr. Rahul Singh (B. Tech, Head of Innovation, R&D Mefyineer) for their assistance and encouragement throughout the process. Thanks to Dr. P. Gopinathan (Scientist, Oxford Nanopore Technologies) for her useful recommendations. The author is also thankful to Besmed Health Business Corp. (Taiwan) for providing the opportunity to understand various challenges in biomedical Regularity Affairs.

References

1. Barrett TH. The woman who invented notepaper: towards a comparative historiography of paper and print. *J R Asiat Soc.* 2011;21:199–210. https://doi.org/10.1017/S1356186311000186.
2. Pávai M, Mihály J, Paszternák A. pH and CO2 sensing by curcumin-coloured cellophane test strip. *Food Anal Methods.* 2015;8:2243–2249. https://doi.org/10.1007/s12161-015-0102-1.
3. Müller RH, Clegg DL. Automatic paper chromatography. *Anal Chem.* 1949;21:192. https://doi.org/10.1021/ac60033a032.
4. Weil H, Williams TI. Early history of chromatography. *Nature.* 1951;167:906–907.
5. Leroy F. *A Century of Nobel Prize Recipients: Chemistry, Physics, and Medicine.* CRC Press; 2003.
6. Convery N, Gadegaard N. 30 years of microfluidics. *Micro Nano Eng.* 2019;2:76–91. https://doi.org/10.1016/j.mne.2019.01.003.
7. Martinez AWW, Phillips STT, Butte MJJ, Whitesides GMM. Patterned paper as a platform for inexpensive, low-volume, portable bioassays. *Angew Chem Int Ed.* 2007; 46:1318–1320. https://doi.org/10.1002/anie.200603817.
8. Gravesen P, Branebjerg J, Jensen OS. Microfluidics—a review. *J Micromech Microeng.* 1993;3:168–182. https://doi.org/10.1088/0960-1317/3/4/002.
9. Lee WG, Kim YG, Chung BG, Demirci U, Khademhosseini A. Nano/microfluidics for diagnosis of infectious diseases in developing countries. *Adv Drug Deliv Rev.* 2010;62:449–457. https://doi.org/10.1016/j.addr.2009.11.016.
10. Peeling RW, Holmes KK, Mabey D, Ronald A. Rapid tests for sexually transmitted infections (STIs): the way forward. *Sex Transm Infect.* 2006;82:1–6. https://doi.org/10.1136/sti.2006.024265.
11. Chin CD, Linder V, Sia SK. Commercialization of microfluidic point-of-care diagnostic devices. *Lab Chip.* 2012;12:2118. https://doi.org/10.1039/c2lc21204h.

12. Companion Diagnostics. *Market Global Forecast to 2025 | MarketsandMarkets*; 2020. MD 3609.
13. Yager P, Edwards T, Fu E, et al. Microfluidic diagnostic technologies for global public health. *Nature*. 2006;442:412–418. https://doi.org/10.1038/nature05064.
14. Sinha A, Gopinathan P, Da Chung Y, et al. An integrated microfluidic platform to perform uninterrupted SELEX cycles to screen affinity reagents specific to cardiovascular biomarkers. *Biosens Bioelectron*. 2018;122:104–112. https://doi.org/10.1016/j.bios.2018.09.040.
15. Sinha A, Gopinathan P, Chung Y-D, Shiesh S-C, Lee G-B. Simultaneous detection of multiple NT-proBNP clinical samples utilizing an aptamer-based sandwich assay on an integrated microfluidic system. *Lab Chip*. 2019;19:1676–1685. https://doi.org/10.1039/c9lc00115h.
16. Sinha A, Tai TY, Li KH, et al. An integrated microfluidic system with field-effect-transistor sensor arrays for detecting multiple cardiovascular biomarkers from clinical samples. *Biosens Bioelectron*. 2019;129:155–163. https://doi.org/10.1016/j.bios.2019.01.001.
17. Clarke SF, Foster JR. A history of blood glucose meters and their role in self-monitoring of diabetes mellitus. *Br J Biomed Sci*. 2012;69:83–93. https://doi.org/10.1080/09674845.2012.12002443.
18. Chard T. Review: pregnancy tests: a review. *Hum Reprod*. 1992;7:701–710. https://doi.org/10.1093/oxfordjournals.humrep.a137722.
19. de Castro LF, de Freitas SV, Duarte LC, de Souza JAC, Paixão TRLC, Coltro WKT. Salivary diagnostics on paper microfluidic devices and their use as wearable sensors for glucose monitoring. *Anal Bioanal Chem*. 2019;411:4919–4928. https://doi.org/10.1007/s00216-019-01788-0.
20. Zhang D, Bi H, Liu B, Qiao L. Detection of pathogenic microorganisms by microfluidics based analytical methods. *Anal Chem*. 2018;90:5512–5520. https://doi.org/10.1021/acs.analchem.8b00399.
21. Parween S, Subudhi PD, Asthana A. An affordable, rapid determination of total lipid profile using paper-based microfluidic device. *Sensors Actuators B Chem*. 2019;285:405–412. https://doi.org/10.1016/j.snb.2019.01.064.
22. Azhar M, Dendukuri D. Microfluidic platforms for point of care (POC) medical diagnostics. In: *Medical Biosensors for Point of Care (POC) Applications*. Woodhead Publishing; 2016:255–273. https://doi.org/10.1016/B978-0-08-100072-4.00011-3.
23. Nosrati R, Gong MM, Gabriel MCS, Pedraza CE, Zini A, Sinton D. Paper-based quantification of male fertility potential. *Clin Chem*. 2016;62:458–465. https://doi.org/10.1373/clinchem.2015.250282.
24. Pandey CM, Augustine S, Kumar S, et al. Microfluidics based point-of-care diagnostics. *Biotechnol J*. 2018;13:1700047. https://doi.org/10.1002/biot.201700047.
25. Wu L, Ma C, Ge L, et al. Paper-based electrochemiluminescence origami cyto-device for multiple cancer cells detection using porous AuPd alloy as catalytically promoted nanolabels. *Biosens Bioelectron*. 2015;63:450–457. https://doi.org/10.1016/j.bios.2014.07.077.
26. Tong X, Ga L, Zhao R, Ai J. Research progress on the applications of paper chips. *RSC Adv*. 2021;11:8793–8820. https://doi.org/10.1039/d0ra10470a.
27. Yetisen AK, Akram MS, Lowe CR. Paper-based microfluidic point-of-care diagnostic devices. *Lab Chip*. 2013;13:2210–2251. https://doi.org/10.1039/c3lc50169h.
28. Sriram G, Bhat MP, Patil P, et al. Paper-based microfluidic analytical devices for colorimetric detection of toxic ions: a review. *TrAC Trends Anal Chem*. 2017;93:212–227. https://doi.org/10.1016/j.trac.2017.06.005.
29. Chen YH, Kuo ZK, Cheng CM. Paper—a potential platform in pharmaceutical development. *Trends Biotechnol*. 2015;33(1):4–9. https://doi.org/10.1016/j.tibtech.2014.11.004.

30. Roberts JC. *The Chemistry of Paper*. Royal Society of Chemistry; 1996.
31. Whatman. *Qualitative Filter Papers*. Whatman; 2011:1–10.
32. Zhong ZW, Wang ZP, Huang GXD. Investigation of wax and paper materials for the fabrication of paper-based microfluidic devices. *Microsyst Technol*. 2012;18:649–659. https://doi.org/10.1007/s00542-012-1469-1.
33. Cate DM, Adkins JA, Mettakoonpitak J, Henry CS. Recent developments in paper-based microfluidic devices. *Anal Chem*. 2015;87:19–41. https://doi.org/10.1021/ac503968p.
34. Camplisson CK, Schilling KM, Pedrotti WL, Stone HA, Martinez AW. Two-ply channels for faster wicking in paper-based microfluidic devices. *Lab Chip*. 2015;15: 4461–4466. https://doi.org/10.1039/c5lc01115a.
35. Masoodi R, Pillai KM. Darcy's law-based model for wicking in paper-like swelling porous media. *AICHE J*. 2012;59:215–228. https://doi.org/10.1002/aic.
36. Guo W, Hansson J, van der Wijngaart W. Capillary pumping independent of the liquid surface energy and viscosity. *Microsyst Nanoeng*. 2018;4:1–7. https://doi.org/10.1038/s41378-018-0002-9.
37. Shin JH, Park J, Kim SH, Park JK. Programmed sample delivery on a pressurized paper. *Biomicrofluidics*. 2014;8:054121. https://doi.org/10.1063/1.4899773.
38. Rattanarat P, Dungchai W, Cate D, Volckens J, Chailapakul O, Henry CS. Multilayer paper-based device for colorimetric and electrochemical quantification of metals. *Anal Chem*. 2014;86:3555–3562. https://doi.org/10.1021/ac5000224.
39. Martinez AW, Phillips ST, Whitesides GM. Three-dimensional microfluidic devices fabricated in layered paper and tape. *Proc Natl Acad Sci U S A*. 2008;105: 19606–19611. https://doi.org/10.1073/pnas.0810903105.
40. Li J, Ge J, Yin Y, Zhong W. Multiplexed affinity-based protein complex purification. *Anal Chem*. 2008;80:7068–7074. https://doi.org/10.1021/ac801251y.
41. Abe K, Suzuki K, Citterio D. Inkjet-printed microfluidic multianalyte chemical sensing paper. *Anal Chem*. 2008;80:6928–6934. https://doi.org/10.1021/ac800604v.
42. Li X, Tian J, Nguyen T, Shen W. Paper-based microfluidic devices by plasma. *Lab Chip*. 2008;80:9131–9134. https://doi.org/10.1039/b811135a.10.1021/ac801729t.
43. Lu R, Shi W, Jiang L, Qin J, Lin B. Rapid prototyping of paper-based microfluidics with wax for low-cost, portable bioassay. *Electrophoresis*. 2009;30:1497–1500. https://doi.org/10.1002/elps.200800563.
44. Fenton EM, Mascarenas MR, López GP, Sibbett SS. Multiplex lateral-flow test strips fabricated by two-dimensional shaping. *ACS Appl Mater Interfaces*. 2009;1:124–129. https://doi.org/10.1021/am800043z.
45. Toley BJ, Das D, Ganar KA, et al. Multidimensional paper networks: a new generation of low-cost pump-free microfluidic devices. *J Indian Inst Sci*. 2018;98:103–136. https://doi.org/10.1007/s41745-018-0077-1.
46. Li X, Tian J, Garnier G, Shen W. Fabrication of paper-based microfluidic sensors by printing. *Colloids Surf B: Biointerfaces*. 2010;76:564–570. https://doi.org/10.1016/j.colsurfb.2009.12.023.
47. Olkkonen J, Lehtinen K, Erho T. Flexographically printed fluidic structures in paper. *Anal Chem*. 2010;82:10246–10250. https://doi.org/10.1021/ac1027066.
48. Chitnis G, Ding Z, Chang CL, Savran CA, Ziaie B. Laser-treated hydrophobic paper: an inexpensive microfluidic platform. *Lab Chip*. 2011;11:1161–1165. https://doi.org/10.1039/c0lc00512f.
49. Dungchai W, Chailapakul O, Henry CS. A low-cost, simple, and rapid fabrication method for paper-based microfluidics using wax screen-printing. *Analyst*. 2011;136: 77–82. https://doi.org/10.1039/c0an00406e.
50. Thuo MM, Martinez RV, Lan WJ, et al. Fabrication of low-cost paper-based microfluidic devices by embossing or cut-and-stack methods. *Chem Mater*. 2014;26: 4230–4237. https://doi.org/10.1021/cm501596s.

51. Mohammadi S, Maeki M, Mohamadi RM, Ishida A, Tani H, Tokeshi M. An instrument-free, screen-printed paper microfluidic device that enables bio and chemical sensing. *Analyst.* 2015;140:6493–6499. https://doi.org/10.1039/c5an00909j.
52. Zhang AL, Zha Y. Fabrication of paper-based microfluidic device using printed circuit technology. *AIP Adv.* 2012;2:022171. https://doi.org/10.1063/1.4733346.
53. Li X, Liu X. Fabrication of three-dimensional microfluidic channels in a single layer of cellulose paper. *Microfluid Nanofluid.* 2014;16:819–827. https://doi.org/10.1007/s10404-014-1340-z.
54. Osann R, Fasching MM. *(12) Patent Application Publication (10) Pub. No.: US 2008/0225123 A1 Patent Application Publication. Priv. Point Contain. Sm Card 1*; 2008:11–14.
55. Ouyang L, Wang C, Du F, Zheng T, Liang H. Electrochromatographic separations of multi-component metal complexes on a microfluidic paper-based device with a simplified photolithography. *RSC Adv.* 2014;4:1093–1101. https://doi.org/10.1039/c3ra43625j.
56. Songjaroen T, Dungchai W, Chailapakul O, Henry CS, Laiwattanapaisal W. Blood separation on microfluidic paper-based analytical devices. *Lab Chip.* 2012;12:3392–3398. https://doi.org/10.1039/c2lc21299d.
57. Curto VF, Lopez-Ruiz N, Capitan-Vallvey LF, Palma AJ, Benito-Lopez F, Diamond D. Fast prototyping of paper-based microfluidic devices by contact stamping using indelible ink. *RSC Adv.* 2013;3:18811–18816. https://doi.org/10.1039/c3ra43825b.
58. Delaney JL, Hogan CF, Tian J, Shen W. Electrogenerated chemiluminescence detection in paper-based microfluidic sensors. *Anal Chem.* 2011;83:1300–1306. https://doi.org/10.1021/ac102392t.
59. Kwong P, Gupta M. Vapor phase deposition of functional polymers onto paper-based microfluidic devices for advanced unit operations. *Anal Chem.* 2012;84:10129–10135. https://doi.org/10.1021/ac302861v.
60. Cai L, Xu C, Lin SH, Luo J, Wu M, Yang F. A simple paper-based sensor fabricated by selective wet etching of silanized filter paper using a paper mask. *Biomicrofluidics.* 2014;8:2–10. https://doi.org/10.1063/1.4898096.
61. Martinez AW, Phillips ST, Wiley BJ, Gupta M, Whitesides GM. FLASH: a rapid method for prototyping paper-based microfluidic devices. *Lab Chip.* 2008;8:2146–2150. https://doi.org/10.1039/b811135a.
62. Li X, Tian J, Nguyen T, Shen W. Paper-based microfluidic devices by plasma treatment. *Anal Chem.* 2008;80:9131–9134. https://doi.org/10.1021/ac801729t.
63. Li X, Ballerini DR, Shen W. A perspective on paper-based microfluidics: current status and future trends. *Biomicrofluidics.* 2012;6:011301–011301-13. https://doi.org/10.1063/1.3687398.
64. Carrilho E, Phillips ST, Vella SJ, Martinez AW, Whitesides GM. Paper microzone plates. *Anal Chem.* 2009;81:5990–5998. https://doi.org/10.1021/ac900847g.
65. He PJW, Katis IN, Eason RW, Sones CL. Laser-based patterning for fluidic devices in nitrocellulose. *Biomicrofluidics.* 2015;9:1–11. https://doi.org/10.1063/1.4919629.
66. Akyazi T, Saez J, Elizalde J, Benito-Lopez F. Fluidic flow delay by ionogel passive pumps in microfluidic paper-based analytical devices. *Sensors Actuators B Chem.* 2016;233:402–408. https://doi.org/10.1016/j.snb.2016.04.116.
67. Wang J, Monton MRN, Zhang X, Filipe CDM, Pelton R, Brennan JD. Hydrophobic sol-gel channel patterning strategies for paper-based microfluidics. *Lab Chip.* 2014;14:691–695. https://doi.org/10.1039/c3lc51313k.
68. Asano H, Shiraishi Y. Development of paper-based microfluidic analytical device for iron assay using photomask printed with 3D printer for fabrication of hydrophilic and hydrophobic zones on paper by photolithography. *Anal Chim Acta.* 2015;883:55–60. https://doi.org/10.1016/j.aca.2015.04.014.

69. Sameenoi Y, Na Nongkai P, Nouanthavong S, Henry CS, Nacaprich D. One-step polymer screen-printing for microfluidic paper-based analytical device (¼PAD) fabrication. *Analyst*. 2014;139:6580–6588. https://doi.org/10.1039/c4an01624f.
70. Yang Y, Noviana E, Nguyen MP, Geiss BJ, Dandy DS, Henry CS. Paper-based microfluidic devices: emerging themes and applications. *Anal Chem*. 2017;89:71–91. https://doi.org/10.1021/acs.analchem.6b04581.
71. Böhm A, Carstens F, Trieb C, Schabel S, Biesalski M. Engineering microfluidic papers: effect of fiber source and paper sheet properties on capillary-driven fluid flow. *Microfluid Nanofluid*. 2014;16:789–799. https://doi.org/10.1007/s10404-013-1324-4.
72. Jeong SG, Kim J, Jin SH, Park KS, Lee CS. Flow control in paper-based microfluidic device for automatic multistep assays: a focused minireview. *Korean J Chem Eng*. 2016;33:2761–2770. https://doi.org/10.1007/s11814-016-0161-z.
73. Schilling KM, Lepore AL, Kurian JA, Martinez AW. Fully enclosed microfluidic paper-based analytical devices. *Anal Chem*. 2012;84:1579–1585. https://doi.org/10.1021/ac202837s.
74. Renault C, Koehne J, Ricco AJ, Crooks RM. Three-dimensional wax patterning of paper fluidic devices. *Langmuir*. 2014;30:7030–7036. https://doi.org/10.1021/la501212b.
75. Fu E, Lutz B, Kauffman P, Yager P. Controlled reagent transport in disposable 2D paper networks. *Lab Chip*. 2010;10:918–920. https://doi.org/10.1039/b919614e.
76. Toley BJ, Mckenzie B, Liang T, Buser JR, Yager P, Fu E. Tunable-delay shunts for paper micro fluidic devices. *Anal Chem*. 2013;85(23):11545–11552.
77. Mendez S, Fenton EM, Gallegos GR, et al. Imbibition in porous membranes of complex shape: quasi-stationary flow in thin rectangular segments. *Langmuir*. 2010;26:1380–1385. https://doi.org/10.1021/la902470b.
78. Houghtaling J, Liang T, Thiessen G, Fu E. Dissolvable bridges for manipulating fluid volumes in paper networks. *Anal Chem*. 2013;85:11201–11204. https://doi.org/10.1021/ac4022677.
79. Lutz B, Liang T, Fu E, Ramachandran S, Kauffman P, Yager P. Dissolvable fluidic time delays for programming multi-step assays in instrument-free paper diagnostics. *Lab Chip*. 2013;13:2840–2847. https://doi.org/10.1039/c3lc50178g.
80. Chen H, Cogswell J, Anagnostopoulos C, Faghri M. A fluidic diode, valves, and a sequential-loading circuit fabricated on layered paper. *Lab Chip*. 2012;12:2909–2913. https://doi.org/10.1039/c2lc20970e.
81. Toley BJ, Wang JA, Gupta M, et al. A versatile valving toolkit for automating fluidic operations in paper microfluidic devices. *Lab Chip*. 2015;15:1432–1444. https://doi.org/10.1039/c4lc01155d.
82. Li X, Zwanenburg P, Liu X. Magnetic timing valves for fluid control in paper-based microfluidics. *Lab Chip*. 2013;13:2609–2614. https://doi.org/10.1039/c3lc00006k.
83. Kong T, Flanigan S, Weinstein M, Kalwa U, Legner C, Pandey S. A fast, reconfigurable flow switch for paper microfluidics based on selective wetting of folded paper actuator strips. *Lab Chip*. 2017;17:3621. https://doi.org/10.1039/c7lc00620a.
84. Lutz BR, Trinh P, Ball C, Fu E, Yager P. Two-dimensional paper networks: programmable fluidic disconnects for multi-step processes in shaped paper. *Lab Chip*. 2011;11:4274–4278. https://doi.org/10.1039/c1lc20758j.
85. Samuel S, Klaus S, Drese S, et al. Lab on a chip lab on a chip. *Lab Chip*. 2014;15:1485–1490.
86. Jahanshahi-Anbuhi S, Henry A, Leung V, et al. Paper-based microfluidics with an erodible polymeric bridge giving controlled release and timed flow shutoff. *Lab Chip*. 2014;14:229–236. https://doi.org/10.1039/c3lc50762a.
87. Noh H, Phillips ST. Metering the capillary-driven flow of fluids in paper-based microfluidic devices. *Anal Chem*. 2010;82:4181–4187. https://doi.org/10.1021/ac100431y.

88. Jagadeesan KK, Kumar S, Sumana G. Application of conducting paper for selective detection of troponin. *Electrochem Commun*. 2012;20:71–74. https://doi.org/10.1016/j.elecom.2012.03.041.
89. Wei X, Tian T, Jia S, et al. Target-responsive DNA hydrogel mediated stop-flow microfluidic paper-based analytic device for rapid, portable and visual detection of multiple targets. *Anal Chem*. 2015;87:4275–4282. https://doi.org/10.1021/acs.analchem.5b00532.
90. Wang W, Wu WY, Wang W, Zhu JJ. Tree-shaped paper strip for semiquantitative colorimetric detection of protein with self-calibration. *J Chromatogr A*. 2010;1217:3896–3899. https://doi.org/10.1016/j.chroma.2010.04.017.
91. Mani V, Kadimisetty K, Malla S, Joshi AA, Rusling JF. Paper-based electrochemiluminescent screening for genotoxic activity in the environment. *Environ Sci Technol*. 2013;47:1937–1944. https://doi.org/10.1021/es304426j.
92. Ferrer IM, Valadez H, Estala L, Gomez FA. Paper microfluidic-based enzyme catalyzed double microreactor. *Electrophoresis*. 2014;35:2417–2419. https://doi.org/10.1002/elps.201400091.
93. Chen Y, Cheng H, Tram K, et al. A paper-based surface-enhanced resonance Raman spectroscopic (SERRS) immunoassay using magnetic separation and enzyme-catalyzed reaction. *Analyst*. 2013;138:2624–2631. https://doi.org/10.1039/c3an36647b.
94. Schonhorn JE, Fernandes SC, Rajaratnam A, Deraney RN, Rollanda JP, Mace CR. A device architecture for three-dimensional, patterned paper immunoassays. *Lab Chip*. 2014;14:4653–4658. https://doi.org/10.1039/c4lc00876f.
95. Yang RJ, Tseng CC, Ju WJ, Fu LM, Syu MP. Integrated microfluidic paper-based system for determination of whole blood albumin. *Sensors Actuators B Chem*. 2018;273:1091–1097. https://doi.org/10.1016/j.snb.2018.07.010.
96. Liu P, Li B, Fu L, et al. Hybrid three dimensionally printed paper-based microfluidic platform for investigating a cell's apoptosis and intracellular cross-talk. *ACS Sens*. 2020;5:464–473. https://doi.org/10.1021/acssensors.9b02205.
97. Yehia AM, Farag MA, Tantawy MA. A novel trimodal system on a paper-based microfluidic device for on-site detection of the date rape drug "ketamine". *Anal Chim Acta*. 2020;1104:95–104. https://doi.org/10.1016/j.aca.2020.01.002.
98. Cai X, Zhang H, Yu X, Wang W. A microfluidic paper-based laser-induced fluorescence sensor based on duplex-specific nuclease amplification for selective and sensitive detection of miRNAs in cancer cells. *Talanta*. 2020;216:120996. https://doi.org/10.1016/j.talanta.2020.120996.
99. Hossain SMZ, Brennan JD. β-Galactosidase-based colorimetric paper sensor for determination of heavy metals. *Anal Chem*. 2011;83:8772–8778. https://doi.org/10.1021/ac202290d.
100. Zhang Y, Ma H, Wu D, Li Y, Du B, Wei Q. Label-free immunosensor based on Au@Ag2S nanoparticles/magnetic chitosan matrix for sensitive determination of ractopamine. *J Electroanal Chem*. 2015;741:14–19. https://doi.org/10.1016/j.jelechem.2015.01.007.
101. Fu LM, Liu CC, Yang CE, Wang YN, Ko CH. A PET/paper chip platform for high resolution sulphur dioxide detection in foods. *Food Chem*. 2019;286:316–321. https://doi.org/10.1016/j.foodchem.2019.02.032.
102. Jin L, Hao Z, Zheng Q, et al. A facile microfluidic paper-based analytical device for acetylcholinesterase inhibition assay utilizing organic solvent extraction in rapid detection of pesticide residues in food. *Anal Chim Acta*. 2020;1100:215–224. https://doi.org/10.1016/j.aca.2019.11.067.
103. Yamada K, Shibata H, Suzuki K, Citterio D. Toward practical application of paper-based microfluidics for medical diagnostics: state-of-the-art and challenges. *Lab Chip*. 2017;17(7):1206–1249. https://doi.org/10.1039/c6lc01577h.

104. Liu S, Su W, Ding X. A review on microfluidic paper-based analytical devices for glucose detection. *Sensors (Basel)*. 2016;16(12):2086. https://doi.org/10.3390/s16122086.
105. Martinez AW, Phillips ST, Carrilho E, Thomas SW, Sindi H, Whitesides GM. Simple telemedicine for developing regions: camera phones and paper-based microfluidic devices for real-time, off-site diagnosis. *Anal Chem*. 2008;80:3699–3707. https://doi.org/10.1021/ac800112r.
106. Martinez AW, Phillips ST, Nie Z, et al. Programmable diagnostic devices made from paper and tape. *Lab Chip*. 2010;10:2499–2504. https://doi.org/10.1039/c0lc00021c.
107. Lepowsky E, Ghaderinezhad F, Knowlton S, Tasoglu S. Paper-based assays for urine analysis. *Biomicrofluidics*. 2017;11:51501. https://doi.org/10.1063/1.4996768.
108. Owen CS. High gradient magnetic separation of erythrocytes. *Biophys J*. 1978;22: 171–178. https://doi.org/10.1016/S0006-3495(78)85482-4.
109. Khan MS, Thouas G, Shen W, Whyte G, Garnier G. Paper diagnostic for instantaneous blood typing. *Anal Chem*. 2010;82:4158–4164. https://doi.org/10.1021/ac100341n.
110. Sharma N, Barstis T, Giri B. Advances in paper-analytical methods for pharmaceutical analysis. *Eur J Pharm Sci*. 2018;111:46–56. https://doi.org/10.1016/j.ejps.2017.09.031.
111. Shiroma LY, Santhiago M, Gobbi AL, Kubota LT. Separation and electrochemical detection of paracetamol and 4-aminophenol in a paper-based microfluidic device. *Anal Chim Acta*. 2012;725:44–50. https://doi.org/10.1016/j.aca.2012.03.011.
112. Weaver AA, Reiser H, Barstis T, et al. Paper analytical devices for fast field screening of beta lactam antibiotics and antituberculosis pharmaceuticals. *Anal Chem*. 2013;85: 6453–6460. https://doi.org/10.1021/ac400989p.
113. Weaver AA, Halweg S, Joyce M, Lieberman M, Goodson HV. Incorporating yeast biosensors into paper-based analytical tools for pharmaceutical analysis. *Anal Bioanal Chem*. 2015;407:615–619. https://doi.org/10.1007/s00216-014-8280-z.
114. Koesdjojo MT, Wu Y, Boonloed A, Dunfield EM, Remcho VT. Low-cost, high-speed identification of counterfeit antimalarial drugs on paper. *Talanta*. 2014;130: 122–127. https://doi.org/10.1016/j.talanta.2014.05.050.
115. Craig D, Mazilu M, Dholakia K. Quantitative detection of pharmaceuticals using a combination of paper microfluidics and wavelength modulated raman spectroscopy. *PLoS One*. 2015;10:e0123334. https://doi.org/10.1371/journal.pone.0123334.
116. Murphy A, Gorey B, De Guzman K, Kelly N, Nesterenko EP, Morrin A. Microfluidic paper analytical device for the chromatographic separation of ascorbic acid and dopamine. *RSC Adv*. 2015;5:93162–93169. https://doi.org/10.1039/c5ra16272f.
117. Su M, Ge L, Ge S, et al. Paper-based electrochemical cyto-device for sensitive detection of cancer cells and in situ anticancer drug screening. *Anal Chim Acta*. 2014;847:1–9. https://doi.org/10.1016/j.aca.2014.08.013.
118. Gomes HIAS, Sales MGF. Development of paper-based color test-strip for drug detection in aquatic environment: application to oxytetracycline. *Biosens Bioelectron*. 2015;65:54–61. https://doi.org/10.1016/j.bios.2014.10.006.
119. Mentele MM, Cunningham J, Koehler K, Volckens J, Henry CS. Microfluidic paper-based analytical device for particulate metals. *Anal Chem*. 2012;84:4474–4480. https://doi.org/10.1021/ac300309c.
120. Jayawardane BM, McKelvie ID, Kolev SD. Development of a gas-diffusion microfluidic paper-based analytical device (μPAD) for the determination of ammonia in wastewater samples. *Anal Chem*. 2015;87:4621–4626. https://doi.org/10.1021/acs.analchem.5b00125.
121. Hossain SMZ, Luckham RE, McFadden MJ, Brennan JD. Reagentless bidirectional lateral flow bioactive paper sensors for detection of pesticides in beverage and food samples. *Anal Chem*. 2009;81:9055–9064. https://doi.org/10.1021/ac901714h.

122. Rattanarat P, Dungchai W, Siangproh W, Chailapakul O, Henry CS. Sodium dodecyl sulfate-modified electrochemical paper-based analytical device for determination of dopamine levels in biological samples. *Anal Chim Acta*. 2012;744:1–7. https://doi.org/10.1016/j.aca.2012.07.003.
123. Vaher M, Kaljurand M. The development of paper microzone-based green analytical chemistry methods for determining the quality of wines. *Anal Bioanal Chem*. 2012;404: 627–633. https://doi.org/10.1007/s00216-012-5930-x.
124. Zhou J, Li B, Qi A, et al. ZnSe quantum dot based ion imprinting technology for fluorescence detecting cadmium and lead ions on a three-dimensional rotary paper-based microfluidic chip. *Sensors Actuators B Chem*. 2020;305:127462. https://doi.org/10.1016/j.snb.2019.127462.
125. Fu E, Kauffman P, Lutz B, Yager P. Chemical signal amplification in two-dimensional paper networks. *Sensors Actuators B Chem*. 2010;149:325–328. https://doi.org/10.1016/j.snb.2010.06.024.
126. Fu E, Liang T, Houghtaling J, et al. Enhanced sensitivity of lateral flow tests using a two-dimensional paper network format. *Anal Chem*. 2011;83:7941–7946. https://doi.org/10.1021/ac201950g.
127. Ramachandran S, Fu E, Lutz B, Yager P. Long-term dry storage of an enzyme-based reagent system for ELISA in point-of-care devices. *Analyst*. 2014;139:1456–1462. https://doi.org/10.1039/c3an02296j.
128. Huang S, Abe K, Bennett S, et al. Disposable autonomous device for swab-to-result diagnosis of influenza. *Anal Chem*. 2017;89:5776–5783. https://doi.org/10.1021/acs.analchem.6b04801.
129. Zhao M, Li H, Liu W, Chu W, Chen Y. Paper-based laser induced fluorescence immunodevice combining with CdTe embedded silica nanoparticles signal enhancement strategy. *Sensors Actuators B Chem*. 2017;242:87–94. https://doi.org/10.1016/j.snb.2016.11.004.
130. Hsu CK, Huang HY, Chen WR, et al. Paper-based ELISA for the detection of autoimmune antibodies in body fluid-the case of bullous pemphigoid. *Anal Chem*. 2014;86:4605–4610. https://doi.org/10.1021/ac500835k.
131. Mazzu-Nascimento T, Gomes Carneiro Leão PA, Catai JR, Morbioli GG, Carrilho E. Towards low-cost bioanalytical tools for sarcosine assays for cancer diagnostics. *Anal Methods*. 2016;8:7312–7318. https://doi.org/10.1039/c6ay01848c.
132. Le S, Zhou H, Nie J, et al. Fabrication of paper devices via laser-heating-wax-printing for high-tech enzyme-linked immunosorbent assays with low-tech pen-type pH meter readout. *Analyst*. 2017;142:511–516. https://doi.org/10.1039/c6an02422j.
133. Badu-Tawiah AK, Lathwal S, Kaastrup K, et al. Polymerization-based signal amplification for paper-based immunoassays. *Lab Chip*. 2015;15:655–659. https://doi.org/10.1039/c4lc01239a.
134. Lathwal S, Sikes HD. Assessment of colorimetric amplification methods in a paper-based immunoassay for diagnosis of malaria. *Lab Chip*. 2016;16:1374–1382. https://doi.org/10.1039/c6lc00058d.
135. Ortega GA, Pérez-Rodríguez S, Reguera E. Magnetic paper-based ELISA for IgM-dengue detection. *RSC Adv*. 2017;7:4921–4932. https://doi.org/10.1039/c6ra25992h.
136. Liu XY, Cheng CM, Martinez AW, et al. A portable microfluidic paper-based device for elisa. In: *Proceedings of the IEEE International Conference on Micro Electro Mechanical Systems (MEMS)*; 2011:75–78. https://doi.org/10.1109/MEMSYS.2011.5734365.
137. Verma MS, Tsaloglou MN, Sisley T, et al. Sliding-strip microfluidic device enables ELISA on paper. *Biosens Bioelectron*. 2018;99:77–84. https://doi.org/10.1016/j.bios.2017.07.034.

138. Han KN, Choi JS, Kwon J. Three-dimensional paper-based slip device for one-step point-of-care testing. *Sci Rep*. 2016;6:25710. https://doi.org/10.1038/srep25710.
139. Rohrman BA, Richards-Kortum RR. A paper and plastic device for performing recombinase polymerase amplification of HIV DNA. *Lab Chip*. 2012;12:3082–3088. https://doi.org/10.1039/c2lc40423k.
140. Connelly JT, Rolland JP, Whitesides GM. "Paper machine" for molecular diagnostics. *Anal Chem*. 2015;87:7595–7601. https://doi.org/10.1021/acs.analchem.5b00411.
141. Dou M, Sanjay ST, Dominguez DC, Liu P, Xu F, Li XJ. Multiplexed instrument-free meningitis diagnosis on a polymer/paper hybrid microfluidic biochip. *Biosens Bioelectron*. 2017;87:865–873. https://doi.org/10.1016/j.bios.2016.09.033.
142. Magro L, Jacquelin B, Escadafal C, et al. Paper-based RNA detection and multiplexed analysis for Ebola virus diagnostics. *Sci Rep*. 2017;7:1–9. https://doi.org/10.1038/s41598-017-00758-9.
143. Shetty P, Ghosh D, Singh M, Tripathi A, Paul D. Rapid amplification of Mycobacterium tuberculosis DNA on a paper substrate. *RSC Adv*. 2016;6:56205–56212. https://doi.org/10.1039/c6ra07529k.
144. Rodriguez NM, Wong WS, Liu L, Dewar R, Klapperich CM. A fully integrated paperfluidic molecular diagnostic chip for the extraction, amplification, and detection of nucleic acids from clinical samples. *Lab Chip*. 2016;16:753–763. https://doi.org/10.1039/c5lc01392e.
145. Linnes JC, Rodriguez NM, Liu L, Klapperich CM. Polyethersulfone improves isothermal nucleic acid amplification compared to current paper-based diagnostics. *Biomed Microdevices*. 2016;18:1–12. https://doi.org/10.1007/s10544-016-0057-z.
146. Rodriguez NM, Linnes JC, Fan A, Ellenson CK, Pollock NR, Klapperich CM. Paper-based RNA extraction, in situ isothermal amplification, and lateral flow detection for low-cost, rapid diagnosis of influenza A (H1N1) from clinical specimens. *Anal Chem*. 2015;87:7872–7879. https://doi.org/10.1021/acs.analchem.5b01594.
147. Choi JR, Hu J, Tang R, et al. An integrated paper-based sample-to-answer biosensor for nucleic acid testing at the point of care. *Lab Chip*. 2016;16:611–621. https://doi.org/10.1039/c5lc01388g.
148. Lafleur LK, Bishop JD, Heiniger EK, et al. A rapid, instrument-free, sample-to-result nucleic acid amplification test. *Lab Chip*. 2016;16:3777–3787. https://doi.org/10.1039/c6lc00677a.
149. Xu G, Nolder D, Reboud J, et al. Paper-origami-based multiplexed malaria diagnostics from whole blood. *Angew Chem Int Ed*. 2016;55:15250–15253. https://doi.org/10.1002/anie.201606060.
150. Yang X, Forouzan O, Brown TP, Shevkoplyas SS. Integrated separation of blood plasma from whole blood for microfluidic paper-based analytical devices. *Lab Chip*. 2012;12:274–280. https://doi.org/10.1039/c1lc20803a.
151. Kar S, Maiti TK, Chakraborty S. Capillarity-driven blood plasma separation on paper-based devices. *Analyst*. 2015;140:6473–6476. https://doi.org/10.1039/c5an00849b.
152. Kim D, Kim S, Kim S. An innovative blood plasma separation method for a paper-based analytical device using chitosan functionalization. *Analyst*. 2020;145:5491–5499. https://doi.org/10.1039/d0an00500b.
153. Derda R, Tang SKY, Laromaine A, et al. Multizone paper platform for 3D cell cultures. *PLoS One*. 2011;6, e18940. https://doi.org/10.1371/journal.pone.0018940.
154. Sapp MC, Fares HJ, Estrada AC, Grande-Allen KJ. Multilayer three-dimensional filter paper constructs for the culture and analysis of aortic valvular interstitial cells. *Acta Biomater*. 2015;13:199–206. https://doi.org/10.1016/j.actbio.2014.11.039.
155. Derda R, Laromaine A, Mammoto A, et al. Paper-supported 3D cell culture for tissue-based bioassays. *Proc Natl Acad Sci U S A*. 2009;106:18457–18462. https://doi.org/10.1073/pnas.0910666106.

156. Kim SH, Lee HR, Yu SJ, et al. Hydrogel-laden paper scaffold system for origami-based tissue engineering. *Proc Natl Acad Sci U S A*. 2015;112:15426–15431. https://doi.org/10.1073/pnas.1504745112.
157. Wang L, Xu C, Zhu Y, et al. Human induced pluripotent stem cell-derived beating cardiac tissues on paper. *Lab Chip*. 2015;15:4283–4290. https://doi.org/10.1039/c5lc00919g.
158. Deiss F, Matochko WL, Govindasamy N, Lin EY, Derda R. Flow-through synthesis on teflon-patterned paper to produce peptide arrays for cell-based assays. *Angew Chem Int Ed*. 2014;53:6374–6377. https://doi.org/10.1002/anie.201402037.
159. Fleischer S, Feiner R, Shapira A, et al. Spring-like fibers for cardiac tissue engineering. *Biomaterials*. 2013;34:8599–8606. https://doi.org/10.1016/j.biomaterials.2013.07.054.
160. Liu Q, Wang J, Wang B, et al. Paper-based plasmonic platform for sensitive, noninvasive, and rapid cancer screening. *Biosens Bioelectron*. 2014;54:128–134. https://doi.org/10.1016/j.bios.2013.10.067.
161. Cramer SM, Larson TS, Lockett MR. Tissue papers: leveraging paper-based microfluidics for the next generation of 3D tissue models. *Anal Chem*. 2019;91:10916–10926. https://doi.org/10.1021/acs.analchem.9b02102.
162. Sharifi F, Ghobadian S, Cavalcanti FR, Hashemi N. Paper-based devices for energy applications. *Renew Sust Energ Rev*. 2015;52:1453–1472. https://doi.org/10.1016/j.rser.2015.08.027.
163. Esquivel JP, Del Campo FJ, Gómez De La Fuente JL, Rojas S, Sabaté N. Microfluidic fuel cells on paper: meeting the power needs of next generation lateral flow devices. *Energy Environ Sci*. 2014;7:1744–1749. https://doi.org/10.1039/c3ee44044c.
164. Pushparaj VL, Shaijumon MM, Kumar A, et al. Flexible energy storage devices based on nanocomposite paper. *Proc Natl Acad Sci U S A*. 2007;104:13574–13577. https://doi.org/10.1073/pnas.0706508104.
165. Rivera JH, Díaz DO, Cruz DMA, et al. A paper-based microfluidic fuel cell using soft drinks as a renewable energy source. *Energies*. 2020;13:1–13. https://doi.org/10.3390/en13102443.
166. Dou M, Dominguez DC, Li X, Sanchez J, Scott G. A versatile PDMS/paper hybrid microfluidic platform for sensitive infectious disease diagnosis. *Anal Chem*. 2014;86:7978–7986. https://doi.org/10.1021/ac5021694.
167. Citron P. Medical devices: lost in regulation. *Issues Sci Technol*. 2011;XXVII(3).
168. European Parliament. *Regulation (EU) 2017/745 of the European Parliament and of the Council—of 5 April 2017—On Medical Devices, Amending Directive 2001/83/EC, Regulation (EC) No 178/2002 and Regulation (EC) No 1223/2009 and Repealing Council Directives 90/385/EEC a, Expert Review of Medical Devices*; 2017.
169. Lim SB, Menon NV, Lim CT. Microfluidic tools for probing micro-culprits. *EMBO Rep*. 2020;21:e49749. https://doi.org/10.15252/embr.201949749.

CHAPTER SIX

Paper based micro/nanofluidics devices for biomedical applications

P.E. Resmi[a], P.V. Suneesh[a,b], T. Ramachandran[b], and T.G. Satheesh Babu[a,b,*]
[a]Amrita Biosensor Research Lab, Amrita School of Engineering Coimbatore, Amrita Vishwa Vidyapeetham, Coimbatore, India
[b]Department of Sciences, Amrita School of Engineering Coimbatore, Amrita Vishwa Vidyapeetham, Coimbatore, India
*Corresponding author: e-mail addresses: tgsatheesh@gmail.com; tg_satheesh@cb.amrita.edu

Contents

1. Importance of paper as a diagnostic device 160
2. Paper as a sample pretreatment platform in clinical diagnosis 161
3. Paper microfluidic diagnostic devices 163
4. Dipstick paper biosensors 164
5. Lateral flow assay (LFA) 164
6. MicroPAD (μPAD) 167
7. Fabrications of paper microfluidics 167
   7.1 Photolithography 168
   7.2 Plotting 169
   7.3 Screen printing 169
   7.4 Inkjet etching 169
   7.5 Wax printing 170
   7.6 Flexographic printing 171
   7.7 Wax dipping 171
   7.8 Inkjet printing 172
   7.9 Plasma treatment 173
   7.10 Ink stamping 173
8. Detection techniques 174
   8.1 Colorimetry 174
   8.2 Quantification in a colorimetric assay 176
   8.3 Chemiluminescence 177
   8.4 Fluorescence detection 177
   8.5 Electrochemical method 177
   8.6 Electrochemiluminescence 178

Progress in Molecular Biology and Translational Science, Volume 186
ISSN 1877-1173
https://doi.org/10.1016/bs.pmbts.2021.07.011

9. Readout devices 178
9.1 Smartphone-based devices 178
9.2 Instrument-free readout devices 180
10. Biomedical applications 180
10.1 Clinical diagnostics 180
10.2 Drug analysis 181
10.3 Detection of biomolecules 181
10.4 Smartphone based applications 182
11. Paper nanofluidics 183
12. Challenges and research gaps 184
13. Conclusions 185
14. Future outlook 185
Acknowledgment 185
References 185

Abstract

This chapter details the significance, fabrication and biomedical applications of paper-based microfluidic devices. The first part of the chapter describes the importance of paper diagnostic devices, highlighting pretreatment, dipsticks, lateral flow assays, and microPADs. Various methods followed for the fabrication of the paper analytical devices are discussed in the second part. The last part is about some of the important biomedical applications of paper analytical devices. Finally, the challenges and research gaps in the paper microfluidics for biomedical applications are presented.

1. Importance of paper as a diagnostic device

Paper is considered one of the best platforms for disease diagnostics due to its relative abundance, low cost, environment friendliness, lightweight, and tunability of detection processes. Cellulose paper is composed of solid fiber which has a pore size of 1–10 μm and diameter in the range of 1–100 μm. Normally paper is made from wood pulp, and it contains lignin fluorophore that affects the optical detection on paper devices. Due to this, specially treated chromatographic paper containing low amounts of lignin and high cellulose are employed for diagnostic purposes. These paper substrates also possess good mechanical and fluid retention properties due to the treatment processes. Like other microfluidic diagnostic platforms, the paper also requires microliter or sub microliter volume of samples, and the device is of a millimeter to centimeter dimensions. Therefore, these devices are well suited for point-of-care needs with rapid, simple, and low-cost operations. Thus, paper devices received attention in environmental monitoring, clinical diagnosis and food safety.[1–3]

Paper is a soft, hydrophilic, flexible, fibrous and porous material with inherent wicking property to absorb solvents due to capillary action.[4] There are mainly two kinds of paper used for analytical purposes. The first one is the regular paper made of cellulose and the second one is the nitrocellulose paper. Cellulose paper is thin, flexible, absorbent, foldable and is composed of glucose moieties. Its chemical and physical modifications are relatively easy compared to other substrates. Nitrocellulose (NC) paper is obtained by reacting nitric acid with cellulose. This results in a white fibrous polymer that has good resistance to different chemical reagents and different oils. Also, nitrocellulose demonstrates better properties than cellulose like controllable polymerization, improved solubility, and better ability to bind proteins and is widely used in analytical fields.[1]

The fabrication and treatment process controls the pore dimension in the paper substrate and, thereby, the wicking properties. When used with clinical samples, this fluid flow controls the analyte movement, and the proper separation and mixing are possible without using an external force. The usual air bubble issue in conventional microfluidic channels is not present in paper channels, and the fluid flow can be achieved without using external pumps. The initial wetting process known as imbibition is the key for fluid transport in porous media like paper. The hydrophilic surface of the cellulose paper has a low contact angle which facilitates the flow of liquids and adhesion. The surface tension of liquid decreases liquid-gas interfacial area, and this supports wicking for capillary-driven fluid flow. The fluid flow results from the pressure difference across the dry area at the liquid front and the wetted area. The wicking of liquids in the paper is usually analyzed by two models derived from Hagen-Poiseuille law and is known as Lucas-Washburn equation and another phenomenologically derived model Darcy's law.[5]

2. Paper as a sample pretreatment platform in clinical diagnosis

The composition of biological samples like blood, saliva, urine, ascitic fluid, etc., is complex and challenging to analyze and requires sample pretreatment to expose the analyte like RNA, DNA, or protein. This necessitates the pretreatment procedures like sample storage, extraction, separation as well as concentration. Traditionally these are performed by storage vials (vacutainer system), centrifugation, ultrafiltration, and dialysis. But these methods are expensive, time-consuming and require bulky equipment, which is not suitable for a point of care diagnosis. Paper can be used as a simple and inexpensive platform for the pretreatment process.[6] Fast technology analysis (FTA) cards from Whatman® can be used to collect and store bioanalytes like blood, cells, and viruses, which are found to be efficient in transportation even in the absence of cold chains. These cards can be used for cell lysis to release nucleic acids, denaturing proteins, and protecting nucleic acids. The hydroxyl groups and surfactants in cellulose paper can help in the efficient binding of analytes without damaging them, making this a suitable substrate for biological sample collection and storage. Specific chemical treatment procedures can tune the function of the paper device. For example, butylated hydroxyl toluene in methanol treatment is done for the paper to collect and store blood. FTA cards are reported for the efficient storing of cancerous cells and pathogens.[7,8] The nucleic acid content from the cards can easily be eluted using a simple procedure, and it is reported to have long storage stability.

Filtration aided by capillary force is employed for the separation of complex biological samples using paper. Commercial cellulose membranes such as LF1, Fusion 5, VF2 and MF1 are commonly used to fabricate POCT devices. The separation of serum from blood has been demonstrated using these devices. The surface of the paper may be modified with antibodies favoring the aggregation of the red blood cells and thereby trapped in the pores on the paper surface. The process is named as agglutination technique. Origami-based blood cell separation and magnetic and electric field aided cell separation are reported on the paper substrate.[9]

Conventional extraction procedures for obtaining the target analyte (DNA or RNA) from complex biological samples include salt precipitation, alkaline lysis, guanidinium thiocyanate method, magnetic particle based method and spin-column method.[10] An efficient method should be having a short extraction time, equipment-free, and can easily be manipulated. FTA elution cards are the commercial system for extracting nucleic acids from the stored cells. Simple sequential folding (origami) of paper is used for the fabrication of the gadolinium thiocyanate method, which otherwise requires complex physical and chemical extraction procedures. In addition, a micro pad filtration isolation nucleic acid (FINA) technology for the extraction of DNA from whole blood is also reported. In this stacked paper device, the alkaline extraction method is adopted to capture DNA using a fusion 5 membrane.[11] This technique is used for HIV detection and is found to be highly efficient while comparing with the traditional methods.

The detection of low concentration analytes requires a concentration step before analysis that improves the detection limit. Traditional methods for this process include ultrafiltration, isotachophoresis, and temperature gradient focusing. Multilayer paper 3D device that has a hydrophobic layer and a more hydrophilic layer is used. The concentration of the sample occurs based on the hydrophilic or hydrophobic nature of the analyte. Nitrocellulose membrane is found to have good efficiency in the concentration of proteins and nucleic acids. Traditional heating-induced evaporation methods using heating plates are also reported for the concentration of sample analytes where the paper is sandwiched between custom-made aluminum heating plates. Thermal damage of the analyte is a hindrance to this method.[12] Electrophoretic and isotachophoretic methods on paper are also employed based on the mass and charge (decides the mobility) of the analyte can be concentrated using an applied electric field.[13] Although there are challenges, these concentration techniques can improve the detection limit of the analytical procedure drastically.

3. Paper microfluidic diagnostic devices

Paper microfluidic diagnostic devices can broadly be classified into two categories as, continuous flow microfluidic devices (p-CMF) and paper digital microfluidic devices (p-DMF). p-CMF devices transport fluid via capillary action while p-DMF uses electrowetting on dielectric (EWOD) technique. In this technique, droplet actuation is achieved by applying an electric field to an array of electrodes coated with a hydrophobic dielectric material layer.[14]

p-CMF was invented in 1949 by Müller and Clegg and came into broad interests after the report by Whitesides' group in 2007. The inherently porous paper matrix provides capillary flow, and various physical and chemical modifications can tune the flow. Standard dipstick assays, micro pads, lateral flow assays are working on this principle.

The first report on p-DMF was by Pollock et al. in 2000 and gained popularity after Shin's group and Wheeler group reports in 2004. These devices have several features, including portability, capability to deliver, delay, split and merge liquid drops. These are fabricated by patterning conductive electrode arrays on a paper substrate and coating the electrode arrays with a hydrophobic dielectric layer. An electronic switching system and programmed control software will be used for fluid control. Details on the engineering design and prototype fabrication are reviewed thoroughly by López-Marzo and Merkoci.[4]

4. Dipstick paper biosensors

It is the simplest form of a paper microfluidic device that incorporates the specific reagent. Specific colorimetric reactions occur when the paper strip is dipped in the sample solution. These are devices valuable tools for investigating, diagnosing, and screening diseases immediately. They are easy to manufacture and convenient to use. In most cases, a color chart is provided for semiquantitative information without any sophisticated equipment. The first paper-based dipstick was devised in the 1950s to determine glucose in urine samples (glycosuria) for diabetic monitoring. This method involves the enzymes glucose oxidase, peroxidase, and orthotoluidine reagent incorporated paper. Enzymatic glucose oxidation produces glucuronic acid and which in turn converts to hydrogen peroxide. This hydrogen peroxide oxidized orthotoluidine to a deep blue chromogen.[15] Similarly, dipstick sensors for ketones, bilirubin, urobilinogen, leukocyte esterase, and nitrite are in practice. Dipsticks are also used to measure urine pH and specific gravity, metabolite products such as protein, and electrolyte content for kidney health. The color changes can be captured with a mobile camera for disease prediction.[16] Paper-based dipsticks are also employed for low-cost pathogen detection.[17,18] Highly sensitive bladder cancer detection using ELISA-based keratin detection and functional DNA-linked gold nanoparticles (AuNPs) based mercury detection[19] are examples in this category.

Although the dipstick method is a simple technology and easy to standardize, it has limitations. For example, the detection is limited to optical methods, nonsuitability for detections involving multiple steps, and qualitative. Some of the commercialized products in this area are Chemstrip1 Micral test strips (human albumin, Roche Diagnostics), Medi-Test Combi 11 (multiparameters, Macherey Nagel), Multistix 10SG reagent strips (UTI, Diabetes, and kidney disorders, Siemens), Unisys 11,001 (multiparameters, Rosche) and Keto Diastix Reagent Test Strip (Urine ketones, Bayer).[20]

5. Lateral flow assay (LFA)

LFAs are simple tools for the detection of different analytes in a quantitative or semiquantitative way. Here the normal laboratory functions like separation and mixing of reagents occur without any external force due to the excellent capillarity action of paper. The ASSURED criteria recommended by WHO are perfectly followed in this case. Different analytes such as infectious viruses, cancer biomarkers, pathogenic bacteria, DNA, or RNA are detected using this equipment-free method.[21] Antibody conjugated gold nanoparticles (AuNPs) are the most common detection probes in conventional colorimetric LFAs. Thus LFA is the technique that combines the biorecognition chemistry to the chromatography to emerge as a simple analytical device. These methods eliminate the conventional incubation and washing steps. LFA is performed on a paper strip in which different parts are assembled on a plastic backing for support. A typical LFA device (Fig. 1) consists of the following components made of nitrocellulose membrane (NC).[23]

1. *Sample pad*: This is where the sample is applied. It is made of either cellulose membrane or glass fiber. The pad contains buffer salts and surfactants that ensures easy interaction of the sample with the detection system. The main role of this component is to transport the sample applied to other parts of the assay in a smooth, continuous and homogeneous manner. This may also have the role of filtering out the unwanted sample contents like blood cells and separate the serum.
2. *Conjugation pad*: This is where the labeled recognition probes are incorporated. Upon the flow of sample from the sample pad through conjugation pad, the pad releases the labeled probes (antibodies, aptamers etc.) easily, and that flows along with the sample. Glass fiber, cellulose, polyesters are also used to make a conjugate pad. During the flow of sample and recognition probes, specific analyte-probe conjugates are formed and flow through this pad and stable throughout the flow. The flow dynamics, proper dispensing of bioreagents, drying, blocking, and conjugate stability can affect the precision of the detection. Also, the nature

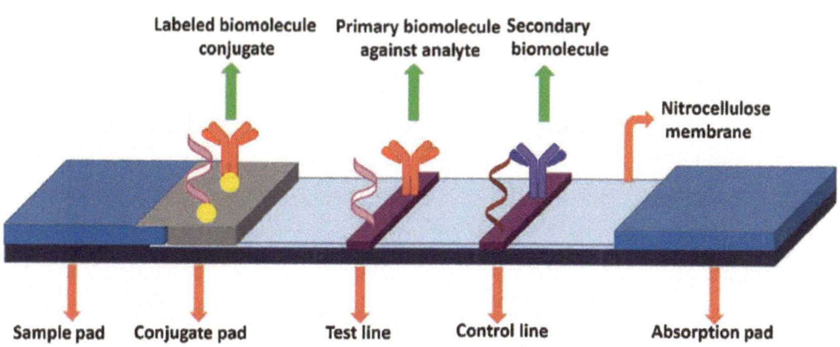

Fig. 1 Schematic presentation of a Lateral Flow Immunoassay (LFA).[22]

of the conjugate membrane to release the conjugates affects the sensitivity of this detection method. This has the specific detection line for the specific binding of the analyte-probe conjugate and the control line for the comparison.

3. *Absorption pad*: It acts as a sink for the sample liquid flowing and restricts the backflow of the liquid. It possesses a very high wicking property that allows the smooth flow of the liquid. The adsorption capacity of the pad to hold the sample can play an important role in the efficiency of the assay.
4. *Backing card*: This provides rigid support for the LFA strip assembly, mainly made of plastic. There are no strict rules in selecting the material for this as it does not have a role in the detection and only acts as a support.

The most known commercial LFA strip is a pregnancy test kit in which the elevated concentration of human chorionic gonadotropin (HCG) hormone during pregnancy is monitored. Most pregnant women respond as positive to the tone week after the first day of a missed menstrual period.[24] But these are qualitative tests and give yes or no answers for pregnancy. Several optical readers are reported for quantification. Another recent example in this area is the COVID-19 antibody test strips.[25,26] Since December 2019, the pathogenic novel coronavirus (SARS-CoV-2) has caused several deaths and was affected worldwide. Therefore, efficient detection methods are essential for the patient's proper isolation and treatment to prevent the disease's spread. The most efficient method used for the detection is nucleic acid-based detection, where the virus RNA is detected using the RT-PCR method. The method is time-consuming, need trained technician and requires costly instrument such as PCR. But because of the specificity, this is considered the gold standard for this disease detection now. Analysis of specific antibodies of SARS-CoV-2 in patient blood is the other method adopted for rapid diagnosis of COVID-19.[27] This involves the detection of an increase in the amount of IgM and IgG antibodies in the blood.[28] This is done via lateral flow immunoassay similar to that of a pregnancy test strip. Raybiotech, Confirm Biosciences and IgG-IgM, and Assay Genie is among the pioneers in the commercialization of this technique. But these methods are less selective to COVID-19, and the IgG and IgM concentrations can increase in other viral attacks. Recently, this has been shown as false positive, and the controversies are still on.[29] But these methods are presently employed as the first-hand screening test for COVID 19.

6. MicroPAD (μPAD)

Micropads are simple paper analytical devices (PADs) appropriately engineered to perform multiple laboratory functions like sequential mixing. This is the outcome of advancements in several fields including, nanotechnology, biotechnology, chemistry, microfluidics, microelectronics, and printing technologies. These devices can be fabricated in both 2D or 3D formats. Specific hydrophobic and hydrophilic channels or points were designed to control the fluid flow on paper that decides the mixing and dilution timings. Materials such as photoresists, wax, and polystyrene formulations are used to fabricate hydrophobic channels, wells and points. Wax printing, lithography and inkjet printing are commonly used to fabricate 2D devices while stalking and vertical folding used for 3D devices.[1,30] Origami is famous for the fabrication of 3D devices and is even used for multiplexed arrays using electrodes or colorimetric reagents. Blood is the most important sample for clinical diagnostics. These devices also do not require an external pumping force for the fluid flow, similar to LFA. However, red blood cells may interfere with the detection and quantification in colorimetric assays, especially assays based on pink, red, and brown dyes, thus necessitate plasma separation.

The first of this kind of lab on a chip device was developed by Martinez et al. from the Whitesides group to detect glucose and proteins on a single platform. Hydrophobic barriers were made using photoresists and in specific hydrophilic zones to observe the characteristic response with the analyte. To introduce the telemedicine concept, they have fabricated four detection zones to detect glucose and total protein with mobile phone camera readout.[31–33]

7. Fabrications of paper microfluidics

One of the novelties behind paper analytical devices (PADs) is their fabrication. Various methods have been employed for the fabrication of paper-based analytical devices. The fabrication approaches consist of increasing solvent compatibility, enhanced flow control designs, deposition of different materials, and functionalities for different applications.[1] Hydrophobic barriers are created to prevent the flow of reagents and samples during fabrication. Fabrication techniques also depend on the type of target and type of paper substrate. The fabrication by patterning includes physical deposition, physical blocking, and chemical modifications of the paper surface. Photolithography and plotting involve the physical blocking of pores with photoresist chemicals. The different methods for physical deposition of reagents on paper include screen printing techniques, inkjet etching, wax printing and flexographic printing. Chemical modification of the paper surface takes place during inkjet printing and plasma treatment.[34]

7.1 Photolithography

The photolithographic method is based on the entire hydrophobization followed by selective dehydrophobization of the paper. In this method, a hydrophobic barrier is created on a chromatography paper with SU-8 photoresist or SC (cyclized poly(isoprene)derivative) photoresist. Next, this photoresist impregnated chromatographic paper is selectively polymerized by UV exposure using a transparency mask. Finally, the unexposed photoresist is removed by washing in a mixture containing isopropyl alcohol and acetone. Fig. 2 depicts the patterning of paper done by Whiteside's group. The steps involved are (i) soaking the chromatographic paper in the

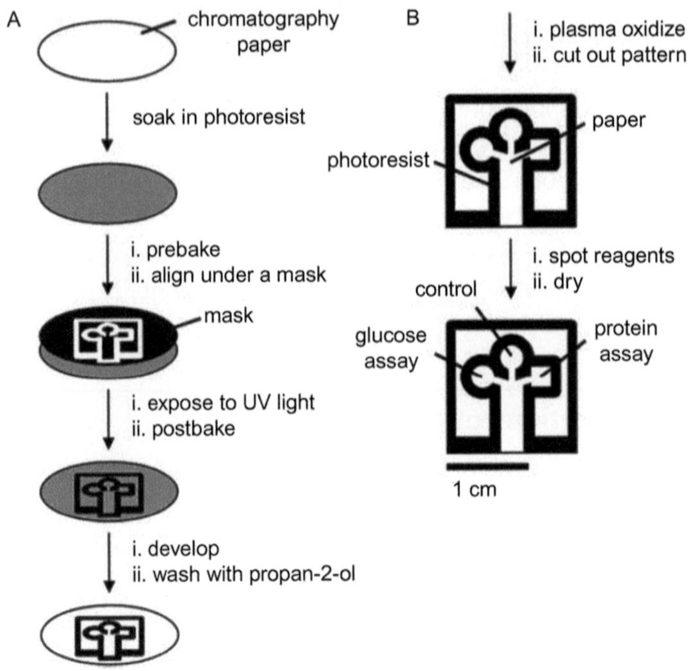

Fig. 2 Schematic presentations of the different steps involved in the photolithography for patterning of paper microfluidic devices.[35]

photoresist, (ii) spinning, (iii) pre-baking, (iv) exposure to UV light using a mask, (v) post-baking and (vi) removal of unpolymerized photoresist. The requirement of expensive photoresists, photolithography equipment and organic solvents to remove unexposed are some disadvantages of this technique. Also, this method leaves a layer of hydrophobic organic precipitate on the paper, which may damage the flexibility of paper.[33]

7.2 Plotting

It is a direct method to create a hydrophobic barrier by printing the hydrophobic polymer. Poly dimethoxy silane (PDMS) is a low-cost, non-toxic, and easily obtainable polymer commonly used for plotting. In the pen plotting method, the hydrophobic inks are added to the pens loaded with PDMS. This pen is then used to draw hydrophobic barriers on paper using an XY plotter. The penetration of the ink depends on the viscosity of the ink and the pressure of the plotter head. The technique is inexpensive but has low resolution and cannot be used for multiple batches of devices.[36] This technique does not harm paper flexibility, but controlled penetration is not possible because of the non-uniformity in the porous substrate paper. It also requires about 1 h at 70 °C to cure the PDMS.[34]

7.3 Screen printing

The screen-printing method is another way of fabricating paper-based devices. For this, commonly available polymer inks can be screen printed on chromatographic paper to create a hydrophobic barrier. This simple screen-printing technique does not require additional post-treatment and heating. It involves only two steps, designing of pattern/mask and screen printing on a paper surface.[37] The electrochemical paper-based device can be fabricated by producing the electrodes directly on the paper. The screen containing the optimized design is set on the paper substrate. The conductive ink is pushed manually or automatically with an appropriate squeegee through the screen to form the printed electrode. The printed paper was cured at room temperature or in an oven. The commercially available inks contain graphite, carbon nanotubes, and other conducting materials.

7.4 Inkjet etching

In inkjet etching, the hydrophilic channel was developed by soaking the paper in hydrophobic chemicals like polystyrene solution for 2 h and drying at room temperature for 15 min. Then, the modified paper was patterned by

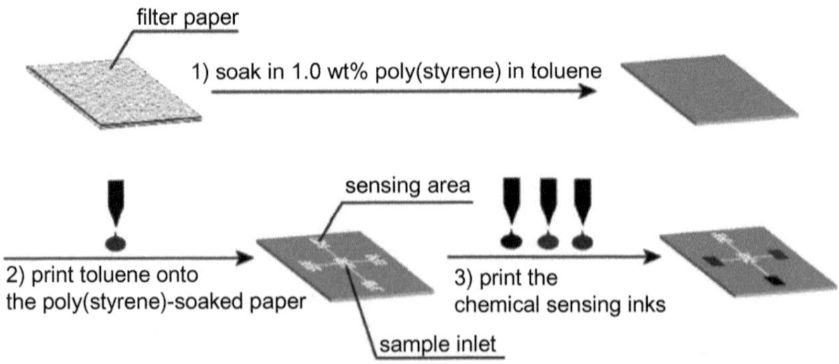

Fig. 3 Schematic representation of the inkjet printing.[39]

inkjet printing of toluene about 10–20 times to dissolve the polymer. This resulted in the formation of a hydrophobic barrier and hydrophilic channel. Finally, the polystyrene is peeled off from the filter paper as etching. The toluene serves as the etchant and the polystyrene on the paper is etched.[38] The schematic presentation of the inkjet printing technique is shown in Fig. 3.

7.5 Wax printing

The wax can be introduced on the chromatographic paper by various methods. The commonly used methods include (i) wax painting using a wax pen, (ii) inkjet printing followed by wax painting and (iii) direct wax printing. The desired pattern was drawn on both sides of the filter paper using a wax pen followed by heating in an oven at 150 °C for 5 min to form a hydrophobic barrier. The complete process takes only 5–10 min. The materials used for this technique are affordable and easily available.[40] Wax patterned paper was also fabricated using a wax pen to create a hydrophobic barrier.

Wax printing is the simplest method for creating a hydrophobic barrier on hydrophilic paper and this can be done using a wax printer. The designs were drawn using Clewin or CAD software on a computer. Then, the designed pattern was printed on a Whatman chromatographic paper and heated on a hotplate at 150 °C to reflow the wax (Fig. 4). When the wax melts, it forms a hydrophobic barrier across the thickness of the paper.[41] It involves the fewest steps, and the largest number of strips can be fabricated in a single batch.[42]

The wax was also introduced on the chromatographic paper by a wax droplet generating system. Here we drop the wax onto the paper through

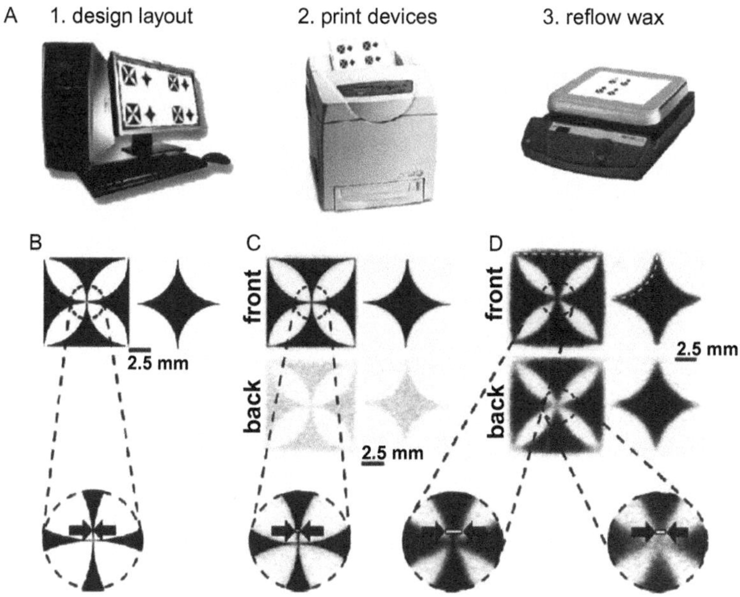

Fig. 4 Schematics of the different steps involved in wax printing.[41]

the nozzles of the droplet generator. Wax patterns were heated to create a hydrophobic barrier and used for the bioassay. Wax dipping is another method for wax printing in which the pattern is obtained by plunging an assembly of iron mold into the melted wax for 1 s.[43] Another method for wax patterning is screen printing, which was performed using traditional screens. Although the fabrication cost is less, this method has a low resolution.[43]

7.6 Flexographic printing

A flexographic printer setup consists of an anilox roll, and the substrate paper is fixed on the impression role. Then, the ink (a mixture of polystyrene solutions in toluene or xylene) was applied to an anilox roll. The ink penetrates the paper due to the rotation of the plate and impression role, which forms a hydrophobic pattern (Fig. 5). The advantages of this method are that heat treatment is not required for polystyrene, which is biocompatible, and the transferring of reagent is possible through a single roll to roll process. But it gives a low-resolution channel.[45]

7.7 Wax dipping

In this technique, wax patterned microfluidic channels are generated on chromatographic paper using an iron mold created by the laser cutting

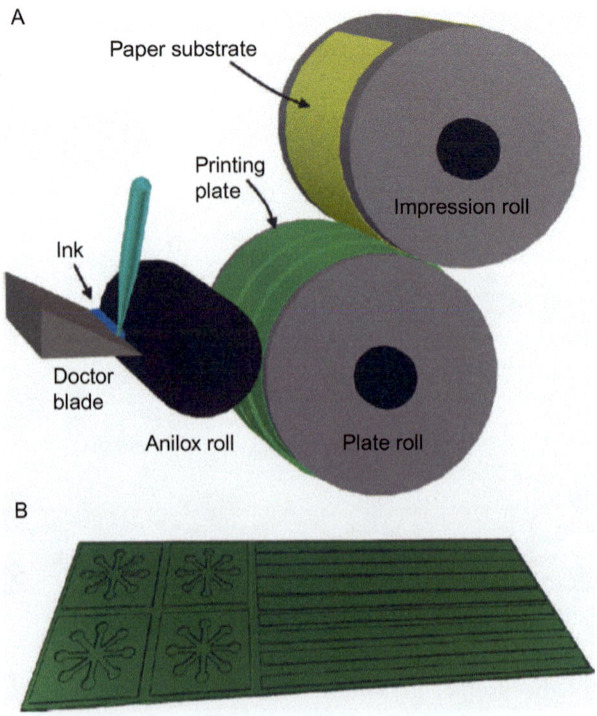

Fig. 5 Schematic presentation of the Flexographic printing.[44]

technique. First, for wax dipping, white beeswax pellets are heated to melt. The temperature is maintained around 120–130 °C and this is monitored using an electronic contact thermometer. Next, the iron mold is placed on a chromatographic paper of a suitable size with the help of a permanent magnet. This setup is then plunged into the wax melt. The paper is then cooled and peeled off from the glass slide. This creates a hydrophobic barrier and forms the wax patterned paper.[43]

7.8 Inkjet printing

Inkjet printing allows precise, rapid, and reproducible deposition of reagents and biomolecules for analytical assays. Based on the application, inkjet printing technology is categorized into a drop-on-demand (DOD) and continuous inkjet (CIJ). In the CIJ mechanism, the continuous flow of ink occurs, whereas the jetting of ink on to substate occurs whenever needed in DOD. Thermal and piezoelectric methods are used for DOD inkjet printing. When bubbles are ejected from the nozzle, it collapses and forms a vacuum that allows automatic filling of ink in the reservoir. Ink is the essential component of inkjet printing. Its physicochemical properties determine the velocity of drop, resolution, and reliability of printing. Inkjet printing can be used for both patterning of a microfluidic channel and the deposition of reagents for multiple analyses. Inkjet etching is another method for the patterning of paper. A filter paper substrate is made hydrophobic by soaking it in polystyrene solution and drying at room temperature in this technique. This hydrophobic paper is then exposed to inkjet printing to create channels having a width of about 550 μm. Desktop inkjet printers with UV curable ink were used for patterning the paper. The ink contains a prepolymer solution composed of a monomer and photoinitiator. The polymerization occurs by exposure to UV light, leading to hydrophobic interpenetrating polymer formation within the cellulose network. For inkjet printing, the commonly available printer can be used in which the printing ink is replaced with hydrophobic materials such as alkyl ketene dimer, resin, polystyrene, and polyacrylate.[46] One of the drawbacks of the method is that it cannot keep the organic solvents and hydrophobic material in the cartridge for a long time because it damages the printer.[45]

7.9 Plasma treatment

For the plasma treatment, the paper is first hydrophobized, followed by treatment with plasma. This is achieved by keeping the paper sample between the metal mask having the desired pattern and then placed into a vacuum plasma reactor. The metal masks are made from stainless steel sheets by cutting mechanically. Water and aqueous solutions were easily transported through the plasma-treated channel through capillary penetration. Plasma treatment increases the wettability of the treated area.[47] The paper devices made by this method can be used for single and multiple tests colorimetric assays as the plasma treatment does not change the color of the paper or its flexibility. Functional elements such as switches, filters, and separators can easily be incorporated into the paper pattern fabricated by plasma treatment.[47]

7.10 Ink stamping

A handheld stamping process was developed for the fabrication of PADs. Before stamping, the paper surface is oxidized to convert the hydroxyl group into aldehyde groups, which helps activate the covalent coupling of enzymes. The oxidized paper is dipped in liquid paraffin at 90 °C for

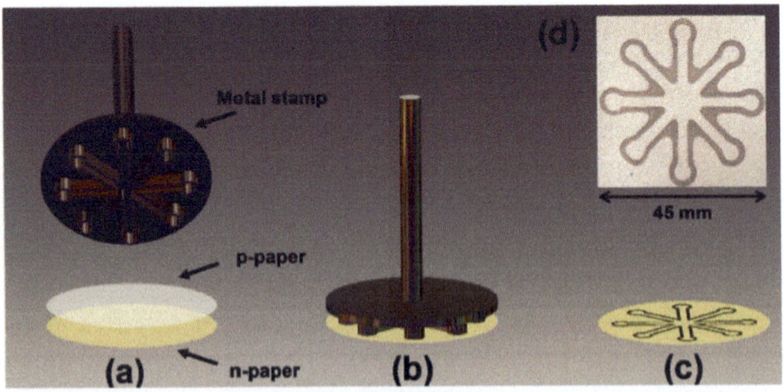

Fig. 6 Schematic presentation of the ink stamping.[48]

1 min, followed by solidification at room temperature. Then, a preheated stainless steel metal stamp is brought in contact with the paraffined paper to develop the pattern. By applying the pressure on the stamp, the thermal transfer of paraffin from p-paper to n-paper takes place.[48] The schematic of the ink stamping is presented in Fig. 6. The drawbacks of this technique like preheating of the stamp and oxidization of the paper surface, make it less suitable for mass production.

8. Detection techniques

The most important applications of PADs are the detection of biomolecules and bioassays. Different techniques employed for the detection include optical and electrochemical. The optical method is the most straightforward and inexpensive and is further classified into colorimetry, chemiluminescence, electrochemiluminescence, and fluorescence.

8.1 Colorimetry

Colorimetry is one of the most commonly used optical method detection of analytes in paper-based devices. The intensity of the color intensity is proportional to the analyte concentration, and the results can be interpreted directly without any equipment. The principle of this method lies in the generation of color by chemical/biochemical reaction between target analyte and reagents. The intensity of the resulting color can be quantified using imaging tools and processing software.

The most significant advantage of the colorimetric assay on paper is its simplicity. Further, the paper substrate is compatible with low detection systems and provides a white background for colorimetric assays. In addition, advantages like faster analysis, lack of complicated instrumentation, ease of application, low-cost detection system, and low sample volume requirement makes colorimetry-based paper analytical devices highly suitable for point of care applications.[49]

The different methods used to produce color on PADs include coupled nanoparticle agglutination, enzymatic assays, redox indicators, and other innovative colorimetric transduction methods. The commonly used enzymatic systems are based on oxidoreductases, transferases and hydrolases. The enzyme oxidase oxidizes the substrate in the presence of oxygen and water to generate hydrogen peroxide along with the oxidized product. The peroxidase catalyzes the oxidation of the redox indicator and the reduction of hydrogen peroxide. The change in the oxidation state of the redox mediator leads to the color change which is proportional to the concentration of H_2O_2. This is the most common enzymatic assay on paper which has been used to detect various biomolecules.[50–53]

Another sensing method used for the colorimetric assay is based on nanoparticles. The nanoparticles can be tagged with antigens, antibodies, oligonucleotides, and their optical properties, such as surface plasmon resonance (SPR) for the naked-eye detection of analytes. Fig. 7 shows the

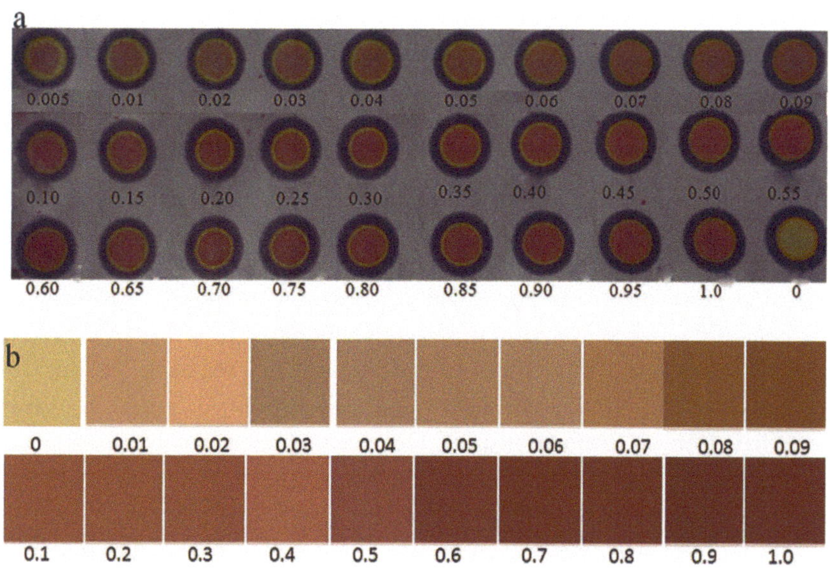

Fig. 7 Colorimetric response obtained on the paper strip with an increase in bilirubin concentrations (A) and the color chart developed for naked-eye detection.[54]

colorimetric response on a paper strip obtained for increasing bilirubin concentrations due to the in situ formation of gold nanoparticles.[54] The commonly used nanoparticles used for detection are gold nanoparticles (AuNPs), silver nanoparticles (AgNPs), latex nanoparticles, and carbon dots.

Another approach is based on light-induced polymerization amplification based on the covalent bonding between the photoinitiator and target analyte. When the monomers are irradiated with visible radiation, photopolymerization occurs in the detection zone, and this leads to a color change.[55] Colored complex formation is another way for colorimetric detection on PADs. The creation or destruction of metallic complexes results in a color change.[56–58] Color changes can be induced due to changes in the chemical environment using chemo responsive dye on PADs.

Color homogeneity is one of the critical factors in the colorimetric paper-based assay. Color homogeneity is affected by factors such as substrate, chemical modification, and nanoparticle incorporation. Whatman Grade 1 chromatographic paper shows the highest color intensity. Color homogeneity can be increased by chemical modification of the chromatography paper. The different method used for chemical modification includes treating the paper with sodium periodate. This enables immobilization via Schiff base formation with the aldehyde group generated by oxidation on the cellulose surface. This eliminates the need for carboxylic coupling agents like 1-Ethyl-3-(3-diethyl aminopropyl) carbodiimide (EDC) and N-hydroxysuccinimide (NHS). The homogeneity can also be increased by modifying using functionalized silica nanoparticles. The design of the pattern also plays a role in the color output.

8.2 Quantification in a colorimetric assay

Colorimetric assay gives either quantitative, semiquantitative, or qualitative assay depending on the reaction. Qualitative assay gives yes or no answers with the help of the naked eye without using instruments and without the help of a trained technician. The color bar code is used as a reference standard for the analysis. Semiquantitative assay results can be estimated by using the color chart based on the calibration plot. Devices such as microscopes, cameras, smartphones, scanners, spectrophotometers, or lasers are incorporated to obtain accurate and precise measurements in quantitative analysis. The quantification of the analyte is carried out by capturing the images and the color intensity values from the different color models, which are obtained with the help of image processing software such as ImageJ and MATLAB.[1]

Reading out the result is another crucial step in the colorimetric analysis on paper. For that, different methods are employed. The analysis can be done with instrumental techniques such as diffuse reflectance spectroscopy or scanners, smartphone cameras, and handheld readers. Smartphone camera and scanners works based on diffuse reflectance spectroscopy. The use of scanners, smartphones, and webcam help in developing point of care testing devices.

Detectors are an essential part of the colorimetric essay on paper to quantify the analytes. Commonly used detectors include complimentary metal-oxide sensors (CMOS) embedded cameras, charge-coupled devices (CCD) and flatbed scanners. The measurement used in smartphone cameras and scanners is based either on light reflectance or light transmittance. In reflectance measurement, the light reflected from the sample after illumination is captured by a photodetector. Conventional flatbed scanners, portable scanners, fiber optics and smartphone-based cameras are based on reflectance-based measurement.[1]

8.3 Chemiluminescence

This is the simplest method to detect the light generated by the chemical or biochemical reaction of the analyte with the reagent. For measuring the chemiluminescence signal, luminescence is incorporated on a paper analytical device. The chemiluminescence-based assay has advantages such as its simplicity, wide calibration range, high sensitivity, and low background.

8.4 Fluorescence detection

Fluorescent detection is based on the interaction of analytes with a fluorescent dye. A light source of appropriate wavelength induces the luminescence of a fluorophore and the emitted light is a measure of the analyte concentration.[59] The fluorescence detection process involves three stages: excitation, excited-state lifetime, and fluorescence emission. The concentration of the analyte can be obtained from the intensity emitted light.[59]

8.5 Electrochemical method

In an electrochemical method, microfluidic channels and electrodes were printed on paper. Carbon and Ag/AgCl inks were used to fabricate the working, counter and reference electrodes, respectively. Finally, a drop of the sample was applied to the detector to read out the measurement. Electrochemical based detection techniques provide low-cost, portable, simple, highly selective and sensitive methods with low electrical power consumption, low cost of electrode fabrication, high compatibility with microfabrication technologies and minimal instrumentation for analysis. The electrochemical systems employed in these PEDs involve microelectrodes, surface-modified electrodes, ion-selective electrodes and ion-exchange membranes. Common techniques used for electrochemical assay include cyclic voltammetry (CV), amperometry, coulometry, pulsed techniques, and impedance spectroscopy.

8.6 Electrochemiluminescence

The detection is based on a signal generated by an electrochemical reaction. Electrochemiluminescence (ECL) biosensors integrate the advantages of both electrochemical and photoluminescence analysis. In electrochemiluminescence, the chemiluminescence is initiated and controlled by the application of electrode potential. ECL is an attractive detection method because it can combine the advantages of luminescence and electrochemical techniques and has good selectivity.[60] A mobile phone-based detection was developed using an inkjet printer and an office laminator with the screen-printed electrode to fabricate the sensor. Herein, the paper strip was fabricated using inkjet printing. Fluids were loaded and dried on the paper and cut into a small strip. This strip is aligned and fixed on to screen-printed carbon electrode through lamination.[61]

9. Readout devices

9.1 Smartphone-based devices

A light source and a detector such as a spectrophotometer or a fluorimeter are required for optical detection. The modern smartphone contains a light source (camera flash) and a detector (digital camera). The bright LED in most smartphones is very bright to offer adequate light intensity as the light source for optical detection.[62] Recent advances in smartphone technology offer an affordable metal oxide semiconductor array (CMOS), which captures the intensity of light that strikes the pixel and converts it into an electrical signal. In addition, color filters are incorporated to increase sensitivity.

Some of the advantages of these devices are that it takes minimum time to familiarize the experimental set up to the user, and the system's total cost is less when we compare other devices. Hence it can be used as a tool for the point of care rapid detection of analytes in resource-limited areas. Using a smartphone-based device, we can interpret the data by sending the results to doctors or technicians after each assay. Based on the interpretation, medical treatment and medical decisions can be taken without delay. A smartphone has the ability to share and store real-time results. This information can be stored in a global database and can be used for further evaluation. Nowadays, most smartphones also have the ability to extract color information from digital images.[63] The smartphone devices may quell the non-homogeneity issue with paper substrate because the light detector used as a digital camera is an imaging device by itself. The non-uniform signal and signal fluctuations obtained from paper are averaged out over a significant area. Also, by using a white LED or ambient light, a range of wavelengths can be used, which is not possible with conventional spectrometric analysis. They are based on the Mie scattering detection, an angle-dependent but less wavelength-dependent method than the other optical detection methods.

The smartphone can also be customized to increase the accuracy of the analysis. A customized mobile phone consists of a built-in camera, an internal microprocessor for image analysis, and a data processing unit. This eliminates the need for external computers and instruments and is capable of providing fast and accurate results. Nowadays, smartphone-based systems can perform analysis based on colorimetry, fluorometry, chemi/bioluminescence, and scattering.[64] Among these techniques, colorimetry is the popular choice.

One of the main limitations of the colorimetric assay is the lack of ability to match the color obtained to a corresponding color chart. This is mainly because of the variations in the identification of the color by the naked eye. Studies are being conducted to overcome this problem. Recently, researchers could develop quick response (QR) codes for paper-based assays and an augmented reality (AR) app to perform colorimetric measurements in real-time.[65]

Smartphone integrated electrochemical paper-based sensors were developed. It contains a low-cost module that is connected to the audio jack of a smartphone and employs screen-printed electrodes (SPEs) to measure different analytes such as antibodies, proteins, or nucleic acids. The module plugs directly into the smartphone through the audio jack. In addition, the module contains a low-power potentiostat that can interface and harvest power from the smartphone. The potentiostat is used to conduct electrochemical experiments to detect and quantify the biomarkers.[64,66]

9.2 Instrument-free readout devices

The instrument-free devices can also be applied for the detection on paper. Distance-based detection is such type of method in which the length of colored product traveled on PAD was measured.[45,67] Another method used for the quantitative assay is based on calibrant loaded paper-based analysis. Before analysis, the multiple sensing areas were preloaded with excess colorimetric reagents required for the assay and a known amount of reagent of the test analyte. The signal obtained from the sensing area relates to the total concentration of the analyte. This total concentration is the sum of the concentration analyte in the tested sample and prestored in the device. The combined signal obtained from each zone is then plotted to form a standard addition calibration curve. This calibration curve is used to calculate the unknown concentration of the analyte in the sample.[68]

10. Biomedical applications

10.1 Clinical diagnostics

Paper-based analytical devices (PAD) are used to detect various biomarkers for tumor, and heart disease, carbohydrate antigen, creatinine kinase-MB, glycogen phosphorylase isoenzyme BB, lipids, low and high-density lipoprotein, triglycerides, and glucose.[69] A PAD was fabricated using the wax printing technique and was used for the colorimetric assay of conjugated bilirubin.[54] Silver nanoparticles were used for detecting dengue, yellow fever, and Ebola. Carbon was used for the colorimetric detection of dengue.[70] A chemiluminescence sensor system was used to detect hydrogen peroxide (H_2O_2) and SARS-CoV-2, L-cysteine.[71–73] Simultaneous detection of three biomarkers of acute myocardial infarction was detected using a 3D microfluidic paper analytical device with chemiluminescence and signal amplification.[74] Multiple tumor markers such as α-fetoprotein (AFP), a cancer antigen, and carcinoembryonic antigen (CEA) were simultaneously detected using a paper-based chemiluminescence ELISA.[75] Multiplexed monitoring of cancer cells was carried out using aptamer modified QDs decorated on mesoporous Si nanoparticles. Transferases such as aspartate aminotransferases and alanine aminotransferases are used for the liver function test.[76] Hydrolase enzymes were used to detect acetylcholinesterase to find out the neurological disorders and organophosphate poisoning.[77] A microfluidic paper-based chemiluminescence biosensor for simultaneous determination of glucose and uric acid was developed. This was based on the reaction between a rhodamine derivative and hydrogen peroxide generated by oxidase enzymes.[78] A wax printed PAD was used for the detection of alkaline phosphatase has also been developed.[79]

10.2 Drug analysis

PAD provides a suitable platform for drug analysis because of advantages like self-driven programmable fluid flow, a versatile substrate for sample pretreatment, low-cost separation and is also an attractive platform for cell culture. One of the essential aspects of drug analysis is fluid manipulation. The fluid transport and constant fluid velocity can be achieved on paper precisely by controlling the geometrical shape of the channels. A wax patterned paper analytical device with AgNPs catalyzed luminol CL system to determine ofloxacin (OFLX) was developed.[59]

10.3 Detection of biomolecules

Paper-based sensors are used to detect various biological molecules such as glucose, lactose, uric acid etc. Standard methods used for the detection are enzymatic and immunoassay. Enzymes are used as the biological catalyst for the detection of biomolecules. The first paper-based enzymatic analytical device was developed for the detection of glucose and protein in the urine. The detection was based on the oxidation of glucose by glucose oxidase into glucuronic acid and hydrogen peroxide. Hydrogen peroxide was reduced by horseradish peroxidase, oxidizes iodide to iodine results in brown color. Simultaneous determination of uric acid, glucose, and lactate in the biological sample was developed based on the electrochemical method. A schematic representation of the paper device for multiple analyte detection is shown in Fig. 8. Nanomaterials of carbon, metal nanoparticles, quantum dots, and functionalized nanoparticles are incorporated on the paper to enhance the sensitivity and selectivity of detection. Nanocomposites of polyaniline/nanohematite/Prussian Blue are incorporated on paper for the detection of cholesterol. Amperometric detection of cholesterol was done by incorporating nanocomposite containing graphene, polyvinyl pyrrolidine, and polyaniline modified electrode.

Simultaneous determination of glucose, lactate and uric acid was colorimetrically detected on paper fabricated by wax dipping method. Biological elements can also be detected using immunoassay method. Electrochemical immunoassay involves the immobilization of recognition elements onto the electrode surface, which forms immunocomplex that leads to a change or

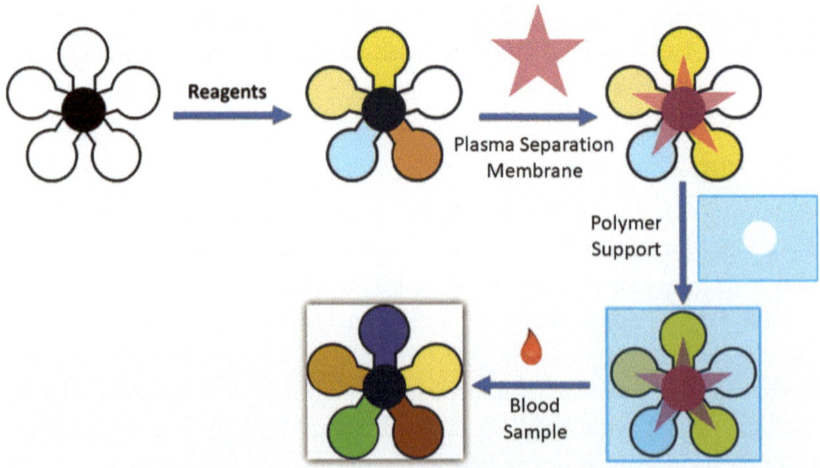

Fig. 8 Schematic representation of the paper device for multiple analyte detection with plasma separation membrane and polymer support.

shift in the response signal. For the development of an electrochemical immunosensor, an electrochemical transducer with a specific design and pattern was constructed on paper. A paper-based immunosensor device was fabricated for the detection of α-fetoprotein, hemoglobin, HbA1c, and human chorionic gonadotropin. DNA sensors for detecting high-risk human papillomavirus, MERS-CoV, MTB and HPV have been developed.

10.4 Smartphone based applications

Combining smartphones with PAD provides a perfect platform for monitoring health problems in remote areas with limited resources. A smartphone-based detection of salmonella on paper microfluidics was developed by preloading microfluidic channel in the paper with *anti-Salmonella typhimurium* and *anti-Escherichia coli* conjugated sub microparticles. The paper devices were dipped into salmonella solutions which led to immunoagglutination of the antibody with the conjugated nanoparticle. The digital images were taken at an optimized angle and distance with a smartphone. The extent of immunoagglutination was quantified by evaluating the Mie scattering from the digital image.[62]

Smartphones were also used for the Quick Response Code-based immunosensors. It involves the superimposition of a transparent QR code on the test paper. In the low concentrations of analyte, the QR code was not disrupted, and the Augmented Reality (AR) app could recognize the QR code and display a symbol indicating negative results. However, the QR code is disrupted with red spots from the gold conjugate at a higher concentration of the analyte. The AR app could not recognize the QR code and this shows a positive result. By this method, *E. coli* was detected in sample during the infections.[65]

Smartphones were also integrated with fluorescent PADs to detect the light emitted from the analyte during excitation. Commonly used fluorescent particles include small fluorophores, quantum dots, and fluorescent nanoparticles.[64] In addition, a quantitative fluorescence assay using a self-powered paper-based microfluidic device and a smartphone was developed. This was achieved by using paper microfluidic device with an internal fluidic battery and a surface-mount LED. The fluorescence image was taken using a smartphone and the images were processed to get the quantitative results.

11. Paper nanofluidics

Nanofluidics deals with the study of fluid transport and manipulation through nano-dimensional pores and channels. The fluid flow in this regime is laminar, and the viscous forces dominate the inertial forces. For a cylinder, the surface-to-volume ratio will be inversely proportional to the diameter. Thus, fluid movement is complex under the pressure gradient because of increased viscous drag on the walls. The electrokinetic properties in the fluid flow through these channels can serve for ionic transport and the separation of charged species like DNA. Due to the non-homogeneity in the number of micro and nanopores in nitrocellulose or cellulose paper substrates, the incorporation of reagents leads to non-reproducible results. Thus, porous anodic aluminum[80] and graphene-based systems[81] are employed mainly for nanofluidic applications. Since most of the pores are in micro dimensions in cellulose paper to generate nanopores, modifications are being carried out. For instance, egg albumin and egg yolk as a nano junction in cellulose paper was used to generate nanofluidic phenomena.[82] Nanofibres with a high aspect ratio from bacterial cellulose are also used for the generation of ion-conducting cellulose aerogel channels that show good strength even after retaining water. The nanofluidic membrane was developed from cellulose derived from the bark of wood after hemicellulose and lignin removal is used as a perm selective membrane.[83]

12. Challenges and research gaps

Though a wide variety of papers have been used to fabricate POCT devices, porous cellulose is the most commonly used. Though a large number of hydroxyl groups are present on the cellulosic matrix, they are not highly reactive for biomolecular binding. Hence functionalization of the surface with reactive groups becomes essential in many cases. However, though the surface modification enhances the binding and fluid flow, it adversely affects the stability of the paper. It is also reported that aggressive chemical treatment for surface modification/functionalization affects the stability and contaminates the paper surface, which negatively affects the subsequent steps. Further, excessive functionalization may reject the biomolecules due to steric repulsion, whereas poor functionalization results in reduced activity. Therefore, precise control over the pretreatment and functionalization is necessary for reproducible results.[84,85]

Paper is a sensitive material to environmental conditions such as temperature and humidity because of its low tensile strength.[86] Unmodified Whatman chromatographic paper exhibits varying stability from a few weeks to a few months, depending on the grade. At elevated temperatures, the stability was found to decrease dramatically. For instance, the sensitivity of the paper device fabricated with Whatman grade 1 was found to decrease by 10 days. In contrast, the devices with Whatman 903 grade paper are highly stable, and thermal degradation occurs only at 300 °C and shows 1-year shelf life at −20 °C. Nitrocellulose and nylon membranes have relatively higher storage stability, and the immunoassays on these membranes were stable, without much reduction of sensitivity, even for a few months.[87,88] However, the treated papers exhibit poor stability compared to the untreated ones. Further, the treatments lose the shape of the paper and lead to wrinkles and folds on the paper. Also, in applications such as immunoassays wherein multiple washing steps are essentially involved, and that affects the storage stability. Therefore, the fabrication process should be carefully planned or designed to involve minimum steps modification or treatment that enhances the shelf life.

Conventionally, paper-based diagnostic devices are used for the naked-eye detection of the analyte. But this is always semiquantitative. In addition, there are camera-enabled smartphones and electronic devices used for the color development and, in turn, the quantity of the analyte to obtain a quantitative output. In the fluorescence method, the white paper can also produce a background signal that causes a bleaching effect.[89]

13. Conclusions

Paper-based point of care devices are well suited for a variety of chemical and biochemical assays. The ease of fabrication, simple fluid handling and eco-friendly nature of the paper substrate promise its applicability for the fabrication of low-cost point of care diagnostic devices. Several devices have been developed and marketed using paper substrates. However, there are challenges in their fabrication and real-time use, including the non-uniform pore size, storage stability, detection of higher concentrations, and reproducibility issues. Suitable functionalization and design of the paper substrates yield diagnostic devices for various biomarkers, including proteins and nucleic acids. In addition, the possibility of developing qualitative, semi-quantitative, and quantitative methods on paper devices are highly suitable for resource-limited settings.

14. Future outlook

Advances in automated fabrication systems, reagent incorporation methods, and reliable detection technologies show the future potential of paper microfluidic devices in diagnostics. The developments in stable color reagents, fluorescent dyes, and electroanalytical techniques for biomarker detection can easily be incorporated in paper devices to develop a point of care diagnostic devices. Improvements in nanofluidic paper devices and the paper shrinkage while dipping in some of the chemical reagents and the development of more sensitive detection reagents can considerably decrease the size of these devices. Low-cost handheld electronic devices and image processing strategies could improve the sensitivity of the detection and which will be a boon for clinicians. Because of the simplicity in fabrication, these devices could serve efficiently as a fast and first-hand screening tool for pandemic conditions such as COVID-19.

Acknowledgment

The authors thank the Department of Biotechnology (DBT), Government of India for financial support (Sanction no. *102/IFD/SAN/2238/2016-17 dated 30-8-2016,* 102/IFD/SAN/1555/2018-2019, August 13, 2018).

References

1. Yang Y, Noviana E, Nguyen MP, Geiss BJ, Dandy DS, Henry CS. Paper-based microfluidic devices: emerging themes and applications. *Anal Chem.* 2017;89(1):71–91. https://doi.org/10.1021/acs.analchem.6b04581.

2. Hosseini S, Vázquez-Villegas P, Martínez-Chapa SO. Paper and fiber-based bio-diagnostic platforms: current challenges and future needs. *Appl Sci*. 2017;7(8):863. https://doi.org/10.3390/app7080863.
3. Cate DM, Adkins JA, Mettakoonpitak J, Henry CS. Recent developments in paper-based microfluidic devices. *Anal Chem*. 2015;87(1):19–41. https://doi.org/10.1021/ac503968p.
4. López-Marzo AM, Merkoçi A. Paper-based sensors and assays: a success of the engineering design and the convergence of knowledge areas. *Lab Chip*. 2016;16(17):3150–3176. https://doi.org/10.1039/c6lc00737f.
5. Cummins BM, Chinthapatla R, Ligler FS, Walker GM. Time-dependent model for fluid flow in porous materials with multiple pore sizes. *Anal Chem*. 2017;89 (8):4377–4381. https://doi.org/10.1021/acs.analchem.6b04717.
6. Tang RH, Yang H, Choi JR, et al. Advances in paper-based sample pretreatment for point-of-care testing. *Crit Rev Biotechnol*. 2017;37(4):411–428. https://doi.org/10.3109/07388551.2016.1164664.
7. Da Cunha Santos G. FTA cards for preservation of nucleic acids for molecular assays a review on the use of cytologic/tissue samples. *Arch Pathol Lab Med*. 2018;142 (3):308–312. https://doi.org/10.5858/arpa.2017-0303-RA.
8. Peluso AL, Cascone AM, Lucchese L, et al. Use of FTA cards for the storage of breast carcinoma nucleic acid on fine-needle aspiration samples. *Cancer Cytopathol*. 2015;123 (10):582–592. https://doi.org/10.1002/cncy.21577.
9. Ge L, Wang S, Song X, Ge S, Yu J. 3D origami-based multifunction-integrated immunodevice: low-cost and multiplexed sandwich chemiluminescence immunoassay on microfluidic paper-based analytical device. *Lab Chip*. 2012;12(17):3150–3158. https://doi.org/10.1039/c2lc40325k.
10. Poh JJ, Gan SKE. Comparison of customized spin-column and salt-precipitation finger-prick blood DNA extraction. *Biosci Rep*. 2014;34(5):629. https://doi.org/10.1042/BSR20140105.
11. McFall SM, Neto MF, Reed JL, Wagner RL. Filtration isolation of nucleic acids: a simple and rapid DNA extraction method. *J Vis Exp*. 2016;2016(114):2–7. https://doi.org/10.3791/54289.
12. Wong SY, Cabodi M, Rolland J, Klapperich CM. Evaporative concentration on a paper-based device to concentrate analytes in a biological fluid. *Anal Chem*. 2014;86 (24):11981–11985. https://doi.org/10.1021/ac503751a.
13. Li X, Luo L, Crooks RM. Low-voltage paper isotachophoresis device for DNA focusing. *Lab Chip*. 2015;15(20):4090–4098. https://doi.org/10.1039/c5lc00875a.
14. Soum V, Park S, Brilian AI, Kwon OS, Shin K. Programmable paper-based microfluidic devices for biomarker detections. *Micromachines*. 2019;10(8):516. https://doi.org/10.3390/mi10080516.
15. Clarke SF, Foster JR. A history of blood glucose meters and their role in self-monitoring of diabetes mellitus. *Br J Biomed Sci*. 2012;69(2):83–93. https://doi.org/10.1080/09674845.2012.12002443.
16. Ginardi RVH, Saikhu A, Sarno R, Sunaryono D, Kholimi AS, Shanty RNT. Intelligent method for dipstick urinalysis using smartphone camera. In: *Lecture Notes in Computer Science (including Subser Lect Notes Artif Intell Lect Notes Bioinformatics)*. vol. 8407. Springer; 2014. https://doi.org/10.1007/978-3-642-55032-4_7. LNCS:66–77.
17. Ostrov N, Jimenez M, Billerbeck S, et al. A modular yeast biosensor for low-cost point-of-care pathogen detection. *Sci Adv*. 2017;3(6):e1603221. https://doi.org/10.1126/sciadv.1603221.
18. Park J, Shin JH, Park JK. Pressed paper-based dipstick for detection of foodborne pathogens with multistep reactions. *Anal Chem*. 2016;88(7):3781–3788. https://doi.org/10.1021/acs.analchem.5b04743.

19. Torabi SF, Lu Y. Small-molecule diagnostics based on functional DNA nanotechnology: a dipstick test for mercury. *Faraday Discuss.* 2011;149:125–135. https://doi.org/10.1039/c005404f.
20. Kuswandi B, Ensafi AA. Perspective—paper-based biosensors: trending topic in clinical diagnostics developments and commercialization. *J Electrochem Soc.* 2020;167(3), 037509. https://doi.org/10.1149/2.0092003jes.
21. Ngom B, Guo Y, Wang X, Bi D. Development and application of lateral flow test strip technology for detection of infectious agents and chemical contaminants: a review. *Anal Bioanal Chem.* 2010;397(3):1113–1135. https://doi.org/10.1007/s00216-010-3661-4.
22. Bahadır EB, Sezgintürk MK. Lateral flow assays: principles, designs and labels. *TrAC Trends Anal Chem.* 2016;82:286–306. https://doi.org/10.1016/j.trac.2016.06.006.
23. Sajid M, Kawde AN, Daud M. Designs, formats and applications of lateral flow assay: a literature review. *J Saudi Chem Soc.* 2015;19(6):689–705. https://doi.org/10.1016/j.jscs.2014.09.001.
24. Ehrenkranz JRL. Home and point-of-care pregnancy tests: a review of the technology. *Epidemiology.* 2002;13(3 SUPPL):S15. https://doi.org/10.1097/00001648-200205001-00003.
25. Liu D, Ju C, Han C, et al. Nanozyme chemiluminescence paper test for rapid and sensitive detection of SARS-CoV-2 antigen. *Biosens Bioelectron.* 2021;173:112817. https://doi.org/10.1016/j.bios.2020.112817.
26. Wang Z, Zheng Z, Hu H, et al. A point-of-care selenium nanoparticle-based test for the combined detection of anti-SARS-CoV-2 IgM and IgG in human serum and blood. *Lab Chip.* 2020;20(22):4255–4261. https://doi.org/10.1039/d0lc00828a.
27. Cui F, Zhou HS. Diagnostic methods and potential portable biosensors for coronavirus disease 2019. *Biosens Bioelectron.* 2020;165(April):112349. https://doi.org/10.1016/j.bios.2020.112349.
28. Azzi L, Baj A, Alberio T, et al. Rapid Salivary Test suitable for a mass screening program to detect SARS-CoV-2: a diagnostic accuracy study. *J Clean Prod.* 2020;81(3):e75. https://doi.org/10.1016/j.jinf.2020.06.042. Published online.
29. Pokhrel P, Hu C, Mao H. Detecting the coronavirus (CoVID-19). *ACS Sens.* 2020;5(8):2283–2297. https://doi.org/10.1021/ACSSENSORS.0C01153.
30. Strong EB, Schultz SA, Martinez AW, Martinez NW. Fabrication of miniaturized paper-based microfluidic devices (MicroPADs). *Sci Rep.* 2019;9(1):1–9. https://doi.org/10.1038/s41598-018-37029-0.
31. Martinez AW, Phillips ST, Whitesides GM. Three-dimensional microfluidic devices fabricated in layered paper and tape. *Proc Natl Acad Sci U S A.* 2008;105(50):19606–19611. https://doi.org/10.1073/pnas.0810903105.
32. Lisowski P, Zarzycki PK. Microfluidic paper-based analytical devices (μPADs) and micro total analysis systems (μTAS): development, applications and future trends. *Chromatographia.* 2013;76(19–20):1201–1214. https://doi.org/10.1007/s10337-013-2413-y.
33. Martinez AW, Phillips ST, Whitesides GM, Carrilho E. Diagnostics for the developing world: microfluidic paper-based analytical devices. *Anal Chem.* 2010;82(1):3–10. https://doi.org/10.1021/ac9013989.
34. Dou M, Sanjay ST, Benhabib M, Xu F, Li XJ. Low-cost bioanalysis on paper-based and its hybrid microfluidic platforms. *Talanta.* 2015;145:43–54. https://doi.org/10.1016/j.talanta.2015.04.068.
35. Martinez AW, Phillips ST, Butte MJ, Whitesides GM. Patterned paper as a platform for inexpensive, low-volume, portable bioassays. *Angew Chem.* 2007;119(8):1340–1342. https://doi.org/10.1002/ange.200603817.
36. Nishat S, Turab A, Martinez AW, Rabbi F. Chemical paper-based microfluidics: simplified fabrication and assay methods. *Sens Actuators B.* 2021;336:129681.

37. Sun JY, Cheng CM, Liao YC. Screen printed paper-based diagnostic devices with polymeric inks. *Anal Sci*. 2015;31(3):145–151. https://doi.org/10.2116/analsci.31.145.
38. Online VA, Wu Y, Fu J, Wu W. *RSC Advances*. Published online, The Royal Society of Chemistry; 2015.
39. Abe K, Suzuki K, Citterio D. Inkjet-printed microfluidic multianalyte chemical sensing paper. *Anal Chem*. 2008;80(18):6928–6934. https://doi.org/10.1021/ac800604v.
40. Lu Y, Shi W, Jiang L, Qin J, Lin B. Rapid prototyping of paper-based microfluidics with wax for low-cost, portable bioassay. *Electrophoresis*. 2009;30:1497–1500. https://doi.org/10.1002/elps.200800563.
41. Carrilho E, Martinez AW, Whitesides GM. Understanding wax printing: a simple micropatterning process for paper-based microfluidics. *Anal Chem*. 2009;81(16):7091–7095. https://doi.org/10.1021/ac901071p.
42. Xia Y, Si J, Li Z. Fabrication techniques for micro fluidic paper-based analytical devices and their applications for biological testing: a review. *Biosens Bioelectron*. 2016;77:774–789. https://doi.org/10.1016/j.bios.2015.10.032.
43. Songjaroen T, Dungchai W, Chailapakul O, Laiwattanapaisal W. Novel, simple and low-cost alternative method for fabrication of paper-based microfluidics by wax dipping. *Talanta*. 2011;85(5):2587–2593. https://doi.org/10.1016/j.talanta.2011.08.024.
44. Olkkonen J, Lehtinen K, Erho T. Flexographically printed fluidic structures in paper. *Anal Chem*. 2010;82(24):10246–10250. https://doi.org/10.1021/ac1027066.
45. Henry A, Ozer T, Mcmahon C, Henry CS. Advances in paper-based analytical devices. *Molecules*. 2020;25(13):2970.
46. Mu X, Zhang Y. Fabrication and applications of paper-based microfluidics. In: *Diagnostic Devices with Microfluidics*. Taylor and Francis; 2017:45–64.
47. Li X, Tian J, Nguyen T, Shen W. Paper-based microfluidic devices by plasma treatment. *Anal Chem*. 2008;80(23):9131–9134. https://doi.org/10.1021/ac801729t.
48. de Tarso Garcia P, Cardoso MG, Garcia CD, Carrilho E, Coltro WKT. A handheld stamping process to fabricate microfluidic paper-based analytical devices with chemically modified surface for clinical assays. *RSC Adv*. 2014;4:37637–37644. https://doi.org/10.1039/C4RA07112C. Published online.
49. Pratiwi R, Henry CS, Noviana E, Blascke D. Emerging applications of paper-based analytical devices for drug analysis: a review. *Anal Chim Acta*. 2020;1116:70–90. https://doi.org/10.1016/j.aca.2020.03.013.
50. Dungchai W, Chailapakul O, Henry CS. Use of multiple colorimetric indicators for paper-based microfluidic devices. *Anal Chim Acta*. 2010;674(2):227–233. https://doi.org/10.1016/j.aca.2010.06.019.
51. Gabriel EF, Garcia PT, Cardoso TM, Lopes FM, Martins FT, Coltro WK. Highly sensitive colorimetric detection of glucose and uric acid in biological fluids using chitosan-modified paper microfluidic devices. *Analyst*. 2016;141:4749–4756. www.rsc.org/analyst. Manuscript A Published online https://doi.org/10.1039/C6AN00430J.
52. Mazzu-Nascimento T, Leão PAGC, Catai JR, Morbioli GG, Carrilho E. Towards low-cost bioanalytical tools for sarcosine assays for cancer diagnostics. *Anal Methods*. 2016;8(40):7312–7318. Manuscript A Published online https://doi.org/10.1039/C6AY01848C.
53. Alien MP, Deliua A, Jeong H. A noninstrumented quantitative test system and its application for determining cholesterol concentration in whole blood. *Clin Chem*. 1990;36(9):1591–1597.
54. Edachana RP, Kumaresan A, Balasubramanian V, Thiagarajan R, Nair BG, Thekkedath Gopalakrishnan SB. Paper-based device for the colorimetric assay of bilirubin based on in-situ formation of gold nanoparticles. *Microchim Acta*. 2020;187(1):60. https://doi.org/10.1007/s00604-019-4051-z.
55. Badu-Tawiah AK, Lathwal S, Kaastrup K, et al. Polymerization-based signal amplification for paper-based immunoassays. *Lab Chip*. 2015;15(3):655–659. https://doi.org/10.1039/C4LC01239A.

56. Mentele MM, Cunningham J, Koehler K, Volckens J, Henry CS. Micro fluidic paper-based analytical device for particulate metals. *Anal Chem*. 2012;84(10):4474–4480. Published online.
57. Hossain SMZ, Brennan JD. β-Galactosidase-based colorimetric paper sensor for determination of heavy metals. *Anal Chem*. 2011;83(22):8772–8778. Published online.
58. Pegalajar MC. Printed disposable colorimetric array for metal ion discrimination. *Anal Chem*. 2014;86(17):8634–8641. Published online.
59. Liu W, Guo Y, Li H, Zhao M, Lai Z, Li B. A paper-based chemiluminescence device for the determination of ofloxacin. *Spectrochim Acta A Mol Biomol Spectrosc*. 2015;137: 1298–1303. https://doi.org/10.1016/j.saa.2014.09.059.
60. Delaney JL, Hogan CF, Tian J, Shen W. Electrogenerated chemiluminescence detection in paper-based microfluidic sensors. *Anal Chem*. 2011;83(4):1300–1306. https://doi.org/10.1021/ac102392t.
61. Rasooly A, Herold KE. *Mobile Health Technologies: Methods and Protocols*. vol. 1256. Springer; 2015:1–496.
62. Park TS, Li W, McCracken KE, Yoon JY. Smartphone quantifies Salmonella from paper microfluidics. *Lab Chip*. 2013;13(24):4832–4840. https://doi.org/10.1039/c3lc50976a.
63. Land KJ, ed. *Paper-Based Diagnostics*. Springer; 2019.
64. Aydindogan E, Guler Celik E, Timur S. Paper-based analytical methods for smartphone sensing with functional nanoparticles: bridges from smart surfaces to Global Health. *Anal Chem*. 2018;90(21):12325–12333. https://doi.org/10.1021/acs.analchem.8b03120.
65. Russell SM, Doménech-Sánchez A, De La Rica R. Augmented reality for real-time detection and interpretation of colorimetric signals generated by paper-based biosensors. *ACS Sens*. 2017;2(6):848–853. https://doi.org/10.1021/acssensors.7b00259.
66. Fujimoto T, Kawahara S, Fuchigami Y, et al. Portable electrochemical sensing system attached to smartphones and its incorporation with paper-based electrochemical glucose sensor. *Int J Electr Comput Eng*. 2017;7(3):1423–1429. https://doi.org/10.11591/ijece.v7i3.pp1423-1429.
67. Soda Y, Citterio D, Bakker E. Equipment-free detection of K + on microfluidic paper-based analytical devices based on exhaustive replacement with ionic dye in ion-selective capillary sensors. *ACS Sens*. 2019;4(3):670–677. https://doi.org/10.1021/acssensors.8b01521.
68. Kappi FA, Tsogas GZ, Christodouleas DC, Giokas DL. Calibrant-loaded paper-based analytical devices for standard addition quantitative assays. *Sens Actuators B*. 2017;253: 860–867. https://doi.org/10.1016/j.snb.2017.07.007.
69. Suntornsuk W, Suntornsuk L. Recent applications of paper-based point-of-care devices for biomarker detection. *Electrophoresis*. 2020;41(5–6):287–305. https://doi.org/10.1002/elps.201900258.
70. Chen G, Chen W, Yen Y, Wang C, Chang H, Chen C. Detection of mercury (II) ions using colorimetric gold nanoparticles on paper-based analytical devices. *Anal Chem*. 2014;86(14):6843–6849.
71. Ito Y, Matsuura T, Murakami M. H 2 0 2. *J Am Chem Soc*. 1988;3693(16):3692–3693.
72. Liu D, Ju C, Han C, et al. Nanozyme chemiluminescence paper test for rapid and sensitive detection of SARS-CoV-2 antigen. *Biosens Bioelectron*. 2021;173(November 2020):112817. https://doi.org/10.1016/j.bios.2020.112817.
73. Liu W, Luo J, Guo Y, Kou J, Li B, Zhang Z. Nanoparticle coated paper-based chemiluminescence device for the determination of L-cysteine. *Talanta*. 2014;120:336–341. https://doi.org/10.1016/j.talanta.2013.12.033.
74. Li F, Guo L, Hu Y, et al. Multiplexed chemiluminescence determination of three acute myocardial infarction biomarkers based on microfluidic paper-based immunodevice dual amplified by multifunctionalized gold nanoparticles. *Talanta*. 2020;207(July 2019): 120346. https://doi.org/10.1016/j.talanta.2019.120346.

75. Wang S, Ge L, Song X, Yu J, Ge S. Paper-based chemiluminescence ELISA: lab-on-paper based on chitosan modified paper device and wax-screen-printing. *Biosens Bioelectron*. 2012;31(1):212–218. https://doi.org/10.1016/j.bios.2011.10.019.
76. Pollock NR, Rolland JP, Kumar S, et al. Paper-based multiplexed transaminase test for low-cost, point-of-care liver function testing. *Sci Transl Med*. 2012;4(152), 152ra129.
77. Courtney KD, Francisco S. A new and rapid colorimetric of acetylcholinesterase determination. *Biochem Pharmacol*. 1961;7:88–95.
78. Yu J, Ge L, Huang J, Wang S, Ge S. Microfluidic paper-based chemiluminescence biosensor for simultaneous determination of glucose and uric acid. *Lab Chip*. 2011;11 (7):1286–1291. https://doi.org/10.1039/c0lc00524j.
79. Resmi PE, Stanley J, Kumar S, Soman KP, Ramachandran T, Satheesh Babu TG. Fabrication of paper microfluidics POCT device for the colorimetric assay of alkaline phosphatase. In: *INDICON 2018—15th IEEE India Council International Conference*; 2018:2018–2021. https://doi.org/10.1109/INDICON45594.2018.8987193 [Ii].
80. Chen R, Du X, Cui Y, et al. Vertical flow assay for inflammatory biomarkers based on nanofluidic channel array and SERS nanotags. *Small*. 2020;16(32):1–13. https://doi.org/10.1002/smll.202002801.
81. Liu M, Weston PJ, Hurt RH. Controlling nanochannel orientation and dimensions in graphene-based nanofluidic membranes. *Nat Commun*. 2021;12(1):1–7. https://doi.org/10.1038/s41467-020-20837-2.
82. Park S, Hong S, Kim J, Son SY, Lee H, Kim SJ. Eco friendly nanofluidic platforms using biodegradable nanoporous materials. *Sci Rep*. 2021;11(1):1–11. https://doi.org/10.1038/s41598-021-83306-w.
83. Li T, Li SX, Kong W, et al. A nanofluidic ion regulation membrane with aligned cellulose nanofibers. *Sci Adv*. 2019;5(2):1–7. https://doi.org/10.1126/sciadv.aau4238.
84. Hosseini S, Ibrahim F, Djordjevic I, Koole LH. Recent advances in surface functionalization techniques on polymethacrylate materials for optical biosensor applications. *Analyst*. 2014;139(12):2933–2943. https://doi.org/10.1039/c3an01789c.
85. Ahmed S, Bui MPN, Abbas A. Paper-based chemical and biological sensors: engineering aspects. *Biosens Bioelectron*. 2016;77:249–263. https://doi.org/10.1016/j.bios.2015.09.038.
86. Nilghaz A, Guan L, Tan W, Shen W. Advances of paper-based microfluidics for diagnostics—the original motivation and current status. *ACS Sens*. 2016;1 (12):1382–1393. https://doi.org/10.1021/acssensors.6b00578.
87. Wang J, Yiu B, Obermeyer J, Filipe CDM, Brennan JD, Pelton R. Effects of temperature and relative humidity on the stability of paper-immobilized antibodies. *Biomacromolecules*. 2012;13(2):559–564. https://doi.org/10.1021/bm2017405.
88. Farahmand E, Ibrahim F, Hosseini S, et al. A novel approach for application of nylon membranes in the biosensing domain. *Appl Surf Sci*. 2015;353(July):1310–1319. https://doi.org/10.1016/j.apsusc.2015.07.004.
89. Li X, Ballerini DR, Shen W. A perspective on paper-based microfluidics: current status and future trends. *Biomicrofluidics*. 2012;6(1):11301–011301-13. https://doi.org/10.1063/1.3687398.

CHAPTER SEVEN

Advances in microfluidics devices and its applications in personalized medicines

Rupesh Maurya[a], Nisarg Gohil[a], Gargi Bhattacharjee[a], Khushal Khambhati[a], Khalid J. Alzahrani[b], Suresh Ramakrishna[c,d], Dinh-Toi Chu[e], and Vijai Singh[a,*]

[a]Department of Biosciences, School of Science, Indrashil University, Rajpur, Mehsana, Gujarat, India
[b]Department of Clinical Laboratories Sciences, College of Applied Medical Sciences, Taif University, Taif, Saudi Arabia
[c]Graduate School of Biomedical Science and Engineering, Hanyang University, Seoul, South Korea
[d]College of Medicine, Hanyang University, Seoul, South Korea
[e]Center for Biomedicine and Community Health, International School, Vietnam National University, Hanoi, Vietnam
*Corresponding author: e-mail addresses: vijaisingh15@gmail.com; vijai.singh@indrashiluniversity.edu.in

Contents

1. Introduction 192
2. Microfluidics for cancer theranostics 193
3. Microfluidics for detection of cardiovascular diseases 194
4. Microfluidics for detection of diabetes 195
5. Microfluidics for detection of pulmonary disease and therapy 196
6. Conclusion and future perspectives 197
Acknowledgments 197
References 198

Abstract

Microfluidics is an exponentially growing area and is being used for numerous applications from basic science to advanced biotechnology and medicines. Microfluidics provides a platform to the research community for studying and building new strategies for the diagnosis and therapeutics applications. In the last decade, microfluidic have enriched the field of diagnostics by providing new solutions which was not possible with conventional detection and treatment methods. Microfluidics has the ability to precisely control and perform high-throughput functions. It has been proven as an efficient and rapid method for biological sample preparation, analysis and controlled drug delivery system. Microfluidics plays significant role in personalized medicine. These personalized medicines are used for medical decisions, practices and other interventions as

Progress in Molecular Biology and Translational Science, Volume 186
ISSN 1877-1173
https://doi.org/10.1016/bs.pmbts.2021.07.012

well as for individual patients based on their predicted response or risk of disease. This chapter highlights microfluidics in developing personalized medical applications for its applications in diseases such as cancer, cardiovascular disease, diabetes, pulmonary disease and several others.

1. Introduction

During the initial stages of disease development, a number of diseases go undetected and later it gets difficult to treat it. In the past few decades, an alarmic increase in the rate of diseases, such as cancer,[1] cardiovascular disease,[2–4] diabetes,[5] autoimmune disease,[6] pulmonary disease[7] have not just left serious impact on healthcare systems, economies, and governments but has also rapidly taken the shape of a global health crisis threatening the lives of millions of people.

Microfluidics devices play vital role in the development of point-of-care (POC) diagnostics,[8–10] as well as for studying biological mechanism or processes.[11,12] Microfluidic devices/chips are composed of fluidic channels and chambers with a linear dimension of order of tens to hundreds of micrometers.[13] This has been for the quantitative measurement of oxygen level,[14] cell adhesion[15] and analyte concentrations.[16]

The progress in the area of developing fabricated micro/nanofluidics devices for its use in the biomedical sciences has been quite significant. Novel nanotechnological approaches have been explored in view to devise an individualized medical approach. Much of the development has been driven by the merits of utilizing nano-scale phenomena to manipulate fluid samples or accomplish drug delivery. Nanochannel technologies have the potential for rapid disease diagnosis and therapeutic applications. In the past decade, personalized medicine has flourished in the field of healthcare and disease management.[17]

Personalized medicine assures prediction, prevention and treatment of a disease by targeting individuals' requirements. Upcoming technologies are initiating profiling of individuals. There is a requirement to develop novel techniques and technologies for collecting and analyzing data. Of several important requirements for the development of personalized medicine, matters such as the requirement of accord between healthcare and science as well as ethical and regulatory framework are considered crucial.[18] Personalized medicines permit matching patients with treatment and also help in lowering treatment cost. It is also emerging as the cause of development of larger biobanks and medical records or health records, thus interlinking biological information to monitoring technologies.[19] World healthcare systems are moving toward personalized medicine against several diseases such as autoimmune diseases,[20] cancer,[21,22] diabetes,[23] cardiovascular diseases,[24] chronic obstructive pulmonary disease[25] and many others. In this chapter, we highlight different microfluidics devices for its applications in personalized medicines including detection and therapy of cancer, cardiovascular disease, diabetes, pulmonary disease and many others.

2. Microfluidics for cancer theranostics

Cancer is a chronic disease which costs the lives of people in large numbers. Chemotherapy or radiation is not effective for every patients due to heterogeneity. Tumors may form due to different hidden genetic causes. A pressing need therefore arises to develop a rapid, sensitive and specific system for detection of cancer at an early stage. Personalized medicines are beneficial due to its linkages with effective drugs.[26] Circulating tumor cells spread from a primary tumor into blood stream and grow in distant organs and metastasizes in all over the body.[27–29] Studies indicate bone marrow is the place of storage for disseminated tumor cells. Circulating tumor cells can be utilized for the judgment and monitoring of response against therapy.[30] Fig. 1 represents microfluidics systems integrated with

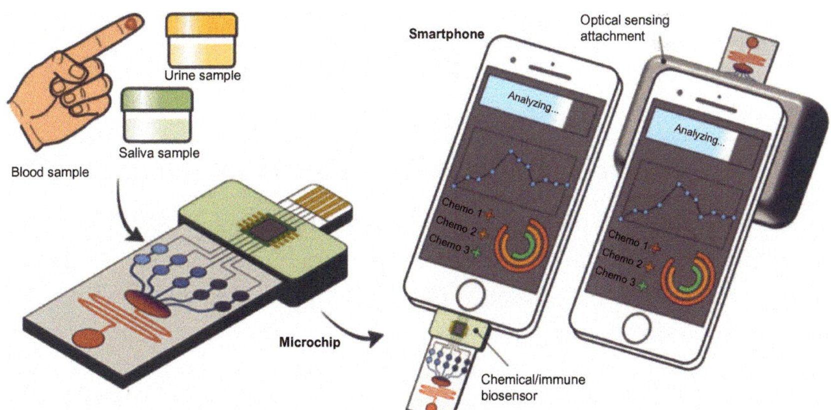

Fig. 1 Microfluidics systems integrated with smartphone to load sample (blood, urine or saliva samples) for cancer detection. *Reproduced with permission from Samandari M, Julia MG, Rice A, Chronopoulos A, Del Rio Hernandez AE. Liquid biopsies for management of pancreatic cancer. Transl Res. 2018;201:98–127, Elsevier.*

smartphone to load biological sample (blood, urine or saliva samples) for cancer detection.[31] In this manner, cancer can be rapidly diagnosed to begin with accurate therapy.

A number of methods have been developed for detection of tumor cells at the earliest. For detection of circulating tumor cells and disseminated tumor cells, techniques such as laser detection,[32] cell search system,[33] EPithelial ImmunoSPOT (EPISPOT),[34] AdnaTest,[35] TelomeScan,[36] aptamers[37] and others have been used. However, these techniques require accessibility and expertise for detecting Circulating Tumor Cells. Microfluidics devices are portable and handy for the detection of circulating tumor cells.[38] Microfluidics have been significantly used for maintaining and culturing circulating tumor cells drawn from patients for personalized drug screening.[39]

Personalized cancer therapies include multiplexed testing in a single chamber and also evaluate drug efficacy within 48 h or less if monitored in real-time,[40] development of mouse model for implantation of patient tumor samples[41] and use of kinase inhibitors.[42] Tumor organoids are the culture of cancer cells which are taken directly from patients. These can be utilized in drug screening and facilitation of drugs.[43] Microfluidics platform used for immunotherapy drug screening in tumor organoids focusses on gene–drug associations, predicting response toward a drug, and thereafter choosing preferred treatment option.[44] Circulating tumor DNA plays an important role of being as a biomarker for liquid biopsy and when used with microfluidics, it can aid the development of personalized cancer chemotherapy.[45]

3. Microfluidics for detection of cardiovascular diseases

Cardiovascular diseases are worldwide known for frequent causes of mortality. Total expenditure on drug development and drug discovery pertaining to diabetes has significantly increased but a decline in drug approvals is witnessed. Microfluidics can be utilized for designing person-centric healthcare. Its applications can improve the effectiveness in detection of biomolecules and monitoring person-centric diagnosis and use in therapeutics. The measurement and visualization of blood cells can be significantly achieved by the use of microfluidic devices. These microfluidic devices exhibit its significance by providing information about POC diagnostics[46] and hydrodynamic characteristics of blood during clinical investigations.[47,48]

Flow visualization techniques can be utilized to access the cardiac risk profile of a specific patient which potentially can support medical staff to decide the appropriate treatment.

New technologies such as organs-on-a-chip can be used to replicate the conditions of cardiovascular system, in particular, heart and general vasculature. Microfluidics systems play a significant role in the development of cardiovascular organs-on-a-chip platforms and have tremendous potential in the area of drug discovery and development.[49–51] Heart-on-a-chip platforms have the potential for heart regeneration, cardiotoxic/cardiotherapeutic drug screening as well as personalized medicine and therapeutics applications.[52] Lactate dehydrogenase and hydrogen peroxide are markers for detection of damaged tissue. Electrochemical measurement and detection of the aforementioned chemicals offer an alternative approach for investigating heart pathophysiology.[53]

4. Microfluidics for detection of diabetes

Microfluidics is highly beneficial for detection of multiple antibodies which can enable immunoprofilling and timely screening of at-risk patients.[54] It can be used for the synthesis of a novel cross-linked carboxymethyl chitosan nanoparticles containing metformin hydrochloride (MET) and used for evaluating its performance. A pressing need has arisen to develop a rapid and sensitive microfluidics device that can easily detect glucose level in a given sample. Fig. 2 depict a smartphone-assisted microfluidics devices for measuring diabetes- and hyperlipidemia-related indexes.[55] Personalized treatment for diabetes patients includes individual behavioral

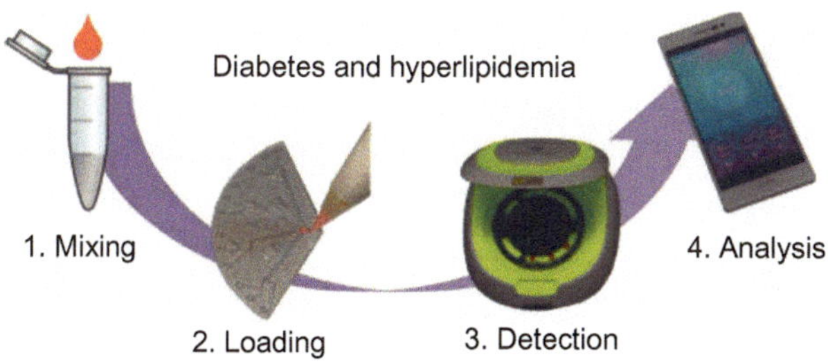

Fig. 2 A smartphone-guided simple and rapid microfluidics devices for measuring diabetes- and hyperlipidemia-related indexes. *Reproduced with permission from Li J, Sun Y, Chen C, Sheng T, Liu P, Zhang G. A smartphone-assisted microfluidic chemistry analyzer using image-based colorimetric assays for multi-index monitoring of diabetes and hyperlipidemia. Anal Chim Acta. 2019;1052:105–112, Elsevier.*

and clinical phenotypic features, standard clinical laboratory findings, and gene sequences and other molecular markers.[56] Next-generation sequencing technologies offer new opportunities for identifying genetic factors and gene products that are linked to diabetes mellitus.[57] Araujo et al.[58] delivered glucagon-like peptide-1 (GLP-1) and dipeptidyl peptidase-4 (DPP4) inhibitors through composites prepared by microfluidics and reported a significant decrease in glucose level, i.e., 44%.

Neutrophils can be important cell-based inflammatory biomarkers for immune health profiling, if speedy purification and measurements is made possible. Neutrophil isolation and functional assay can be done by using microfluidics technologies. This has already been developed and integrated into a POC testing for identification of effects of diabetes, hypertension and hyperlipidemia. Microfluidics can also be used for identification of vulnerable patients and reducing the risks in different aspects of endothelial activation pathway.[59] Islet perifusion systems are used for monitoring highly dynamic insulin release of pancreatic islets in glucose-stimulated insulin secretion (GSIS) assays. Wu Jin et al.[60] introduced a self-regulating drop-height mechanism that enables continuous flow and maintains a constant liquid volume in the chip, thereby enabling simple and robust operation. Upon glucose stimulation, reproducible biphasic insulin release was simultaneously observed from all islets in the system.

5. Microfluidics for detection of pulmonary disease and therapy

Asthma is a chronic obstructive pulmonary disease and is often mortal.[61] Conventional methods such as spirometry,[62] cell migration assay, gene expression analysis,[63] cell culture assay,[64] and interferon gamma release assay[65,66] are routinely used for diagnosis of asthma and chronic obstructive pulmonary disease. Recent advances in different therapies and interventions techniques cannot fulfill the expectations of clinical requirements. This is limited by two major factors, significant heterogeneity within each of these conditions and additionally significant overlap in many of the clinical and inflammatory features. Asthma is grouped into at least two molecular phenotypes defined by degree of Th2 inflammation mediated by IL-4, IL-5, and IL-13. Th2 cytokines have emerged as a significant and important therapeutic target in only a subset of patients with asthma. Results also prove that asthma is heterogeneous.[67]

Biomarkers are useful for phenotyping and endotyping of asthma. Clinical and immunological biomarkers such as blood eosinophil count,[68] eosinophil cationic protein,[69] sputum eosinophils[70,71] offer better therapeutic management and identification of patients who can respond to new therapies. A non-invasive approach of using volatile organic compounds such as combination of nonanal, 1-propanol, and hexane play a vital role for diagnosis of neutrophilic asthma.[72] There is an issue with performing diagnosis of asthma in older patients.[73] Delivery of precision medicine in airways disease is now feasible and is a core component of a personalized healthcare delivery in asthma. Microfluidics is used for preparation of drug carriers which are more effective therapeutic responses. These techniques are also being used for production of mono dispersed and multifunctional drug carriers having the potential to be used for drug transportation and controlled release.[74] Microfluidics can sort neutrophils from a drop of blood and discriminate between asthmatic and nonasthmatic patients within a minute.[75] Ezrin, a membrane cytoskeleton protein[76] and chitinases like proteins[77] are novel biomarkers, and such biomarkers can be targeted, identified and used against various subtypes for future detection and therapies.

6. Conclusion and future perspectives

Microfluidic devices have great potential for being employed in personalized or precision medicine for medical diagnosis and therapeutics. Microfluidics have both benefits and risks for the purpose of personalized medicine. It also provides a platform for rapid and accurate screening of larger samples and patients in a short span of time. In future, microfluidics has the potential to replace the expensive and bulky equipment as well as time consuming diagnostic systems. The microfluidic system has revolutionized drug identification and testing of drug formulations for patients. This can be used for future medical healthcare monitoring in patients from sitting at home or rural areas.

Acknowledgments

N.G. acknowledges the Indian Council of Medical Research, Government of India for financial assistance as Senior Research Fellowship (File No. 5/3/8/63/ITR-F/2020). The financial support from Gujarat State Biotechnology Mission (GSBTM) (Project ID: 5LY45F), Gujarat, India to G.B. and V.S. is duly acknowledged. The financial assistance from National Research Foundation of Korea (NRF) (2017M3A9C6061361) to S.R. is also duly acknowledged. R.M., N.G., G.B., K.K. and V.S. thank Indrashil University, Rajpur, Mehsana, India for providing the infrastructure facility.

References

1. Ngoma T. World Health Organization cancer priorities in developing countries. *Ann Oncol.* 2006;17:viii9–viii14.
2. Kaptoge S, Pennells L, De Bacquer D, et al. World Health Organization cardiovascular disease risk charts: revised models to estimate risk in 21 global regions. *Lancet Glob Health.* 2019;7:e1332–e1345.
3. Nabel EG. Cardiovascular disease. *N Engl J Med.* 2003;349:60–72.
4. Kishore SP, Blank E, Heller DJ, et al. Modernizing the world health organization list of essential medicines for preventing and controlling cardiovascular diseases. *J Am Coll Cardiol.* 2018;71:564–574.
5. Kononenko IV, Smirnova OM, Mayorov AY, Shestakova MV. *Classification of Diabetes.* vol. 23. Geneva: World Health Organization 2019; 2020:329–339.
6. Velikova T, Georgiev T. SARS-CoV-2 vaccines and autoimmune diseases amidst the COVID-19 crisis. *Rheumatol Int.* 2021;41:509–518.
7. Bafadhel M, Criner G, Dransfield MT, et al. Exacerbations of chronic obstructive pulmonary disease: time to rename. *Lancet Respir Med.* 2020;8:133–135.
8. Sia SK, Kricka LJ. Microfluidics and point-of-care testing. *Lab Chip.* 2008;8:1982–1983.
9. Pandey CM, Augustine S, Kumar S, et al. Microfluidics based point-of-care diagnostics. *Biotechnol J.* 2018;13:1700047.
10. Yetisen AK, Akram MS, Lowe CR. Paper-based microfluidic point-of-care diagnostic devices. *Lab Chip.* 2013;13:2210–2251.
11. Breslauer DN, Lee PJ, Lee LP. Microfluidics-based systems biology. *Mol BioSyst.* 2006;2:97–112.
12. Tian WC, Finehout E, eds. *Microfluidics for Biological Applications.* vol. 16. US: Springer-Verlag; 2019.
13. Streets AM, Huang Y. Chip in a lab: microfluidics for next generation life science research. *Biomicrofluidics.* 2013;7:11302.
14. Mehta G, Mehta K, Sud D, et al. Quantitative measurement and control of oxygen levels in microfluidic poly (dimethylsiloxane) bioreactors during cell culture. *Biomed Microdevices.* 2007;9:123–134.
15. Lu H, Koo LY, Wang WM, Lauffenburger DA, Griffith LG, Jensen KF. Microfluidic shear devices for quantitative analysis of cell adhesion. *Anal Chem.* 2004;76:5257–5264.
16. Kamholz AE, Weigl BH, Finlayson BA, Yager P. Quantitative analysis of molecular interaction in a microfluidic channel: the T-sensor. *Anal Chem.* 1999;71:5340–5347.
17. Chan IS, Ginsburg GS. Personalized medicine: progress and promise. *Annu Rev Genomics Hum Genet.* 2011;12:217–244.
18. Harvey A, Brand A, Holgate ST, et al. The future of technologies for personalised medicine. *New Biotechnol.* 2012;29:625–633.
19. Joyner MJ, Paneth N. Seven questions for personalized medicine. *JAMA.* 2015;314:999–1000.
20. Tavakolpour S. Towards personalized medicine for patients with autoimmune diseases: opportunities and challenges. *Immunol Lett.* 2017;190:130–138.
21. Anderson JE, Hansen LL, Mooren FC, et al. Methods and biomarkers for the diagnosis and prognosis of cancer and other diseases: towards personalized medicine. *Drug Resist Updat.* 2006;9:198–210.
22. Jackson SE, Chester JD. Personalised cancer medicine. *Int J Cancer.* 2015;137:262–266.
23. McCarthy MI. Painting a new picture of personalised medicine for diabetes. *Diabetologia.* 2017;60:793–799.
24. Lee MS, Flammer AJ, Lerman LO, Lerman A. Personalized medicine in cardiovascular diseases. *Korean Circ J.* 2012;42:583–591.

25. Agusti A, MacNee W. The COPD control panel: towards personalised medicine in COPD. *Thorax*. 2013;68:687–690.
26. Krzyszczyk P, Acevedo A, Davidoff EJ, et al. The growing role of precision and personalized medicine for cancer treatment. *Dent Tech*. 2018;6:79–100.
27. Alunni-Fabbroni M, Sandri MT. Circulating tumour cells in clinical practice: methods of detection and possible characterization. *Methods*. 2010;50:289–297.
28. Massagué J, Obenauf AC. Metastatic colonization by circulating tumour cells. *Nature*. 2016;529:298–306.
29. Alix-Panabières C, Pantel K. Circulating tumor cells: liquid biopsy of cancer. *Clin Chem*. 2013;59:110–118.
30. Pantel K, Alix-Panabières C. Circulating tumour cells in cancer patients: challenges and perspectives. *Trends Mol Med*. 2010;16:398–406.
31. Samandari M, Julia MG, Rice A, Chronopoulos A, Del Rio Hernandez AE. Liquid biopsies for management of pancreatic cancer. *Transl Res*. 2018;201:98–127.
32. Hsieh HB, Marrinucci D, Bethel K, et al. High speed detection of circulating tumor cells. *Biosens Bioelectron*. 2006;21:1893–1899.
33. Riethdorf S, Fritsche H, Müller V, et al. Detection of circulating tumor cells in peripheral blood of patients with metastatic breast cancer: a validation study of the CellSearch system. *Clin Cancer Res*. 2007;13:920–928.
34. Ramirez JM, Fehm T, Orsini M, et al. Prognostic relevance of viable circulating tumor cells detected by EPISPOT in metastatic breast cancer patients. *Clin Chem*. 2014;60: 214–221.
35. Andreopoulou E, Yang LY, Rangel KM, et al. Comparison of assay methods for detection of circulating tumor cells in metastatic breast cancer: AdnaGenAdna TestBreastCancer Select/Detect™ versus VeridexCellSearch™ system. *Int J Cancer*. 2012;130:1590–1597.
36. Danova M, Torchio M, Mazzini G. Isolation of rare circulating tumor cells in cancer patients: technical aspects and clinical implications. *Expert Rev Mol Diagn*. 2011;11: 473–485.
37. Hassan EM, Willmore WG, DeRosa MC. Aptamers: promising tools for the detection of circulating tumor cells. *Nucleic Acid Ther*. 2016;26:335–347.
38. Abate MF, Jia S, Ahmed MG, et al. Visual quantitative detection of circulating tumor cells with single-cell sensitivity using a portable microfluidic device. *Small*. 2019;15: 1804890.
39. Menon NV, Lim SB, Lim CT. Microfluidics for personalized drug screening of cancer. *Curr Opin Pharmacol*. 2019;48:155–161.
40. Moon HS, Yoo CE, Kim S, Lee JE, Park WY. Application of an open-chamber multi-channel microfluidic device to test chemotherapy drugs. *Sci Rep*. 2020;10:20343.
41. Malaney P, Nicosia SV, Davé V. One mouse, one patient paradigm: new avatars of personalized cancer therapy. *Cancer Lett*. 2014;344:1–12.
42. McDermott U, Settleman J. Personalized cancer therapy with selective kinase inhibitors: an emerging paradigm in medical oncology. *J Clin Oncol*. 2009;27:5650–5659.
43. Weeber F, Ooft SN, Dijkstra KK, Voest EE. Tumor organoids as a pre-clinical cancer model for drug discovery. *Cell Chem Biol*. 2017;24:1092–1100.
44. Es HA, Montazeri L, Aref AR, Vosough M, Baharvand H. Personalized cancer medicine: an organoid approach. *Trends Biotechnol*. 2018;36:358–371.
45. Gauri S, Ahmad MR. ctDNA detection in microfluidic platform: a promising biomarker for personalized cancer chemotherapy. *J Sens*. 2020;2020:8353674.
46. Gomez FA. The future of microfluidic point-of-care diagnostic devices. *Bioanalysis*. 2013;5:1–3.

47. Tripathi S, Kumar YBV, Prabhakar A, Joshi SS, Agrawal A. Performance study of microfluidic devices for blood plasma separation—a designer's perspective. *J Micromech Microeng*. 2015;25, 084004.
48. Catarino SO, Rodrigues RO, Pinho D, Miranda JM, Minas G, Lima R. Blood cells separation and sorting techniques of passive microfluidic devices: from fabrication to applications. *Micromachines*. 2019;10:593.
49. Ribas J, Sadeghi H, Manbachi A, et al. Cardiovascular organ-on-a-chip platforms for drug discovery and development. *Appl In Vitro Toxicol*. 2016;2:82–96.
50. Beverung S, Wu J, Steward R. Lab-on-a-chip for cardiovascular physiology and pathology. *Micromachines*. 2020;11:898.
51. Zhu J. Application of organ-on-chip in drug discovery. *J Biosci Med*. 2020;8:119–134.
52. Zhang YS, Aleman J, Arneri A, et al. From cardiac tissue engineering to heart-on-a-chip: beating challenges. *Biomed Mater*. 2015;10, 034006.
53. Cheah LT, Dou YH, Seymour AML, et al. Microfluidic perfusion system for maintaining viable heart tissue with real-time electrochemical monitoring of reactive oxygen species. *Lab Chip*. 2010;10:2720–2726.
54. Duan K. *Microfluidic Assays to Enhance Biodetection and Diabetes Research*. Doctoral dissertation; 2018.
55. Li J, Sun Y, Chen C, Sheng T, Liu P, Zhang G. A smartphone-assisted microfluidic chemistry analyzer using image-based colorimetric assays for multi-index monitoring of diabetes and hyperlipidemia. *Anal Chim Acta*. 2019;1052:105–112.
56. Malandrino N, Smith RJ. Personalized medicine in diabetes. *Clin Chem*. 2011;57: 231–240.
57. Ginsburg GS, Willard HF. Genomic and personalized medicine: foundations and applications. *Transl Res*. 2009;154:277–287.
58. Araujo F, Shrestha N, Gomes MJ, et al. *In vivo* dual-delivery of glucagon-like peptide-1 (GLP-1) and dipeptidyl peptidase-4 (DPP4) inhibitor through composites prepared by microfluidics for diabetes therapy. *Nanoscale*. 2016;8:10706–10713.
59. Hou HW, Tay HM, Dalan R, Boehm BO. Advances in neutrophil testing in type 2 diabetes mellitus. *Curr Res Diabetes Obes J*. 2017;1:555572.
60. Wu Jin P, Rousset N, Hierlemann A, Misun PM. A microfluidic hanging-drop-based islet perifusion system for studying glucose-stimulated insulin secretion from multiple individual pancreatic islets. *Front Bioeng Biotechnol*. 2021;9:352.
61. Diaz-Guzman E, Khosravi M, Mannino DM. Asthma, chronic obstructive pulmonary disease, and mortality in the US population. *COPD: J Chron Obstruct Pulmon Dis*. 2011;8:400–407.
62. Lee TA, Bartle B, Weiss KB. Spirometry use in clinical practice following diagnosis of COPD. *Chest*. 2006;129:1509–1515.
63. Chen ZH, Kim HP, Ryter SW, Choi AM. Identifying targets for COPD treatment through gene expression analyses. *Int J Chron Obstruct Pulmon Dis*. 2008;3:359.
64. Müller KC, Welker L, Paasch K, et al. Lung fibroblasts from patients with emphysema show markers of senescence *in vitro*. *Respir Res*. 2006;7:32.
65. Taki-Eddin L, Monem F. Utility of an interferon-gamma release assay as a potential diagnostic aid for active pulmonary tuberculosis. *J Infect Dev Ctries*. 2012;6:67–72.
66. Kobashi Y, Abe M, Mouri K, Obase Y, Miyashita N, Oka M. Usefulness of tuberculin skin test and three interferon-gamma release assays for the differential diagnosis of pulmonary tuberculosis. *Intern Med*. 2012;51:1199–1205.
67. Woodruff PG, Modrek B, Choy DF, et al. T-helper type 2–driven inflammation defines major subphenotypes of asthma. *Am J Respir Crit Care Med*. 2009;180:388–395.
68. Katz LE, Gleich GJ, Hartley BF, Yancey SW, Ortega HG. Blood eosinophil count is a useful biomarker to identify patients with severe eosinophilic asthma. *Ann Am Thorac Soc*. 2014;11:531–536.

69. Vanto T, Koskinen P. Serum eosinophil cationic protein in the evaluation of asthma severity in children. *Allergy*. 1998;53:415–419.
70. Jatakanon ANON, Lim SAM, Barnes PJ. Changes in sputum eosinophils predict loss of asthma control. *Am J Respir Crit Care Med*. 2000;161:64–72.
71. Wagener AH, de Nijs SB, Lutter R, et al. External validation of blood eosinophils, FENO and serum periostin as surrogates for sputum eosinophils in asthma. *Thorax*. 2015;70:115–120.
72. Schleich FN, Zanella D, Stefanuto PH, et al. Exhaled volatile organic compounds are able to discriminate between neutrophilic and eosinophilic asthma. *Am J Respir Crit Care Med*. 2019;200:444–453.
73. Busse WW. Asthma diagnosis and treatment: filling in the information gaps. *J Allergy Clin Immunol*. 2011;128:740–750.
74. Riahi R, Tamayol A, Shaegh SAM, Ghaemmaghami AM, Dokmeci MR, Khademhosseini A. Microfluidics for advanced drug delivery systems. *Curr Opin Chem Eng*. 2015;7:101–112.
75. Sackmann EK, Berthier E, Young EW, et al. Microfluidic kit-on-a-lid: a versatile platform for neutrophil chemotaxis assays. *Blood*. 2012;120:e45–e53.
76. Jia M, Yan X, Jiang X, et al. Ezrin, a membrane cytoskeleton cross-linker protein, as a marker of epithelial damage in asthma. *Am J Respir Crit Care Med*. 2019;199:496–507.
77. Hartl D, Lee CG, Da Silva CA, Chupp GL, Elias JA. Novel biomarkers in asthma: chemokines and chitinase-like proteins. *Curr Opin Allergy Clin Immunol*. 2009;9:60–66.

CHAPTER EIGHT

Microfluidics for single cell analysis

Rupesh Maurya[a], Nisarg Gohil[a], Gargi Bhattacharjee[a], Khalid J. Alzahrani[b], Suresh Ramakrishna[c,d], and Vijai Singh[a,]*

[a]Department of Biosciences, School of Science, Indrashil University, Rajpur, Mehsana, Gujarat, India
[b]Department of Clinical Laboratories Sciences, College of Applied Medical Sciences, Taif University, Taif, Saudi Arabia
[c]Graduate School of Biomedical Science and Engineering, Hanyang University, Seoul, South Korea
[d]College of Medicine, Hanyang University, Seoul, South Korea
*Corresponding author: e-mail address: vijaisingh15@gmail.com

Contents

1. Introduction 203
2. Design and fabrication of microfluidics devices and chips 204
3. Droplets-based single cell analysis and their applications 207
4. Use of microfluidics devices for single cell analysis 208
5. Conclusion and future remarks 210
Acknowledgments 211
References 211

Abstract

Cells have several internal molecules that are present in low amounts and any fluctuation in its number drives a change in cell behavior. These molecules present inside the cells are continuously fluctuating, thus producing noises in the intrinsic environment and thereby directly affecting the cellular behavior. Single-cell analysis using microfluidics is an important tool for monitoring cell behavior by analyzing internal molecules. Several gene circuits have been designed for this purpose that are labeled with fluorescence encoding genes for monitoring cell dynamics and behavior. We discuss herewith designed and fabricated microfluidics devices that are used for trapping and tracking cells under controlled environmental conditions. This chapter highlights microfluidics chip for monitoring cells to promote their basic understanding.

1. Introduction

Microfluidics chips or devices play a key role in several areas of biotechnology. It has recently gained considerable scientific attention due to its ease to design and uses. It is an important tool which has been used in various

Progress in Molecular Biology and Translational Science, Volume 186
ISSN 1877-1173
https://doi.org/10.1016/bs.pmbts.2021.07.013

fields such as biochemistry,[1] cell signaling,[2,3] drug discovery and development,[4,5] drug testing,[6] drug delivery,[7] immunology,[8] genomics and proteomics,[9] and can bring revolution in the food and agriculture industry,[10] and pharmaceutical[11] sector as well.

Laser capture microdissection is a very useful tool for isolation of single cells from sections of solid tissues.[12] Fluorescence activated cell sorting systems have the advantage of sorting capability but it is costly and harmful to cells. Micromanipulator-assisted cell picking is comparatively slow but allows control over individual cells. Microfluidics technology helps for integration of entire application specific workflows and only requires very small amounts of samples and reagents.[13–15] It provides accessibility to single cells from a bulk cell suspension. Due to its flexible design, microfluidics devices fulfill the need of individual cell types.[16,17]

It plays a significant role in separation and purification of biomacromolecules,[18] nanoparticles,[19] micro algal cells,[20] proteins,[21] radionuclides,[22] nucleic acids,[23] particle separation,[24] viruses,[25] platelets,[26] etc.

Analysis of single cells has become popular due to heterogeneity in the population. Several questions such as function, transformation and reaction toward stimuli can be answered by studying heterogeneity.[27–30] It helps in analyzing very important parameters such as growth, productivity, heterogeneity of single cells,[31] and ionophore-mediated intracellular Ca^{2+} flux measurements.[32] In this chapter, we highlight design, fabrication and use of microfluidics chips for single cell analysis for better understanding of biological process and cellular dynamics.

2. Design and fabrication of microfluidics devices and chips

AutoCAD is used for designing patterns in order to create a chip. Several companies are currently working on printing silicon wafers that have been used for fabrication of chips. Polydimethylsiloxane (PDMS) and curing agent in 10:1 ratio is first mixed properly. Air bubbles are removed by centrifugation at 4000 rpm for 15–20 min. Subsequently, it is poured on top of wafer and allowed to settle properly. Then, it is kept at 65 °C for 1 h in hot air oven. Once it is dried, the wafer is carefully removed with polymers of solid PDMS embossed on it. PDMS is cut with a sharp knife and made a syringe hole for connecting with tubing for port for cell, media and waste. Oxygen plasma cleaner is used for activating the surface of solid PDMS and a

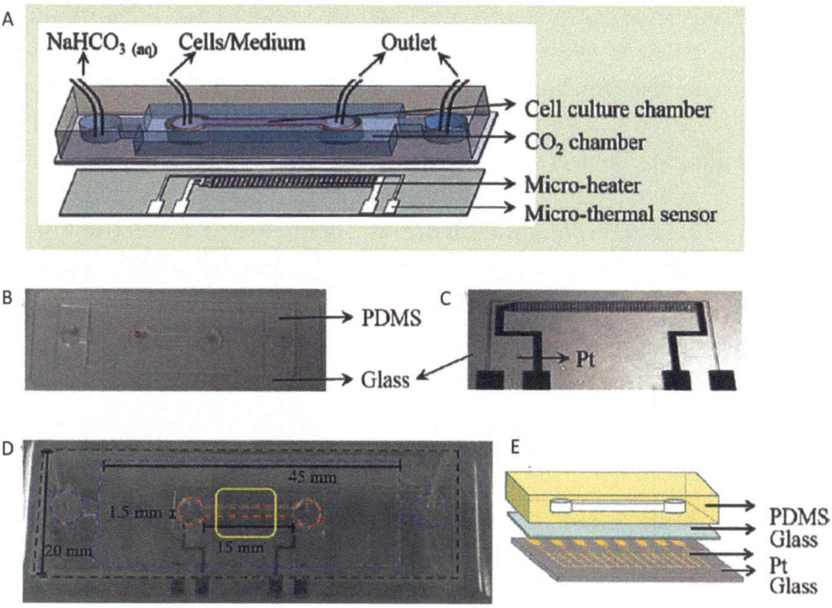

Fig. 1 Microfluidics chip design and fabrication. (A) Micro cell-culture chip. (B) CO_2 and cell-culture chambers. (C) Micro-heater. (D) Assembled chip and (E) layer of chip design. *Figure reproduced with permission from Yu, IF, Yu, YH, Chen, LY, Fan, SK, Chou, HYE, Yang, JT. A portable microfluidic device for the rapid diagnosis of cancer metastatic potential which is programmable for temperature and CO_2.* Lab Chip. *2014;14:3621–3628.*

glass slide for 30 s and then bonded together with a small press. Then, it is kept at 80 °C for overnight to facilitate proper attachment of the two. Prior to the experiment, the chip is cleaned with acetone, isopropanol and air dried to remove dust particles.[33] Fig. 1 depicts the design and fabrication of microfluidics chip.[34]

Escherichia coli cells are first cultured overnight at 37 °C and subcultured into a 100 mL of fresh Luria-Bertani (LB) media with 100 μL of overnight grown culture. It is then incubated at 37 °C for 3–4 h and when the optical density at 600 nm reached to 0.3–0.4, the culture is then centrifuged at 37 °C for 5 min at 5000 rpm. Supernatant is discarded and 2 mL of fresh LB media with 0.1%. Tween 20 as surfactant is added into the tubes and the resuspended cells are loaded onto the chip that has tubing connected to media port and waste ports. Once a cell gets trapped in a channel, fresh media (200 μL/min) is supplied and incubated for 4 h on chip. All extra cells and excess media is moved into waste. Once the monolayer of cells is completely formed, one can begin to take phase contrast and fluorescent images using time lapse fluorescent microscope (TLFM). Phase contrast images are taken at every 30 s and fluorescent images are taken at every 5 min of interval. ImageJ is used to analyze a single cell by creating a movie.[33,35]

Microfluidics devices consist of several important components such as pumps, valves, filters, and mixers which can be manipulated, and in addition, it also incorporates analytical techniques such as electrophoresis, chromatography, fluorescence and electrochemical detection. Since the beginning, construction of microfluidics is achieved on glass material, an calcium-assisted glass-to-glass bonding.[36] Channels are very important part of microfluidics devices which allow flow of different chemicals in the device. Several characteristics are to be ensured for a better channel system, such as resistance to chemicals, capability to sustain high temperature and biocompatibility.[37]

Later other methods such as hot embossing[38] and soft lithography[39] started being devised for channel fabrication and creation of micropatterns. Use of PDMS or microprinting and creating patterns on the surface is recommended. Pre-doping PDMS with acrylic acid, extraction of uncured PDMS (oligomer), and monomer grafting using 2-hydroxyethyl methacrylate (HEMA) and monomer grafting using polyethylene glycol (PEG) should also be considered. Extraction method improves zeta potential in modified PDMS surfaces as compared to other methods.[40] In a quest to discover alternatives to PDMS devices, new thermoset polyester (TPE) microfluidic devices were developed. After a series of improvements such as thin bottom layer in the chips, it was able to provide better sensitivity.[41] Other methods which have brought revolution in design and fabrication include polymer embossing,[42] hot roller embossing,[43] cut and stack method,[44] embossing using PDMS molds,[45] and hot embossing and fabrication in poly methyl methacrylate.[46]

Injection molding technique is used in the plastic industry for creating cavities on objects by using temperature and pressure.[47] Rapid injection molding[48] and injection molding PMMA-based microfluidic chips[49] are more productive for diagnosis and research purpose. Other polymer based, i.e., off-stoichiometric thiol–ene epoxy (OSTE+) thermosetting polymers[50] and magnetic field assisted finishing of mold[51] are also being utilized in injection molding. Laser ablation has also been utilized for fabrication of microchips in order to remove materials from thermoplastic sheets. Femtosecond laser,[52] ArF-excimer laser,[53] excimer laser,[54] and optofluidic dye lasers[55] have been tested for fabrication of microchips. Currently, a wide range of microfluidics chips and devices are being designed and fabricated for several biotechnological applications.

3. Droplets-based single cell analysis and their applications

Droplet-based method for single cell analysis is extensively used. In this method, a droplet generator is used to generate picoliter to nanoliter droplets at very high frequencies. It is used for its low-cost and high performance. Fig. 2 shows, droplet for screening of single cells.[56] More than thousands of droplets per second can be generated. Each single cell is separated in droplets and each droplet acts as a microchamber where the cell can undergo lysis. Each droplet can be analyzed.[57] In this manner, about one million cells can be screened within 5–6 h.[58] Surfactants play a significant role in stabilization of droplet interfaces. Fluorosurfactants are synthesized by coupling oligomeric perfluorinated polyethers (PFPE) with polyethyleneglycol (PEG) for better throughput.[59] Other biocompatible surfactants that have been reported include anionic tris(hydroxymethyl)methyl (Tris) polar head groups[60] and hydrogel–solid dimer particles.[61]

Through droplet-based method, cell growth, metabolism and cell death can be measured as well as DNA, RNA, proteins can be characterized. It offers detection of rare mutation,[62] DNA sequence analysis,[63] single cell hybridoma screening,[64] artificial cell generation,[65] detecting cytokine-secreting cells,[66] detection of different organism, tissues and chemicals,[67] analysis of protein-protein interactions,[68] single digital cell analysis,[69] also used in bioprospecting of cellulolytic organism,[70] investigation of host-microbe interaction,[71] detection of circulating tumor cells,[72] elucidation of cellular metabolism mechanism,[73] detection of fluorescent signals,[74] and for recreation of living cell functions in their mechanical, morphological, or metabolic characteristics.[75] Droplet-based microfluidics shows high sensitivity and it is a potential tool for optimization of computer designed enzymes. It may also be helpful in quantitative assessment of evolutionary models.[76] It can significantly improve catalytic properties of RNA.[77] Food industry requires improved strains which can be utilized in fermentation, and microfluidics may help in identification and screening of improved phenotypes.[78]

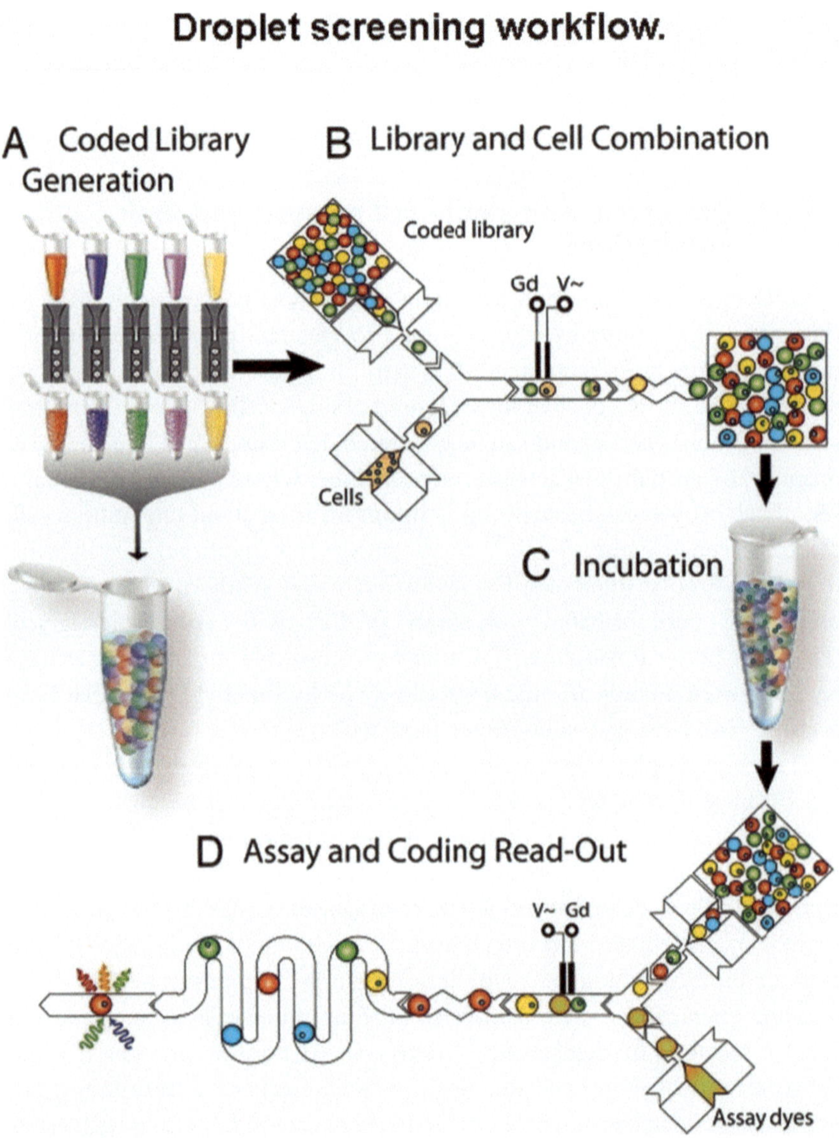

Fig. 2 Workflow for droplet screening. *Adopted from Brouzes, E, Medkova, M, Savenelli, N, Marran, D, Twardowski, M, Hutchison, JB, et al. Droplet microfluidic technology for single-cell high-throughput screening.* Proc Natl Acad Sci U S A. *2009;106:14195–14200.*

4. Use of microfluidics devices for single cell analysis

A single cell analysis is important for monitoring cellular activity and behavior. In the past decades, microfluidics chips were designed and fabricated for trapping the monolayer of either prokaryotic or eukaryotic cells and TLFM was used for capturing phase contrast and fluorescent image. Both images were superimposed and movies were created that were utilized for analyzing gene networks or other artificial circuit behavior along with cell division in different cell culture conditions such as media compositions, temperature, sugar and others. Microfluidics play a key role in single-cell analysis and has recently gained a lot of attention due to accurate measurement of cellular behavior.[79] Fig. 3 depicts, microfluidics for analysis of gene expression at single-cell level.[80] In a study, a single cell of *E. coli* monolayer was trapped and tracked in a microfluidics chip by providing controlled environmental conditions for growth. This helped to obtain high resolution time lapse images for monitoring cell morphology and cell multiplication.[81]

In a prior study, Stricker et al.[82] designed a synthetic oscillator[83,84] based on activator and repressor. Oscillator was designed for *E. coli* which was fast, robust and highly tunable. It was tested using microfluidics chips to support

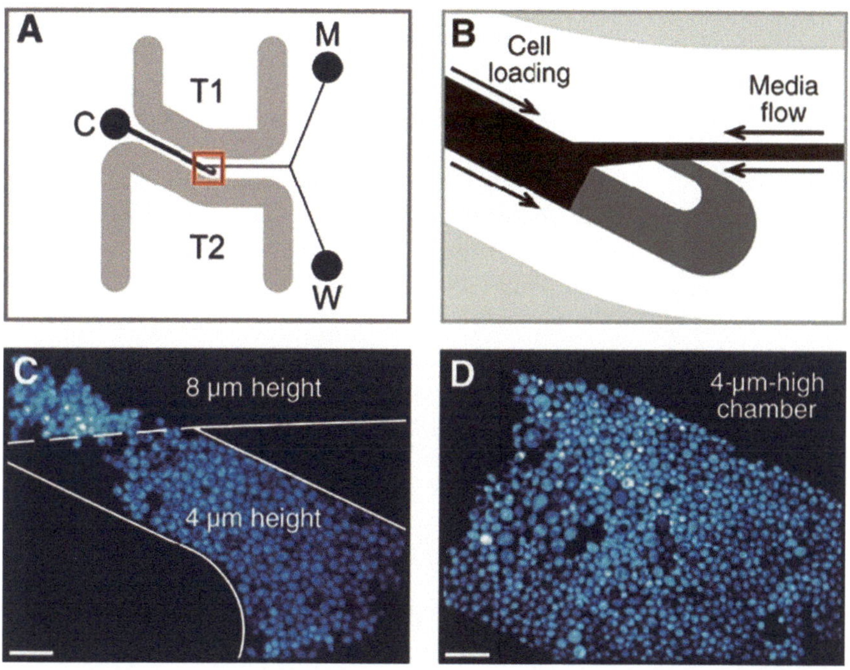

Fig. 3 Microfluidics for single-cell gene expression over multiple cell division. *Adopted from Cookson, S, Ostroff, N, Pang, WL, Volfson, D, Hasty, J. Monitoring dynamics of single-cell gene expression over multiple cell cycles.* Mol Syst Biol. *2005;1:2005.0024.*

monolayer growth of *E. coli* under constant nutrient flow. Different parts designed on the chip included traps for cell division, media flow and wastes. In that chip, 4 ports were connected, one for cell loading, one for media supply and the remaining 2 for wastes. In the experiment, microfluidics media was mixed with a solution of 0.1% Tween 20 surfactant and reported the optimal growth at 37 °C. This study provided better insights for long term monitoring of single cells that could help to measure its oscillatory behavior along with cell division time.

Microfluidics devices nowadays are inexpensive, easy to use and can be designed and fabricated as per experimental needs. It can be used for trapping cells for TLFM for studying cells by changing environmental conditions and response. Balagaddé et al.[85] designed microfluidics devices for long term analysis of cells in a continuous chemostat. In this study, microfluidics bioreactor was used for long term culture and measurement of small bacterial (*E. coli*) population with single cell resolution. They have studied *E. coli* population control that regulates cell density via quorum sensing[86] mechanism. Similarly, other groups have designed microfluidics chips for monitoring dynamic gene regulation and also intracellular signaling by trapping and tracking cells through different time interval. It was either used for single cell analysis or small cell population for long term imaging.[87,88]

Mondragón-Palomino et al.[89] engineered a quorum sensing-based synthetic oscillator. It was tested using a microfluidics chip for trapping a single cell of *E. coli* that allows it to reach to a certain concentration of quorum sensing molecules[86] and then begin to oscillate. The monolayer of cells was constantly supplied with a fresh media with an inducer. In another work, Zare and Kim[17] used microfluidics for studying and control of fluidics behavior at single cell level. Small volume of liquid was used to perform cell lysis and chemical reaction. Chromatography and electrophoresis was integrated within microchannel to analyze several intracellular components. It has shown the potential for rapid, accurate and cost-effective analysis of single cells.

5. Conclusion and future remarks

Microfluidics is a rapidly growing area and have shown a wide range of applications including disease diagnosis, sensing of RNA, DNA, protein, metabolites, chemicals, high-throughput screening and many such applications. Droplets-based microfluidics are also being used in detection of biomolecules. Microfluidics have shown major applications in single-cell trapping and analysis in microchannel with constant supply of fresh media for growth of organisms. Microfluidics can also be useful for drug toxicity testing, growth of cells or organs on chip, delivery and monitoring of effects of drugs. In the near future, microfluidics can be extended in many more areas to solve major issues of slow process to high-throughput processing.

Acknowledgments

N.G. acknowledges the Indian Council of Medical Research, Government of India for financial assistance as Senior Research Fellowship (File No. 5/3/8/63/ITR-F/2020). The financial support from Gujarat State Biotechnology Mission (GSBTM) (Project ID: 5LY45F), Gujarat, India to G.B. and V.S. is duly acknowledged. The financial support by National Research Foundation of Korea (NRF) (2021M3A9H3015389) to S.R. is also acknowledged. R.M., N.G., G.B. and V.S. thank Indrashil University, Rajpur, Mehsana, India for providing the infrastructure facility.

References

1. Ohno KI, Tachikawa K, Manz A. Microfluidics: applications for analytical purposes in chemistry and biochemistry. *Electrophoresis*. 2008;29:4443–4453.
2. Sun J, Masterman-Smith MD, Graham NA, et al. A microfluidic platform for systems pathology: multiparameter single-cell signaling measurements of clinical brain tumor specimens. *Cancer Res*. 2010;70:6128–6138.
3. Zhou Q, Patel D, Kwa T, et al. Liver injury-on-a-chip: microfluidic co-cultures with integrated biosensors for monitoring liver cell signaling during injury. *Lab Chip*. 2015;15:4467–4478.
4. Kang L, Chung BG, Langer R, Khademhosseini A. Microfluidics for drug discovery and development: from target selection to product lifecycle management. *Drug Discov Today*. 2008;13:1–13.
5. Lombardi D, Dittrich PS. Advances in microfluidics for drug discovery. *Expert Opin Drug Discovery*. 2010;5:1081–1094.
6. Patra B, Peng CC, Liao WH, Lee CH, Tung YC. Drug testing and flow cytometry analysis on a large number of uniform sized tumor spheroids using a microfluidic device. *Sci Rep*. 2016;6, 21061.
7. Kleinstreuer C, Li J, Koo J. Microfluidics of nano-drug delivery. *Int J Heat Mass Transf*. 2008;51:5590–5597.
8. Jammes FC, Maerkl SJ. How single-cell immunology is benefiting from microfluidic technologies. *Microsyst Nanoeng*. 2020;6:1–14.
9. Kotz KT, Xiao W, Miller-Graziano C, et al. Clinical microfluidics for neutrophil genomics and proteomics. *Nat Med*. 2010;16:1042–1047.
10. Neethirajan S, Kobayashi I, Nakajima M, Wu D, Nandagopal S, Lin F. Microfluidics for food, agriculture and biosystems industries. *Lab Chip*. 2011;11:1574–1586.
11. Maeki M. Microfluidics for pharmaceutical applications. In: Santos H, Liu D, Zhang H, eds. *Microfluidics for Pharmaceutical Applications*. William Andrew Publishing; 2019: 101–119.
12. Espina V, Heiby M, Pierobon M, Liotta LA. Laser capture microdissection technology. *Expert Rev Mol Diagn*. 2007;7:647–657.
13. Gross A, Schoendube J, Zimmermann S, Steeb M, Zengerle R, Koltay P. Technologies for single-cell isolation. *Int J Mol Sci*. 2015;16:16897–16919.
14. Whitesides GM. The origins and the future of microfluidics. *Nature*. 2006;442:368–373.

15. Sackmann EK, Fulton AL, Beebe DJ. The present and future role of microfluidics in biomedical research. *Nature*. 2014;507:181–189.
16. Yin H, Marshall D. Microfluidics for single cell analysis. *Curr Opin Biotechnol*. 2012;23:110–119.
17. Zare RN, Kim S. Microfluidic platforms for single-cell analysis. *Annu Rev Biomed Eng*. 2010;12:187–201.
18. Vicente FA, Plazl I, Ventura SP, Žnidaršič-Plazl P. Separation and purification of biomacromolecules based on microfluidics. *Green Chem*. 2020;22:4391–4410.
19. Salafi T, Zeming KK, Zhang Y. Advancements in microfluidics for nanoparticle separation. *Lab Chip*. 2017;17:11–33.
20. Syed MS, Rafeie M, Vandamme D, et al. Selective separation of microalgae cells using inertial microfluidics. *Bioresour Technol*. 2018;252:91–99.
21. Sarkar A, Hou HW, Mahan AE, Han J, Alter G. Multiplexed affinity-based separation of proteins and cells using inertial microfluidics. *Sci Rep*. 2016;6, 23589.
22. Mariet C, Vansteene A, Losno M, et al. Microfluidics devices applied to radionuclides separation in acidic media for the nuclear fuel cycle. *Micro Nano Eng*. 2019;3:7–14.
23. Wen J, Legendre LA, Bienvenue JM, Landers JP. Purification of nucleic acids in microfluidic devices. *Anal Chem*. 2008;80:6472–6479.
24. Sajeesh P, Sen AK. Particle separation and sorting in microfluidic devices: a review. *Microfluid Nanofluid*. 2014;17:1–52.
25. Chen GD, Alberts CJ, Rodriguez W, Toner M. Concentration and purification of human immunodeficiency virus type 1 virions by microfluidic separation of superparamagnetic nanoparticles. *Anal Chem*. 2010;82:723–728.
26. Pommer MS, Zhang Y, Keerthi N, et al. Dielectrophoretic separation of platelets from diluted whole blood in microfluidic channels. *Electrophoresis*. 2008;29:1213–1218.
27. Bocci F, Gearhart-Serna L, Boareto M, et al. Toward understanding cancer stem cell heterogeneity in the tumor microenvironment. *Proc Natl Acad Sci U S A*. 2019;116: 148–157.
28. Goldman SL, MacKay M, Afshinnekoo E, Melnick AM, Wu S, Mason CE. The impact of heterogeneity on single-cell sequencing. *Front Genet*. 2019;10:8.
29. Irish JM, Kotecha N, Nolan GP. Mapping normal and cancer cell signalling networks: towards single-cell proteomics. *Nat Rev Cancer*. 2006;6(2):146–155.
30. Katz E. Heterogeneity in transformation efficiency of NIH 3T3 cells as determined by transfection with oncogenes. *Eur J Cell Biol*. 1986;42:231–235.
31. Grünberger A, Wiechert W, Kohlheyer D. Single-cell microfluidics: opportunity for bioprocess development. *Curr Opin Biotechnol*. 2014;29:15–23.
32. Wheeler AR, Throndset WR, Whelan RJ, et al. Microfluidic device for single-cell analysis. *Anal Chem*. 2003;75:3581–3586.
33. Carrera J, Rodrigo G, Singh V, Kirov B, Jaramillo A. Empirical model and in vivo characterization of the bacterial response to synthetic gene expression show that ribosome allocation limits growth rate. *Biotechnol J*. 2011;6:773–783.
34. Yu IF, Yu YH, Chen LY, Fan SK, Chou HYE, Yang JT. A portable microfluidic device for the rapid diagnosis of cancer metastatic potential which is programmable for temperature and CO_2. *Lab Chip*. 2014;14:3621–3628.
35. Shen S, Rodrigo G, Prakash S, et al. Dynamic signal processing by ribozyme-mediated RNA circuits to control gene expression. *Nucleic Acids Res*. 2015;43:5158–5170.
36. Allen PB, Chiu DT. Calcium-assisted glass-to-glass bonding for fabrication of glass microfluidic devices. *Anal Chem*. 2008;80:7153–7157.
37. Hwang J, Cho YH, Park MS, Kim BH. Microchannel fabrication on glass materials for microfluidic devices. *Int J Precis Eng Manuf*. 2019;20:479–495.
38. Jeon JS, Chung S, Kamm RD, Charest JL. Hot embossing for fabrication of a microfluidic 3D cell culture platform. *Biomed Microdevices*. 2011;13:325–333.

39. Kim P, Kwon KW, Park MC, Lee SH, Kim SM, Suh KY. Soft lithography for microfluidics: a review. *Biochip J*. 2008;2:1–11.
40. Almutairi Z, Ren CL, Simon L. Evaluation of polydimethylsiloxane (PDMS) surface modification approaches for microfluidic applications. *Colloids Surf A*. 2012;415: 406–412.
41. Fiorini GS, Yim M, Jeffries GD, et al. Fabrication improvements for thermoset polyester (TPE) microfluidic devices. *Lab Chip*. 2007;7:923–926.
42. Narasimhan J, Papautsky I. Polymer embossing tools for rapid prototyping of plastic microfluidic devices. *J Micromech Microeng*. 2003;14:96.
43. Yeo LP, Ng SH, Wang ZF, et al. Investigation of hot roller embossing for microfluidic devices. *J Micromech Microeng*. 2009;20, 015017.
44. Thuo MM, Martinez RV, Lan WJ, et al. Fabrication of low-cost paper-based microfluidic devices by embossing or cut-and-stack methods. *Chem Mater*. 2014;26: 4230–4237.
45. Goral VN, Hsieh YC, Petzold ON, Faris RA, Yuen PK. Hot embossing of plastic microfluidic devices using poly (dimethylsiloxane) molds. *J Micromech Microeng*. 2010;21: 017002.
46. Qi S, Liu X, Ford S, et al. Microfluidic devices fabricated in poly (methyl methacrylate) using hot-embossing with integrated sampling capillary and fiber optics for fluorescence detection. *Lab Chip*. 2002;2:88–95.
47. Rosato DV, Rosato MG. *Injection Molding Handbook*. Springer Nature; 2012.
48. Lee UN, Su X, Guckenberger DJ, et al. Fundamentals of rapid injection molding for microfluidic cell-based assays. *Lab Chip*. 2018;18:496–504.
49. Ma X, Li R, Jin Z, Fan Y, Zhou X, Zhang Y. Injection molding and characterization of PMMA-based microfluidic devices. *Microsyst Technol*. 2020;26:1317–1324.
50. Sandström N, Shafagh RZ, Vastesson A, Carlborg CF, van der Wijngaart W, Haraldsson T. Reaction injection molding and direct covalent bonding of OSTE+ polymer microfluidic devices. *J Micromech Microeng*. 2015;25, 075002.
51. Guo J, Liu K, Wang Z, Tnay GL. Magnetic field-assisted finishing of a mold insert with curved microstructures for injection molding of microfluidic chips. *Tribol Int*. 2017;114:306–314.
52. Gomez D, Goenaga I, Lizuain I, Ozaita M. Femtosecond laser ablation for microfluidics. *Optim Eng*. 2005;44, 051105.
53. Sabbert D, Landsiedel J, Bauer HD, Ehrfeld W. ArF-excimer laser ablation experiments on cycloolefin copolymer (COC). *Appl Surf Sci*. 1999;150:185–189.
54. Kim J, Xu X. Excimer laser fabrication of polymer microfluidic devices. *J Laser Appl*. 2003;15:255.
55. Li Z, Psaltis D. Optofluidic dye lasers. *Microfluid Nanofluid*. 2008;4:145–158.
56. Brouzes E, Medkova M, Savenelli N, et al. Droplet microfluidic technology for single-cell high-throughput screening. *Proc Natl Acad Sci U S A*. 2009;106:14195–14200.
57. Seemann R, Brinkmann M, Pfohl T, Herminghaus S. Droplet based microfluidics. *Rep Prog Phys*. 2011;75, 016601.
58. Mazutis L, Gilbert J, Ung WL, Weitz DA, Griffiths AD, Heyman JA. Single-cell analysis and sorting using droplet-based microfluidics. *Nat Protoc*. 2013;8:870–891.
59. Holtze C, Rowat AC, Agresti JJ, et al. Biocompatible surfactants for water-in-fluorocarbon emulsions. *Lab Chip*. 2008;8:1632–1639.
60. Chiu YL, Chan HF, Phua KK, et al. Synthesis of fluorosurfactants for emulsion-based biological applications. *ACS Nano*. 2014;8:3913–3920.
61. Chen D, Amstad E, Zhao CX, et al. Biocompatible amphiphilic hydrogel–solid dimer particles as colloidal surfactants. *ACS Nano*. 2017;11:11978–11985.
62. Pekin D, Skhiri Y, Baret JC, et al. Quantitative and sensitive detection of rare mutations using droplet-based microfluidics. *Lab Chip*. 2011;11:2156–2166.

63. Abate AR, Hung T, Sperling RA, et al. DNA sequence analysis with droplet-based microfluidics. *Lab Chip*. 2013;13:4864–4869.
64. El Debs B, Utharala R, Balyasnikova IV, Griffiths AD, Merten CA. Functional single-cell hybridoma screening using droplet-based microfluidics. *Proc Natl Acad Sci U S A*. 2012;109:11570–11575.
65. Martino C, deMello AJ. Droplet-based microfluidics for artificial cell generation: a brief review. *Interf Focus*. 2016;6, 20160011.
66. Chokkalingam V, Tel J, Wimmers F, et al. Probing cellular heterogeneity in cytokine-secreting immune cells using droplet-based microfluidics. *Lab Chip*. 2013;13: 4740–4744.
67. Cao J, Köhler JM. Droplet-based microfluidics for microtoxicological studies. *Eng Life Sci*. 2015;15:306–317.
68. Srisa-Art M, Kang DK, Hong J, et al. Analysis of protein–protein interactions by using droplet-based microfluidics. *Chembiochem*. 2009;10:1605–1611.
69. Najah M, Griffiths AD, Ryckelynck M. Teaching single-cell digital analysis using droplet-based microfluidics. *Anal Chem*. 2012;84:1202–1209.
70. Najah M, Calbrix R, Mahendra-Wijaya IP, Beneyton T, Griffiths AD, Drevelle A. Droplet-based microfluidics platform for ultra-high-throughput bioprospecting of cellulolytic microorganisms. *Chem Biol*. 2014;21:1722–1732.
71. Tauzin AS, Pereira MR, Van Vliet LD, et al. Investigating host-microbiome interactions by droplet based microfluidics. *Microbiome*. 2020;8:141.
72. Del Ben F, Turetta M, Celetti G, et al. A method for detecting circulating tumor cells based on the measurement of single-cell metabolism in droplet-based microfluidics. *Angew Chem Int Ed Engl*. 2016;128:8723–8726.
73. Fan W, Chen X, Ge Y, Jin Y, Jin Q, Zhao J. Single-cell impedance analysis of osteogenic differentiation by droplet-based microfluidics. *Biosens Bioelectron*. 2019;145, 111730.
74. Lim J, Gruner P, Konrad M, Baret JC. Micro-optical lens array for fluorescence detection in droplet-based microfluidics. *Lab Chip*. 2013;13:1472–1475.
75. Staufer O, Schröter M, Platzman I, Spatz JP. Bottom-up assembly of functional intracellular synthetic organelles by droplet-based microfluidics. *Small*. 2020;16, 1906424.
76. Chiu FW, Stavrakis S. High-throughput droplet-based microfluidics for directed evolution of enzymes. *Electrophoresis*. 2019;40:2860–2872.
77. Ryckelynck M, Baudrey S, Rick C, et al. Using droplet-based microfluidics to improve the catalytic properties of RNA under multiple-turnover conditions. *RNA*. 2015; 21:458–469.
78. Chen J, Vestergaard M, Shen J, Solem C, Dufva M, Jensen PR. Droplet-based microfluidics as a future tool for strain improvement in lactic acid bacteria. *FEMS Microbiol Lett*. 2018;365, fny258.
79. Bennett MR, Hasty J. Microfluidic devices for measuring gene network dynamics in single cells. *Nat Rev Genet*. 2009;10:628–638.
80. Cookson S, Ostroff N, Pang WL, Volfson D, Hasty J. Monitoring dynamics of single-cell gene expression over multiple cell cycles. *Mol Syst Biol*. 2005;1:2005.0024.
81. Probst C, Grünberger A, Wiechert W, Kohlheyer D. Polydimethylsiloxane (PDMS) sub-Micron traps for single-cell analysis of Bacteria. *Micromachines*. 2013;4:357–369.
82. Stricker J, Cookson S, Bennett MR, Mather WH, Tsimring LS, Hasty J. A fast, robust and tunable synthetic gene oscillator. *Nature*. 2008;456:516–519.
83. Bhattacharjee G, Gohil N, Singh V. An introduction to design of microbial strain using synthetic biology toolboxes for production of biomolecules. In: Singh V, Singh A, Bhargava P, Joshi M, Joshi C, eds. *Engineering of Microbial Biosynthetic Pathways*. Singapore: Springer; 2020:1–10.
84. Patel S, Panchasara H, Braddick D, Gohil N, Singh V. Synthetic small RNAs: current status, challenges, and opportunities. *J Cell Biochem*. 2018;119:9619–9639.

85. Balagaddé FK, You L, Hansen CL, Arnold FH, Quake SR. Long-term monitoring of bacteria undergoing programmed population control in a microchemostat. *Science*. 2005;309:137–140.
86. Gohil N, Ramírez-García R, Panchasara H, Patel S, Bhattacharjee G, Singh V. Book review: quorum sensing vs. quorum quenching: a battle with no end in sight. *Front Cell Infect Microbiol*. 2018;8:106.
87. Di Carlo D, Aghdam N, Lee LP. Single-cell enzyme concentrations, kinetics, and inhibition analysis using high-density hydrodynamic cell isolation arrays. *Anal Chem*. 2006;78:4925–4930.
88. Prokop A, Prokop Z, Schaffer D, et al. NanoLiterBioReactor: long-term mammalian cell culture at nanofabricated scale. *Biomed Microdevices*. 2004;6:325–339.
89. Mondragón-Palomino O, Danino T, Selimkhanov J, Tsimring L, Hasty J. Entrainment of a population of synthetic genetic oscillators. *Science*. 2011;333:1315–1319.

CHAPTER NINE

Fluorescence based miniaturized microfluidic and nanofluidic systems for biomedical applications

Aishwarya P. Waghchoure, J. Prakasha Reddy*, and Rajesh S. Bhosale*

Department of Chemistry, Indrashil University, Rajpur, Mehsana, Gujarat, India
*Corresponding authors: e-mail address: j.prakashareddy@gmail.com; rajeshbhosale24@gmail.com

Contents

1. Introduction 218
2. Microfluidic systems for biomedical applications 219
   2.1 Nucleic acid sensing and analysis 219
   2.2 Protein sensing and analysis 225
   2.3 Microfluidic systems to generate aggregation induced emission (AIE) droplets/dots for biological applications 228
3. Nanofluidic systems for biomedical applications 232
   3.1 Nucleic acid sequencing and sensing 232
   3.2 Protein binding kinetics and DNA interaction study 235
4. Conclusion and future prospects 239
Acknowledgments 239
References 240

Abstract

Over the last two decades miniaturized microfluidic and nanofluidic systems with fluorescence setup emerged as a powerful technological platform for diverse biomedical applications. Bio-macromolecules such as nucleic acids and proteins are the core cellular components, their single molecule analysis allow us to understand biological processes, disease creation and progression, and development of novel treatment policies. Design and development of foolproof treatment methods requires rigorously analysis of nucleic acids and proteins such as length quantifications, sequence profiling, sequence mapping, analysis of conformational changes, analysis and recognition of epigenetic changes, and their interactions with other biomolecules. Miniaturized microfluidic and nanofluidic systems with fluorescence spectroscopy enable worldwide researchers to perform nucleic acids and proteins extractions and single molecule analysis from the trace amount of biological samples. In the present chapter we mostly

Progress in Molecular Biology and Translational Science, Volume 186
ISSN 1877-1173
https://doi.org/10.1016/bs.pmbts.2021.07.029

highlighted over one decade applications of microfluidic and nanofluidic systems for single cell micro ribonucleic acid (miRNA) isolation and detection, deoxyribonucleic acid (DNA) mapping, DNA barcoding, identification of epigenetic mark on single DNA molecule, DNA-protein interactions study, protein sensing, protein sequencing, protein binding kinetics and many other applications. We also presented the recently reported microfluidic platform for the preparation of reproducible unisize aggregation induced emission (AIE) active nanomaterials and their biological applications.

1. Introduction

Microfluidic and nanofluidic systems are miniaturized engineered devices where non-turbulent fluid flow is manipulated in highly ordered channels.[1–3] From last two decades these systems gained enormous recognition in the diverse scientific fields including energy, defense, diagnosis, chemo/bio-sensing & biomedical applications to compete and even surpass the conventional macro scale analytical techniques in terms of enhanced analysis sensitivity and selectivity, minimized analytical sample volume, shorten analysis time and lowered analysis price.[4] Operational length scale for microfluidic devices is ranging from 1 to 100 μm, whereas for nanofluidic devices it is ranging from 1 to 100 nm. Owing to such small operational length microfluidic & nanofludic technologies enable worldwide researchers to perform scientific research at single molecule/cell levels.[5,6] Recent literature revealed that technological advancement in the fabrication of micro/nano scale device permit researchers to fabricate various well-defined/ordered nanochannels/pores network whose dimensions are competing with the dimension of biological cell objects, for example DNA strands.[7] These devices under the influence of applied electric fields at low voltages perform several biological applications such as separation and concentration of biological molecules, detection and sequencing of single DNA biomolecules, protein binding kinetic study and many mores. Such devices have enormous potential in a wide range of biomedical applications, which makes them a prominent tool for bioengineers, biophysicists, biochemists & medical doctors. Mostly the sensing readout mechanism used in these miniaturized engineered devices is primarily based on mechanical, electrochemical, and optical methods.[8–10]

In the last two decades optical based sensing technology is emerged as a most promising method over the conventional technologies owing to its robustness, highly sensitive and easy operational.[11,12] Various optical methods including absorption, fluorescence (emission of light) and chemiluminescence were reported for many different applications in biology, chemistry, environmental and medical fields. Among these optical methods fluorescence based biomolecule detection in solution phase is employed enormously for distinct biological & biomedical assays because fluorescence methods have several advantages such as easy to operate, high sensitivity and results reproducibility.[13] Fluorescence is a photophysical phenomenon in which fluorophore molecules such as small π-conjugated molecule, protein, carbon dot and quantum dot emits longer wavelength of light upon the excitation by specific wavelength of light. In the field of bio-sensing and biomedical these fluorophore are extensively utilized for the labeling of biomolecules such as lipids, nucleic acids, peptides and proteins. Brandner and coworkers recent review report on microfluidic and nanofluidic systems revealed that full integration of fluorescence based sensing techniques within lab-on-a-chip devices is a challenging task because of complexity.[12] In the past few years important technological developments were reported in the field of integration of fluorescence sensing techniques within microfluidic and nanofluidic lab-on-a-chip devices.[14–18] This Most demanding development of fluorescence based microfluidic and nanofluidic lab-on-a-chip devices provides an opportunity to worldwide researchers to employ this technology for diverse biomedical applications. In this chapter we mainly discussed the published reports over last one decade on bio-macromolecules such as nucleic acid and protein analysis by employing microfluidic and nanofluidic devices based on fluorescence technique.

2. Microfluidic systems for biomedical applications

Miniaturized microfluidic devices with the assistance of fluorescence technology has been greatly utilizing from the last two decade for several biomedical applications such as nucleic acids analysis & sensing, proteins analysis, imaging & concentration detection study, gene transfection study, single cell detections & analysis.

2.1 Nucleic acid sensing and analysis

Nucleic acids made by various nucleotides are vital for all diverse life present on our planet.[19,20] Nucleic acids structure is basically constructed by pairs of organic bases with distinct sequence and lengths. They are generally classified into two types such as ribonucleic acid (RNA) and deoxyribonucleic acid (DNA). RNA utilized for the biosynthesis of protein whereas DNA utilized for the genetic information transfer, storage and replication.[21–23]

In the modern scientific days the knowledge of nucleobases sequence in nucleic acid strands become crucial for various scientific fields such as molecular biology, biotechnology, virology, medicine, forensics and many more.[24,25] Vast advancement in sequencing technologies allow researchers to get valuable insights into complex biological systems including cancer genomics to distinct microbial colonies.[26] Modern sequencing technologies are providing an opportunity to characterize nucleic acids in the single cells for the study of genomics, transcriptomics, and epigenomics. Owing to very low abundance of nucleic acids in living biological cells the specific recognition and signal amplification are becoming challenging tasks. For the advancement of detection and sensitivity of nucleic acids in single cells, distinct DNA amplification methods reported including rolling circle amplification (RCA), hybridization chain reaction (HCR), strand displacement amplification (SDA), polymerase chain reaction (PCR) and many more.[27] Li and coworkers reported a quantitative assay for miRNA detection in a single cell by employing ligation-based droplet digital polymerase chain reaction (dd-PCR) method (Fig. 1).[28] Single cell lysis releases the desire miRNAs were binding with partially complementary oligonucleotide

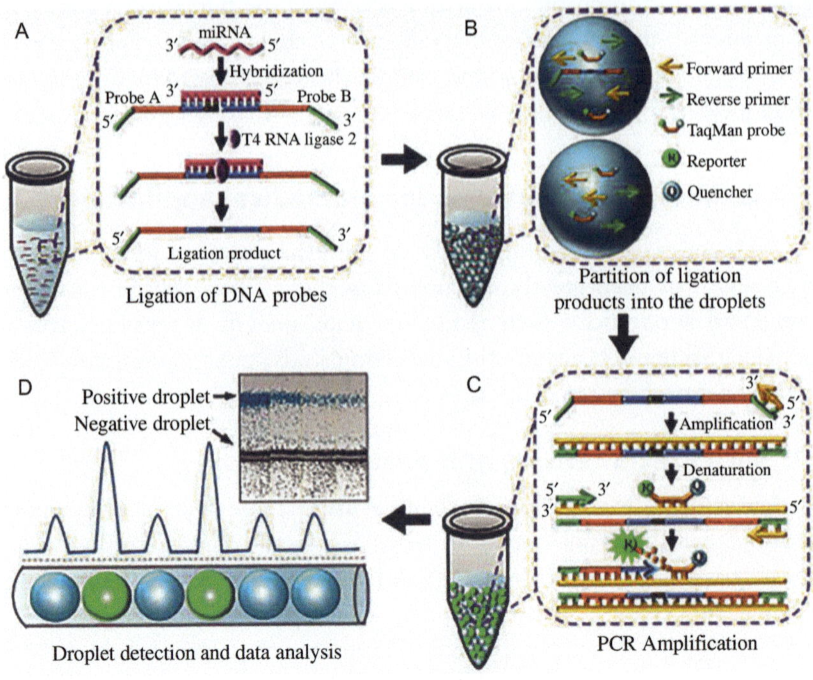

Fig. 1 Detection assay of miRNA based on ddPCR and fluorescence techniques.

probe A and B. Which further catalyze by T4 RNA ligase 2. The resultant ligation products and the PCR mixture (universal primers, TaqMan probes, dNTPs, DNA polymerase and PCR buffer) were partitioned into water-in-oil droplets. The droplets contains ligation product & PCR mixture illustrate PCR amplification of ligation product and cleavage of TaqMan probes owing to 5′-3′ exonuclease activity of DNA polymerase and exhibits fluorescence signals. Whereas the droplets without the ligation products were became fluorescence inactive. The miRNAs detection in single cell was achieved by the counting number of fluorescence-positive droplets. Similarly based on dd-PCR method Zhao and coworkers was recently reported quantification of miRNA related to lungs cancer.[29]

In the early year of last decade an elegant work of target RNA detection based on reverse transcription polymerase chain reaction (RT-PCR) has been reported by Yang and his coworkers.[30] Fig. 2 illustrate the operating

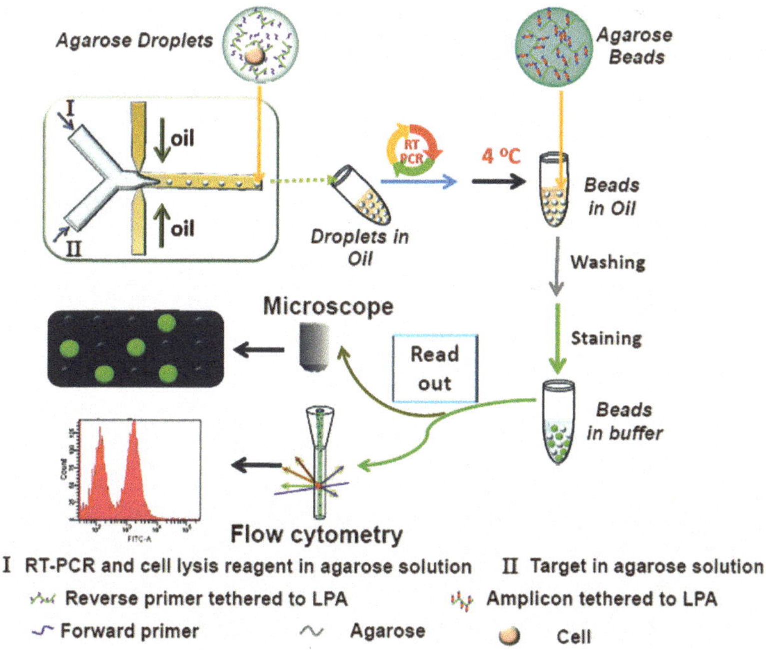

Fig. 2 Microfluidic platform operating overview. Monodispersed droplets synthesized by mixing aqueous solution-I (2% agarose solution contain RTPCR and cell lysis reagents) and solution-II (2% agarose solution contain RNA or cells). The obtained droplets undergo sequential process like cell lysis, reverse transcription reaction, PCR operation, and agarose beads generation via cooling. The resultant beads stained with SYBR Green by removing oil, and put for further analysis by employing flow cytometry/ fluorescence microscopy.

principle utilized in the presented research work. At first unisize well dispersed picoliter agarose droplets with encapsulated RT-PCR reagent, cell lysis buffer and template RNA or cell were prepared by employing microfluidic droplet generator. In the resultant droplets sequence of events were takes place such as cell lysis followed by reverse transcription reaction and finally PCR. Later the droplets were cooled at 4 °C and generate agarose beads. Further the agarose beads were washed and stained with small molecule SYBR Green cyanine dye for analysis via fluorescence microscopy or flow cytometry. In this study they utilized 5′-acrydite primer to graft it to the polymer matrix for trapping of RT-PCR products in agarose. Successful single-molecule RT-PCR in agarose matrix revealed gene expression level of EpCAM at the single-cell level of various cancer cells. Microfluidic device also served as an efficient platform for the enhancement of very low abundant circulating tumor cells (CTCs) from blood samples and further the gene expression of obtained CTCs analyzed. In this field of exciting research Haber and coworkers performed considerable amount work.[31] They have utilized microfluidic devices for successful sorting of pancreatic CTCs from bold sample and further performed single-cell RNA sequencing. The obtained results illustrate three different populations & revealed various paths in the metastatic cascade.

In 2010, Hancen and coworkers reported an elegant example of fully integrated microfluidic platform in which they performed all the steps such as single cell processing and capturing, lysis of captured single cell, reverse transcription and further quantitative PCR (qPCR).[32] Basically the reported microfluidic system split into four units such as (i) reagent and cell loaded unit, (ii) Single cell capture and lysis unit, (iii) Reverse transcription (RT) unit of 10 nL, and (iv) 50 nL PCR unit for the processing of amplification and detection (Fig. 3). This well designed integrated microfluidic platform utilized for the detection of miRNA and mRNA in the single embryonic stem cells by employing TaqMan probe as a fluorescence indicator. In the year 2017 an integrated dd-PCR technology with fluoresce color-coded Luminex® beads was reported by Joensson and coworkers for the detection of several pathogen DNA biomarkers.[33] In this study they developed an assay for the detection DNA of three pathogens infecting to poultry such as *avian influenza*, infectious *laryngotracheitis* virus and *Campylobacter jejuni*. The amplification of target DNA in mono disperse picoliter droplets with fluorescent labeled primers was achieved by PCR technology.

RT-PCR is a vastly utilized technique allows researcher transcriptional profiling as well as sequencing analysis on bulk populations of cells. Many

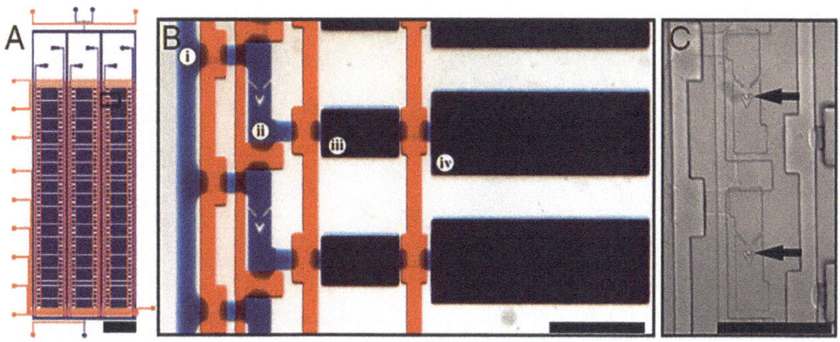

Fig. 3 Microfluidic platform for analysis of single-cell. (A) Microfluidic cheap. Rectangular box zoomed in B. (B) Optical image of array unit. Every unit contains (i) reagent inlet, (ii) cell capture and trap chamber, (iii) RT chamber, and (iv) PCR chamber. (C) Optical image of two cell capture chambers with trapped single cells.

researchers reported that the cell lysate in microdroplets has inhibition effect on RT-PCR method. To overcome this barrier an elegant strategy reported by Abate and his coworkers, in which they studied the merging of the micro drops into larger drop under the effect of applied electric field and achieve the dilution of cell lysate concentration.[34] The diluted droplet subsequently undergoes PCR addition, collection, theromocycling and finally achieves TaqMan RT-PCR study. By employing the developed protocol they illustrated specific cells identification from a mixed human cell population via the detection of multiplex target genes (Fig. 4A).

Mathies and his coworkers in 2011 reported a robust microfluidic agarose-droplet platform for single cell based detecting and sequencing of multiple genetic targets. It was illustrates that firstly the microfluidic encapsulation of single cells with primer-functionalized beads in agarose gel droplets.[35] Further the DNA of encapsulated single cells was released in the agarose gel droplets upon lysis by sodium dodecyl sulfate (SDS) and proteinase K. Later the resultant genomic DNA containing agarose gel droplets were incubated with PCR reagents containing fluorescent primers, followed re-emulsification with dispersing oil via mechanical agitation. Subsequently multiplex PCR amplification performed in droplets. The amplicon-attached beads were recovered via rupturing the emulsion and melting the agarose. The fluorescent beads were quickly quantified by flow cytometry. Epifluorescence micrograph of the RL cell revealed that primer-labeled beads become fluorescent if chromosomal translocation t(14;18) + of RL cell is present in the same droplet, but otherwise remain dark (Fig. 4B).

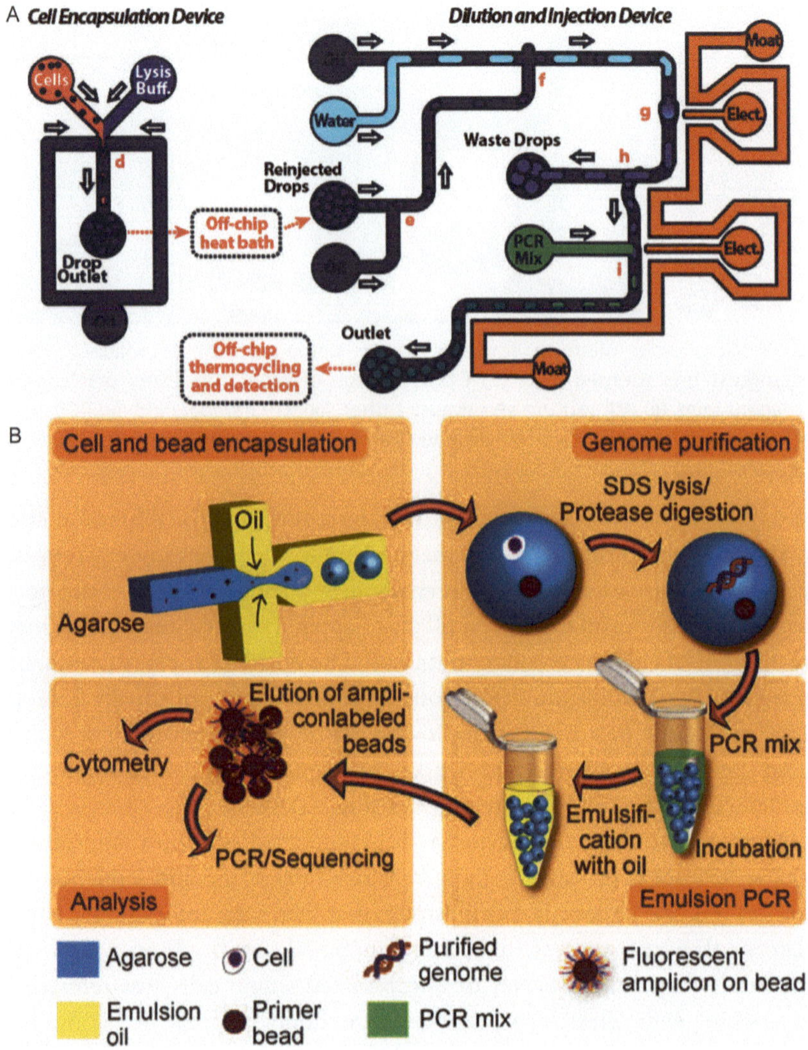

Fig. 4 Microfluidic platform for the preparation of (A) larger drops from micro droplets, (B) agarose-emulsion droplets and further used for the genomic analysis.

Ryckelynck and his coworkers in 2009 presented a microfluidic system based on hyper branched rolling circle amplification (HRCA) technique for high-throughput single cell DNA molecules quantification.[36] Microfluidic system firstly produced large number of single cells packed microdroplets. This further followed lysis of single cells, HRCA mix and intercalate of fluorophore. Subsequently the droplets get collected to complete off-chip

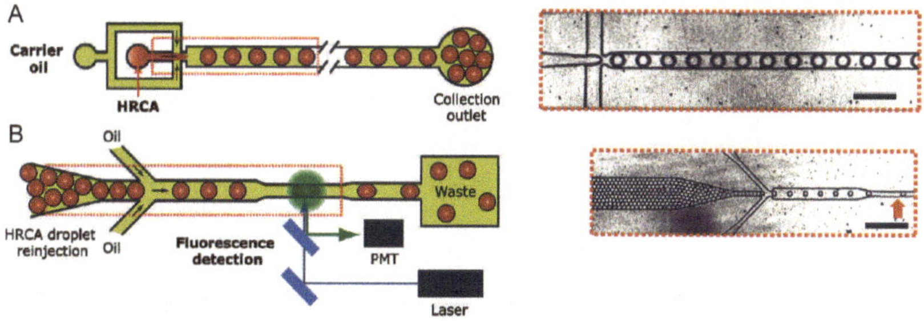

Fig. 5 Microfluidic platform for the preparation of (A) HRCA droplets, (B) HRCA analysis channels.

HRCA reaction. Finally fluoresce detection was done by the reinjection of resultant droplets into the analysis chip (Fig. 5). Authors further employed the developed protocol for β-galactosidase activity detection. Quantification of β-galactosidase activity was achieved by in vitro transcription/translation (IVT) reaction.

2.2 Protein sensing and analysis

Proteins are as important as other biomolecules such as nucleic acids for all diverse life present on our planet. Proteins are macromolecules basically made up of 20 naturally occurring amino acids. Roughly 1,00,000 distinct proteins are present in the cells, whereas their contents are greatly varied. They play a vital role in almost every operation taking place in the living organisms. Intracellular proteins including receptor proteins, transporter proteins, channel proteins, enzyme proteins and structural proteins play significant roles in cell structure stabilization, signal transduction and control cell functions.[37] For modulating biological functions of cells and its practical applications the recognition and quantification of proteins are important issues. Usually several natural proteins expressed in very low quantities such as some structural proteins are expressed up to about 10 molecules, whereas some receptor proteins expressed less than 100 molecules.[38] Expression of proteins in low concentration makes its analysis challenging in trace amount. For single cell protein sensing and analysis the designing of protein specified probes and signals amplification protocol is quite interesting and demanding task.[39] The protein analysis and recognition were achieved by several methods such as labeling of small fluorophore molecules, fluorescent protein binding to target proteins via gene transfection technique, target proteins linking to specificity labeled-antibodies using immunofluorescence protocol.[40–44] Current literature study of single-cell protein analysis on microfluidic device exhibits that immunofluorescence was most commonly used protocol. In this technique specific antibodies chemically linked to fluorophore, the resultant antibodies further bind with cellular antigens and exhibit information of target proteins qualitatively and quantitatively.[45] Preckel and coworkers reported a fluorescence cytometric analysis assays based on microfluidic platform for apoptosis in human primary cells and protein expression.[46] This assay illustrate systematic working principle, at first single cells along with antibodies targeting fluorescence labeled surface proteins or fluorescence labeled cytosolic proteins were moved inside the array of microfluidic channels. Next step was the evaluation of fluorescence intensity of labeled single cells with the help of integrated fluorescence laser and fluorescence detection unit. Further the quantification of low-copy number targeted proteins by employing calibration curves derived by calibration beads. Huang and his coworkers published interesting research work on quantification of low copy number proteins in a single cell via single molecule fluorescence counting method.[47] They have outlined a high efficiency microfluidic device in which following events manipulated such as (i) single cell capture, (ii) lysis of captured cell, (iii) the released protein labeling with fluorophore tagged antibodies, (iv) electrophoresis based separation of protein-antibody complexes and free antibodies, and (v) finally based on single-molecule fluorescence counting technique the recognition of low–copy number proteins. In this work they have performed microfluidic device based following two experiments (i) quantification of β_2 adrenergic receptors expression in insect cells, and (ii) Analyzed phycobiliprotein contents in single cyanobacteria cells.

Worldwide researchers utilizing droplet-based microfluidic flow-cytometry system for the quantification of various specific proteins secreted by single cells.[48–51] In the droplets forming event the ultimate goal was single-cell capture and isolation with simultaneously encapsulation of distinct reagents basically needed for single-cell analysis. The resultant droplets would perform as an independent micro-reactor for the immunoreactions such as expedite the reaction rate and enhance the signals. Huck and his coworkers in 2013 reported the analysis of multiple cytokine secretion of single T-cells by using droplet based microfluidic system.[51] As illustrated in the Fig. 6, at first step the single T-cells along with functionalized capture beads were incorporated in agarose droplets, in which the functionalized capture beads can bind to secreted cytokines (L-2, IFN-γ, TNF-α) by single T-cells. In the following steps, the droplets under low temperature were gelled out and washed for breaking the emulsion which further allows

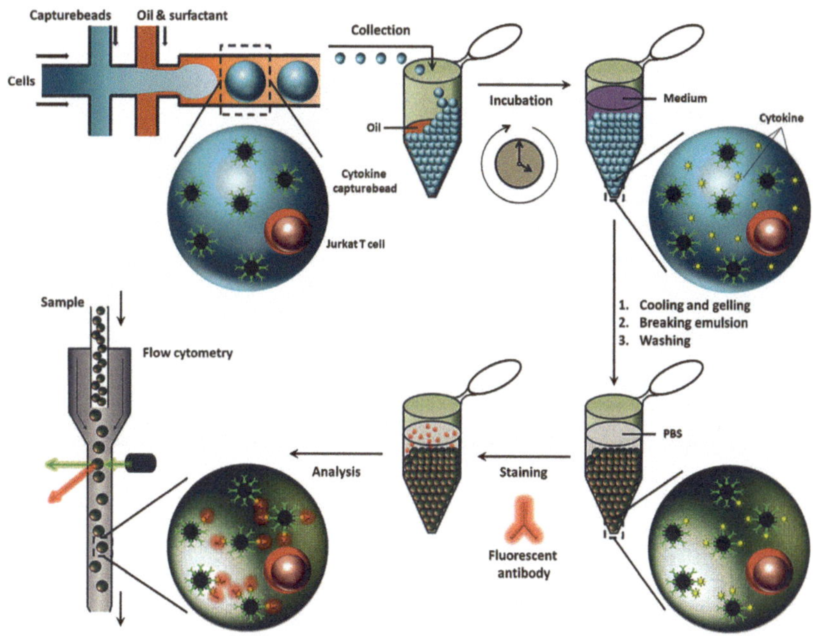

Fig. 6 Microfluidic platform for the preparation of single-cell encapsulated droplets and identification of secreted proteins such as cytokines.

entrancing of multicolor fluorophore-labeled antibodies intending cytokines coupled on the beads. At the end of experiment the emission intensities of the signals in the beads were quantified by employing flow-cytometry. 7415 single cells experimental analysis data revealed that (i) eight distinct cellular subpopulations were present, and (ii) 85% of single T-cells were secreted one or more cytokines.

Heyman and his coworkers presented single cells secreted proteins quantification protocol having the combination of capture beads and droplet based microfluidic platform.[52] Fig. 7 demonstrate that single cells, antibody capture beads and fluorophore labeled antibodies were encapsulated in microfluidic droplets. In the resultant droplets forms sandwich assay, at first the single cells secreted proteins were binds to the exterior of antibody capture beads, afterwards the fluorophore labeled antibodies were binding to secreted proteins. Finally the droplets containing fluorescent beads were sorted out from the emulsion by employing droplet sorter bearing fluorescence-activator and detector.

Microfluidic platforms also successfully employed for the detection of low abundant single cells membrane protein.[53,54] The cells in droplets were

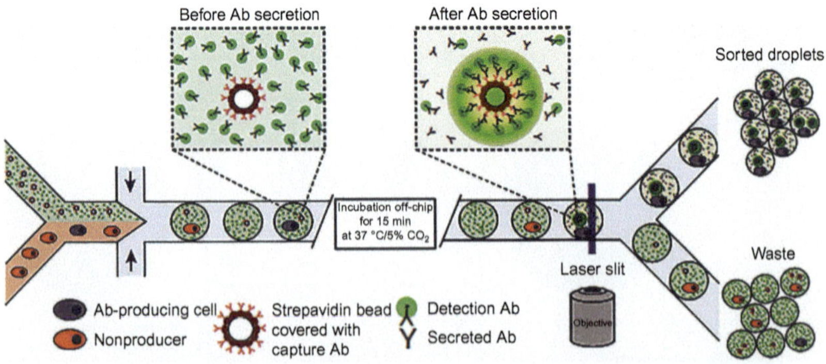

Fig. 7 Microfluidic platform for the preparation of micro droplets containing mixture of single cells, streptavidin beads with capture antibodies, and green fluorescence labeled antibodies. The incubated droplets reinjection and detection of secreted proteins via fluorescence activated droplet sorter.

utilized as the solid phase of the immunoreaction owing to non-essential cell lysis for cell membrane protein detection. The resultant protein antibody complex was bound directly on the surface of cells. Yarmuch and his coworkers reported microfluidic nano-liter system for ultrasensitive detection of less abundant cell membrane proteins via isothermal rolling circle amplification (RCA) technique (Fig. 8).[53] In this study PC3 human prostate cancer cell using as solid phase for detecting epithelial cell adhesion molecule (EpCAM). Primarily in the nano-liter microfluidic reaction droplets following essential items were encapsulated including the PC3 prostate cancer cells, the biotinylated DNA primers, the biotinylated-antibodies, and the RCA reagents. Initially the biotinylated-antibodies were complexes with cell membrane proteins which further linked with DNA primer by biotin-avidin bridging. Later RCA was activated and yielded one-dimensional long DNA strand. Cell membrane proteins (EpCAM) information conceivable reflects via fluorophore-labeled nucleotides generated on the DNA strand. This protocol employed immunoreaction for discrete identification of intended proteins, and the design of RCA might considerably improve the test sensitivity, which especially suitable for quantification of less abundant membrane proteins in single cells.

2.3 Microfluidic systems to generate aggregation induced emission (AIE) droplets/dots for biological applications

Aggregation induced emission (AIE) a photophysical phenomenon of organic/inorganic compounds is well explored in the distinct fields such

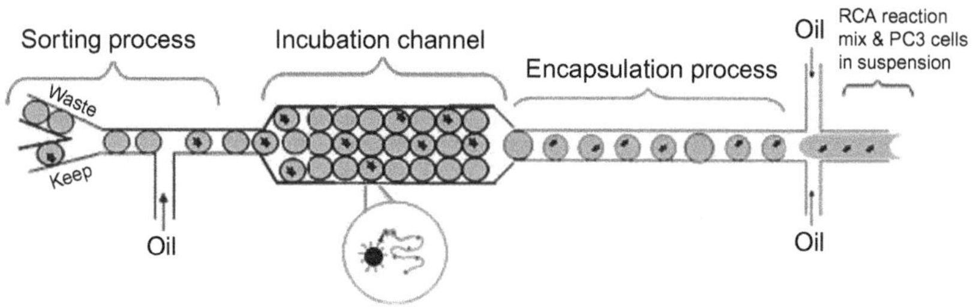

Fig. 8 Microfluidic platform for the preparation of nano-RCA droplets.

as diagnosis, biomedical, organic electronics and many mores.[55,56] AIE molecules for example tetraphenylethylene (TPE) and hexaphenylsilole (HPS) are special class of compounds which under excitation with specific wavelength of light emits very weak light or become non-emissive in the solution states where as in the aggregate states they emits high intensity light. However, conventional fluorophore molecules such as naphthalene, phenanthrene, pyrene, perylene, fluorescein, cyanine and many more featured with planar large conjugated structures exhibits good fluorescence emission in dilute solution. Whereas the conventional fluorophore molecules in the aggregate or solid state exhibit strong π-π stacking interactions and become non-emissive via non-radiative decay of excited-state energy. This phenomenon popularly referred to as aggregation-caused quenching (ACQ). ACQ is the main obstacle for many real-world applications of conventional fluorophore molecules in the solid/aggregate state. In the last two decades several AIE active molecules emerged as prominent choice for many real-world applications and tackle ACQ issue.[55] AIE molecules emits light strongly in its nano-aggregates form and exhibits promising biomedical applications including bioimaging, therapeutic functions such as formation of reactive oxygen species (ROS), photodynamic therapy (PDT) and many more.[57,58] AIE nanoparticles namingly AIE dots are manly fabricated by sonication mixing methods such as encapsulation of AIE molecules in matrix including bovine serum albumin (BSA), poly(DL-lactide-*co*-glycolide) (PLGA), amphiphile copolymers 1,2-distearoyl-sn-glycero-3-phosphoethanolamine-N-[methoxy(poly(ethylene glycol))] (DSPE-PEG) and may more.[59–61] The challenges in the fabrication of AIE dots are precisely control size and shape, and reproducibility in the large scale fabrication. To tackle these issues of AIE dots fabrication via automated microfluidic mixing platform is an elegant choice.[56,62] The development of automated millifluidic/microfluidic mixing platform illustrated in Fig. 9 produce

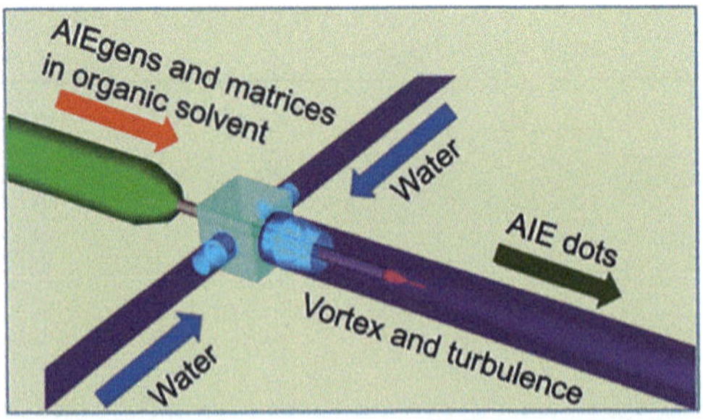

Fig. 9 Millifludic/microfluidic platform for the preparation of uniform AIE dots.

continues uniform AIE dots via jet mixing method. In the jet mixing method firstly the organic solvent containing AIE active material and matrix from one channel and water from another channel mixed in a small tube and form vortex and turbulence to induce AIE dot synthesis. The desire AIE dots with precise shape and size would achieve via modulating Reynolds numbers, flow rates, and solvent ratios.[56,63]

Till date very little research work on the fabrication of AIE dots via microfluidic mixing system for biomedical applications has been reported. Very recently Jiang X. and coworkers reported synthesis of sub-10 nm AIE quantum dots by employing microfluidic mixing system having three inlets, one outlet, one straight mixing channel and a double spiral mixing channel.[64] In this study four different AIE active molecules referred as AIEgens namingly 4,7-bis[4-(1,2,2-triphenylvinyl)phenyl]benzo-2,1,3-thiadiazole (BT), 4,7-bis(4- (1,2,2-triphenylvinyl)phenyl)-spiro[benzo[*d*] imidazole-2,1′-cyclohexane] (BI), 2,3-bis(4-(phenyl(4-(1,2,2-triphenylvinyl) phenyl)amino)phenyl)fumaronitrile (FN), and 4,7-(bis4-(4-octylthiophen-2-yl)-*N*,*N*-diphenylaniline)benzo[1,2-c:4,5-c']bis([1,2,5]thiadiazole) (TTB) were prepared and utilized for the synthesis of small sized organic AIE quantum dots. Fig. 10 illustrate that firstly the organic solvent of AIEgen (BT, BI, FN and TTB) and amphipathic 1,2-distearoyl-sn-glycero-3-phosphoethanolamine-N-[methoxy(polyethylene glycol)-2000](DSPE-PEG2k) matrix injected from one inlet and water injected from other two inlets, subsequently mixed in straight mixing channel and a double spiral mixing channel. Rapid and efficient mixing in the microfluidic channels forces the encapsulation of small hydrophobic AIEgen in amphipathic DSPE-PEG2k matrix resulted ~10 nm sized AIE quantum dots (BT-QDs, BI-QDs, FN-QDs and TTB-QDs) with good reproducibility. The synthesized AIE quantum dots BT-QDs, BI-QDs, FN-QDs and

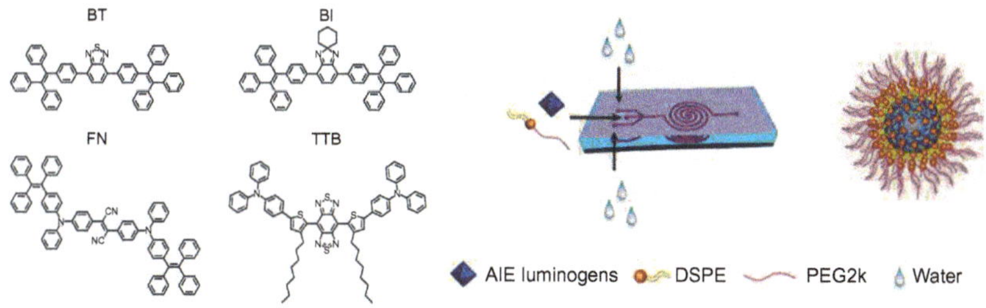

Fig. 10 Chemical structures of AIE molecules used in experiment and microfluidic platform for the preparation of AIE quantum dots.

TTB-QDs exhibits emissions ranging from green to the second near-infrared (NIR-II). They efficiently undergo cellular uptake process and shows good imaging properties. The resultant quantum dots of AIEgens with excellent NIR-II emission illustrate efficient tumor detection and evasion from the liver. Comparative to conventionally synthesized AIE dots the AIE quantum dots synthesized by microfluidic mixing method showed excellent in vitro and in vivo imaging performance.

Li and coworkers in the few years back developed an elegant fluoresce based sensing platform in which AIEgen-doped poly(ionic liquid) (PIL) photonic spheres were used for the sensing biomolecules such as amino acids.[65] Fig. 11 illustrate systematic procedures utilized for the synthesis of AIEgen-doped poly(ionic liquid) (PIL) photonic spheres. At first droplet-based microfluidics system was employed for the preparation of mono-disperse silica nanoparticles of ~170 nm diameter. The resultant silica nanoparticles further self-assembled into colloidal crystal spheres. The obtained colloidal silica crystal spheres were infiltrated with the solutions of ionic liquid (IL) monomer, IL cross-linker, AIEgen, and photo-initiator. This subsequently underwent to photo-initiated polymerization and generates AIEgen-doped PIL SiO_2 photonic spheres. The obtained AIEgen-doped PIL SiO_2 photonic spheres on treated with hydrofluoric acid (HF) remove the SiO_2, producing the AIE-doped PIL photonic spheres. The resultant AIE-doped PIL spheres with macro-pores multi-channel sensing structure showed excellent fluorescence and photonic properties. The AIE-doped PIL single sphere was able to distinguish wide group of analytes with excellent selectivity and sensitivity. These AIE-doped PIL photonic spheres exhibits successful discriminate sensing of naturally occurring 20 amino acids in the real urine sample of human along with their 26 unprecedented complexes.

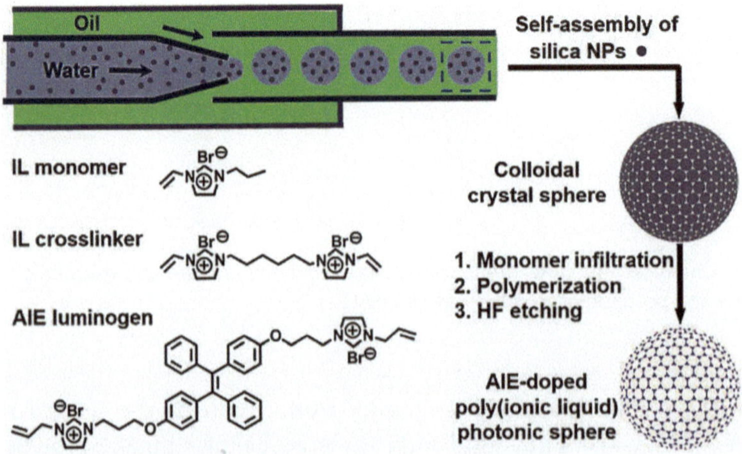

Fig. 11 Microfluidic platform for the preparation of uniform silica nano particles. Further silica nano particles transformed into colloidal particles which subsequently used as template for AIE-doped photonic sphere formation via polymerization.

3. Nanofluidic systems for biomedical applications

Nanofluidic technology emerged not only as an extension/advancement of the available microfluidic technologies, but also providing an opportunity to discover a new range of phenomena that generally do not present at the microscale level. Recent nano-technological advancement provides an opportunity to fabricate device with nano-channel networks having channel diameters ranging from 1 to 100 nm.[7,66] Extensive literature examinations reviled that fluorescence assisted miniaturized nanofluidic devices have been greatly utilizing for various biomedical applications including biomolecules separation, biomolecules sieving, biomolecules pre-concentration, single biomolecule analysis, bio-sensing and many mores.[67–70] In this section we mostly discussed & presented recently developed fluorescence assisted nanofluidic devices for biomedical applications.

3.1 Nucleic acid sequencing and sensing

Nucleic acids, i.e., RNA & DNA research is the centre of biomedical investigations. Specially DNA purification, size separation, sequencing, and single DNA molecule visualizations are the key research areas. Conventional bioanalytical techniques including gel-electrophoresis, PCR, RT-PCR, southern blot analysis, and DNA fingerprinting enables research for DNA analysis, however these techniques has its own advantages and disadvantages. Fluorescent based PCR method is appeared as an elegant approach for nucleic acid analysis, in which the RNA or DNA-based probe having fluorescent tag at one end and quencher at another end. The close proximity of quencher and fluorescent units breaks down during the process of PCR reaction and exhibits fluorescence turn-ons. Prompt literature study revealed that the combination of fluorescent technique and nanofluidic nano-channel platform providing an opportunity to researchers for significant analysis of nucleic acids including single DNA molecules stretching and sequencing.[66] Study of single DNA molecules stretching and sequencing via nanofluidic platform provides structural insights such as genome structures and their variations at a chromosome level along base pair level accuracy. Basically two methods have been reported for single DNA molecule stretching and visualizing via nanofluidic platform. In the first method spontaneous stretching of DNA molecules achieved while passing through nanofludic channel of decreasing dimension based on gradual increasing of flow velocity due to funneling effect.[71,72] Whereas in the second method the stretching of DNA molecules achieved via immobilization through nanofludic channel due to entropic confinement.[73,74] Examination of literatures demonstrate that stretching of single DNA molecules in nano channels open up gateway for identifying specific base pare (AT or GC) sequencing/mapping. Austin and his coworkers reported DNA sequencing by employing restriction reaction approach in the nanochannel.[75] Over the last decade sequencing of DNA were successively achieved via nanofludic nanochannel platform by employing optical mapping technology. This was mainly performed by sequence specific labeling using two labeling strategies such as enzymatic based labeling and affinity based labeling. Both strategies reveal DNA sequence information via single DNA molecule stretching and fluorescent read-out. In the enzymatic labeling strategy the fluorescent marks covalently binds to DNA molecule at specific sites. Whereas, in the affinity based labeling strategy fluorophores non-covalently interacted with DNA molecule and at lower denaturation temperature they dissociate from AT-rich regions ad remains at GC-rich regions.

Schwartz and his coworkers was firstly reported the optical method single DNA molecule mapping in nanofluidic nanoslit platform via nick-translation labeling DNA barcoding (Fig. 12).[76] Firstly off-chip homogeneous DNA labeling were performed and followed full length stretching in large-scale nanoslits. In this system sequence specific endonucleases

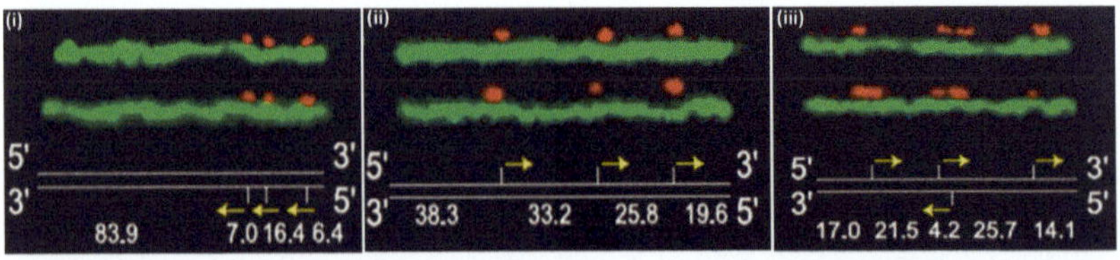

Fig. 12 DNA molecules labeled by nicking and barcode calculated (in kilobasepairs) from DNA sequence for three bacterial artificial chromosomes (i) BAC79 (113.7 kbp), (ii) BAC150 (116.8 kbp) and BAC614 (82.5 kb).

enzyme firstly nicked one of the DNA strands and followed incorporation of nucleotides labeled with another fluorophore at the nicking site. The position of nicking sites of DNA molecule was readout by the experiment know as frequency resonance energy transfer (FRET). FRET takes place between the DNA labeling donor dye and the nucleotides labeling acceptor dye at nicking sites. Barcodes are calculated from three bacterial artificial chromosomes such as BAC79 (113.7 kbp), BAC150 (116.8 kbp), and BAC614 (82.5 kb).

Xiao and coworkers reported the extension of nick labeling strategy for the short DNA sequencing in nanochannel (Fig. 13).[73] In this study it illustrate that the labeling strategy could recognize the site of specific sequences simultaneously DNA molecule starching. At first the formation of single strand DNA flaps nicked on dsDNA by using strand displacement polymerase enzyme. The obtained single strand DNA flaps were further hybridized with sequence specific fluorophore labeled probes. The resultant DNA molecules imaging in nanofludic nanochannel platform provides quantitative recognition of the probes sties. This would allow researcher to control the probe sequence length and frequency of hybridization sites on the DNA. This methodology further extended to different fluorophore labeling on nicking sites to resolve the complex structural variations in the DNA molecule.

To label a longer sequence and recognize a specific gene using regular nicking enzyme is challenging task due to the smaller nicking site (4–7 bp). Recently Xiao and coworkers illustrate the application of CRISPR/Cas9 mutant as similar as the nicking enzymes.[77] This strategy was able to recognize 23 bp stretch, which was large enough to recognize target specific genes. Flyvbjerg and coworkers reported denaturation mapping of single DNA molecule using nanofluidic nanochannel platform.[78] In this strategy, the partial desaturations of YOYO-labeled DNA with combine effect of local heating and formamide generates sequence-dependent barcode. The generated barcode corresponds with a series of emission intensity off and

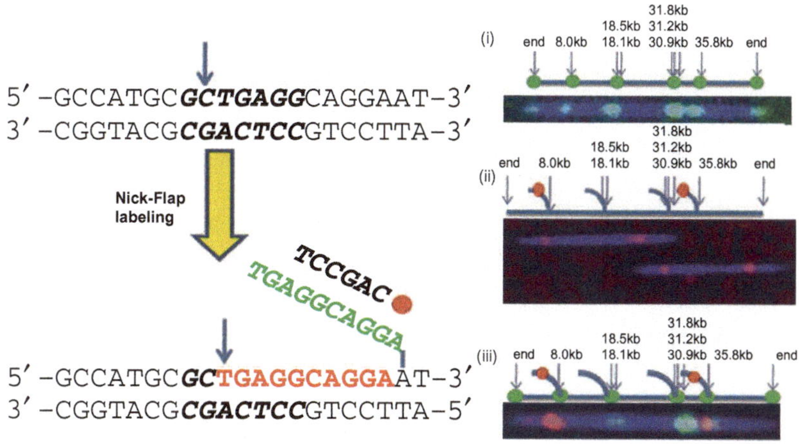

Fig. 13 Recognition sequence of nicking endonuclease and labeled by nick-flap.

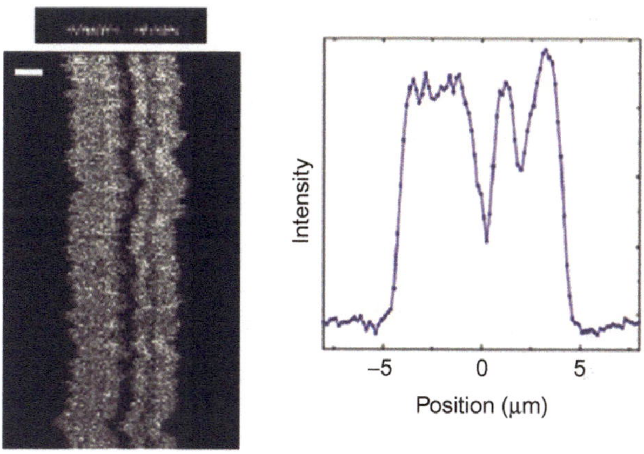

Fig. 14 Emission pattern revealed by partial melting of λ-DNA.

on along the DNA molecule (Fig. 14). It was noted that in the partial denaturation (partly melted) of DNA the labeled fluorophore mainly dissociate from AT-rich regions and turned as dark whereas GC-rich regions remain bright. The developed barcoding system showed many applications such as optical genome mapping, molecules location through the genome, base pare sequencing, sequence specific proteins binding and many mores.[66,70]

3.2 Protein binding kinetics and DNA interaction study

Nanofludic nanochannel platforms and fluorescence microscopy have been extensively employed for protein study such as purification, concentration detection, sequencing, fluorophore labeling, DNA-protein interaction/binding and many more. To tackle cellular permeability issue of microscale protein hydrogels in the current year Knowles and coworkers reported the synthesis of attoliter protein nano gels by using droplet nanofluidic platform.[79] In the study they first prepared nanosized waterin-oil emulsions. The formed nanoemulsions further employed as templates to synthesize supramolecular protein nanogels which stabilized by fibrils proteins. The resultant protein nanogels exhibits biocompatibility and used as intracellular delivery system. In the recent year Xia and coworkers employed micro/nanofluidic platform bearing intergraded nanochannel for the detection of ultrasensitive protein concentration using emission quenching technique, i.e., FRET.[80] Fig. 15A illustrate that at first in the reservoir 1 the bovine serum albumin (BSA) sample protein is adsorbed on gold nanoparticles (AuNPs). Subsequently the fixed quantity of fluorescein-labeled dog serum albumin (FITC – DSA) protein is added to the mixture of BSA and AuNPs. Few of FITC – DSA are absorbed on residual active surface of AuNPs and

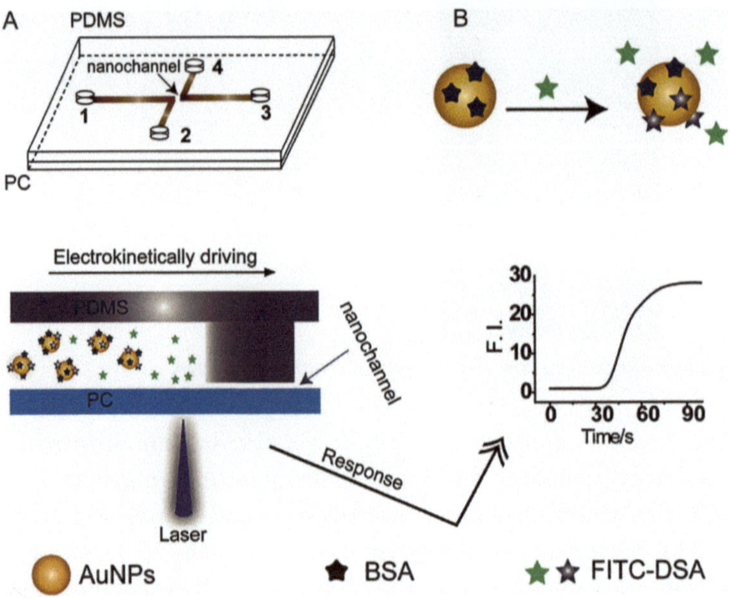

Fig. 15 Micro/nanofluidic platform for protein detection using the FRET method. (A) Micro/nanofluidic device in which nanochannel is presented between two V-shaped micro channels; at bottom the sides view of device where the enrichment and detection of protein BSA in front of the nanochannel. (B) FRET detection scheme for the protein detection; at bottom the corresponding fluorescence signal of the free FITC-DSA molecules in solution.

rest FITC – DSA present in a free state. The adsorbed FITC-DSA on BSA-AuNPs is inducing FRET due to the close proximity of fluorescein dye with AuNPs surface and develops FRET based detection system (Fig. 15B). Trace concentration of BSA leads large residual active surface of AuNPs for adsorption of FITC – DSA and remains less FITC – DSA in the free states. In front of nanochannel in micro/nanofluidic device free states FITC – DSA protein is preconcentrated and generate efficient emission signal. Using resultant emission signal one can detect the concentration of sample BSA protein. Xia and coworkers successfully detected trace concentration of sample BSA protein is 0.1 ng/mL.

Few years early Leichle and coworkers presented protein kinetic study by employing fluorescence based nanofluidic system.[81] The real-time protein binding kinetic is studied by using fluorescence labeled protein binding. In single experiment one can measure association and dissociation kinetics of protein just by recording emission and employing reversed buffer flow. For protein binding kinetic study two pairs of protein-receptor are selected based on their different affinities for example high affinity pair streptavidin/biotin and medium affinity pair mouse IgG/anti-mouse IgG selected. Fig. 16A shows the experimental setup of nanofludic device with fluorescence microscope used in the study. At first the probe biomolecules such as thiol-OEG-biotin conjugates immobilized on gold strip and then encapsulated at bottom of nanofluidic nanochannel chip. Under pressure-driven flow the fluorophore labeled protein analytes such as fluorophore-streptavidin conjugates are pass through the nanochannels. Emissions signal related with streptavidin adsorption on biotin and desorption from biotin

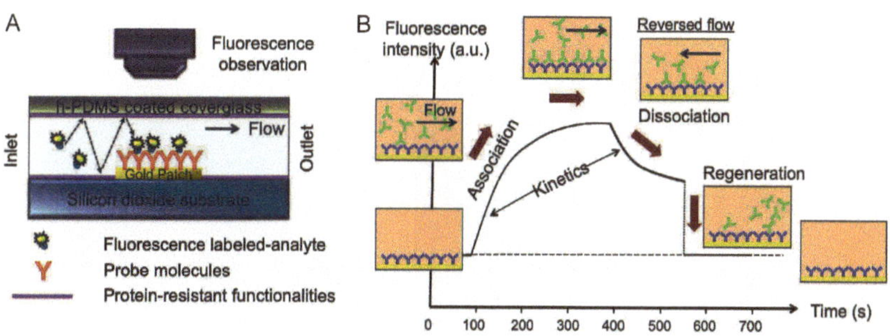

Fig. 16 (A) Bio-functional nanoslit used in protein kinetic study. Nanochannel with gold sensor surface bearing immobilized receptor probe molecules. Under pressure-driven flow the fluorophore labeled target molecules are introduced. Real-time protein-ligand binding events are monitored by fluorescence microscopy. (B) Typical protein kinetic (binding and dissociation) analysis spectrum performed in nanoslits.

sensor surface are recorded in real-time. Fig. 16B illustrate the kinetic sensing graph plotted the change of emission signal intensity versus time. Firstly the injected analyte streptavidin is binding to the immobilized biotin receptor, i.e., association phase leads to an increase of emission signal. Later equilibrium is settled and emission signal gain a plateau. Under reversed buffer flows the dissociation of streptavidin from biotin receptor, i.e., dissociation phase leads to a decrease of emission signal. Association and dissociation phases of real-time sensing graph are used for the determination of kinetic rate constant. The average binding association constant (k_a) and binding dissociation constant (k_d) of streptavidin/biotin kinetic experiment are recorded as $7.1 \times 10^5\ M^{-1}\ s^{-1}$ and $8.0 \times 10^{-7}\ s^{-1}$, respectively. Whereas, the average k_a and k_d of mouse IgG/anti-mouse IgG kinetic experiment are recorded as $8.0 \times 10^5\ M^{-1}\ s^{-1}$ and $6.2 \times 10^{-4}\ s^{-1}$, respectively.

Nanofludic nanochannel platforms with fluorescence microscopy are extensively used for the study of DNA-protein interactions.[82] DNA-protein interactions are the key aspect of cellular processes and its understanding is highly desirable for biomedical research such as gene regulation. Epigenetic information on the DNA molecule is required for gene regulation can studied via DNA-protein interactions.[83] Riehna and coworkers reported direct visualization of epigenetic mark such as methylation of cytosine bases (5mC) on the DNA molecule via DNA starching by employing nanofludic nanochannel platform.[84] In this study location of 5mC base in DNA molecule is identified via binding of fluorophore (Alexa-Fluor-568) labeled methyl-binding-domain (MBD) protein to 5mC base (Fig. 17). Firstly they generated hybrid concatamers, i.e., engineered DNA via mixing of 5mC

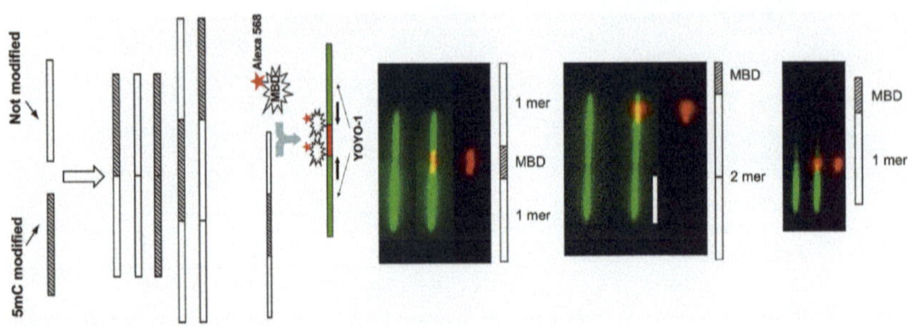

Fig. 17 Detection of DNA methylation sites by use of MBD proteins which labeled by Alexa-Fluor 568, can visualized on λ-DNA concatamers in red. DNA stained with YOYO-1 visualized in green.

and non-methylated segments. Further hybrid concatamer sequentially interact with green stain YOYO-1 and MBD protein. Later the monitoring of emission signals using a fluorescence microscope reveals the methylated pattern versus non-methylated. The same research group in the following years successfully employed nanochannels platform for the detection of histone modifications on chromatin by using fluorophore dye labeled antibodies.[85]

4. Conclusion and future prospects

In this chapter we mostly highlighted biomedical applications of fluorescence based miniaturized microfluidic and nanofluidic systems specially the analysis and sensing of biomolecules such as nucleic acids and proteins over the last two decades. Development of fluorescence based miniaturized microfluidic and nanofluidic platform for diverse medical application is rapidly growing and promising research filed. In the last decade fluorescence based microfluidic/nanofluidic platforms were greatly used by worldwide researchers for single cell miRNA detection, DNA mapping, identification of epigenetic mark on single DNA molecule, DNA barcoding, DNA-protein interaction study, protein sensing, protein sequencing, protein binding kinetics and many more applications. Microfluidic/nanofluidic platforms are also used for the synthesis of confined nanomaterial from biomolecules for biomedical applications. We also presented the applications of microfluidic platform for the construction of biologically important AIE active nanomaterials. In point of view of design and operation of fluorescence based microfluidic/nanofluidic systems still there are some key questions need to be addressed by worldwide researchers for its full potential applications. Most important challenge is the incorporation of fluorescence microscopy in miniaturized device for real-time analysis. Development of such advanced fluorescence based microfluidic/nanofluidic platform would useful to broader scientific community for tackle advance scientific challenges.

Acknowledgments

APW is acknowledges research SHODH fellowship from Education Department, Gujarat Government. JPR and RSB are thankful to Dr. J. S. Yadav Provost & Research Director, Indrashil University for providing research facilities and scientific environment.

References

1. Chiu DT, de Mello AJ, Di Carlo D, et al. Small but perfectly formed? Successes, challenges, and opportunities for microfluidics in the chemical and biological sciences. *Chem.* 2017;2:201–223.
2. Sparreboom W, Van den Berg A, Eijkel JCT. Principles and applications of nanofluidic transport. *Nat Nanotechnol.* 2009;4:713–720.
3. Piruska A, Gong M, Sweedler JV, Bohn PW. Nanofluidics in chemical analysis. *Chem Soc Rev.* 2010;39:1060–1072.
4. Prakash S, Yeom J. *Nanofluidics and Microfluidics: Systems and Applications.* 1st ed. Waltham: William Andrew; 2014.
5. Easley CJ, Karlinsey JM, Bienvenue JM, et al. A fully integrated microfluidic genetic analysis system with sample-in–answer-out capability. *Proc Natl Acad Sci U S A.* 2006;103:19272–19277.
6. Whitesides GM. What comes next? *Lab Chip.* 2011;11:191–193.
7. Abgrall P, Nguyen NT. Nanofluidic devices and their applications. *Anal Chem.* 2008;80:2326–2341.
8. Raiteri R, Grattarola M, Butt HJ, Skládal P. Micromechanical cantilever-based biosensors. *Sensors Actuators B Chem.* 2001;79:115–126.
9. Wang J. Electrochemical biosensors: Towards point-of-care cancer diagnostics. *Biosens Bioelectron.* 2006;21:1887–1892.
10. Kuswandi B, Huskens J, Verboom W. Optical sensing systems for microfluidic devices: a review. *Anal Chim Acta.* 2007;601:141–155.
11. Pires NMM, Dong T, Hanke U, Hoivik N. Recent developments in optical detection technologies in lab-on-a-chip devices for biosensing applications. *Sensors (Basel).* 2014;14:15458–15479.
12. Măriuţa D, Colin S, Barrot-Lattes C, et al. Miniaturization of fluorescence sensing in optofluidic devices. *Microfluid Nanofluid.* 2020;24:65.
13. Nawrot W, Drzozga K, Baluta S, Cabaj J, Malecha K. A fluorescent biosensors for detection vital body fluids' agents. *Sensors (Basel).* 2018;18:2357.
14. Yang X, Yuan T, Yue G, Li E, Yuan L. Optofluidic integrated in-fiber fluorescence online optical fiber sensor. *Sensors Actuators B Chem.* 2015;215:345–349.
15. Hong L, Li H, Yang H, Sengupta K. Fully integrated fluorescence biosensors on-chip employing multi-functional nanoplasmonic optical structures in CMOS. *IEEE J Solid State Circuits.* 2017;52:2388–2406.
16. Wei L, Yan W, Ho D. Recent advances in fluorescence lifetime analytical microsystems: contact optics and CMOS time-resolved electronics. *Sensors (Basel).* 2017;17:2800.
17. Babikian S, Li GP, Bachman M. A digital signal processing assisted microfluidic PCB for on-chip fluorescence detection. *IEEE Trans Compon Packag Manuf Technol.* 2017;7:846–854.
18. Ryu G, Huang J, Hofmann O, et al. deMello JC, deMello AJ, Bradley DDC. Highly sensitive fluorescence detection system for microfluidic lab-on-a-chip. *Lab Chip.* 2011;11:1664–1670.
19. Watson JD, Crick FHC. Molecular structure of nucleic acids: a structure for deoxyribose nucleic acid. *Nature.* 1953;171:737–738.
20. Zhao Y, Chen F, Li Q, Wang L, Fan C. Isothermal amplification of nucleic acids. *Chem Rev.* 2015;115:12491–12545.
21. Alberts B, Johnson A, Lewis J. *Molecular Biology of the Cell.* 4th ed. New York, NY: Garland Science; 2002.
22. Salmon L, Yang S, Al-Hashimi HM. Advances in the determination of nucleic acid conformational ensembles. *Annu Rev Phys Chem.* 2014;65:293–316.

23. Sun H, Sheng J, Hassan AEA, Jiang S, Gan J, Huang Z. Novel RNA base pair with higher specificity using single selenium atom. *Nucleic Acids Res.* 2012;40:5171–5179.
24. Graham CA, Hill AJM. *DNA Sequencing Protocols.* 2nd ed. Springer; 2001.
25. Janitz M. *Next Generation Genome Sequencing: Towards Personalized Medicine.* 1st ed. Wiley; 2008.
26. Hwang B, Lee JH, Bang D. Single-cell RNA sequencing technologies and bioinformatics pipelines. *Exp Mol Med.* 2018;50:96.
27. Fan Y, Dong D, Li Q, et al. Fluorescent analysis of bioactive molecules in single cells based on microfluidic chips. *Lab Chip.* 2018;18:1151–1173.
28. Tian H, Sun Y, Liu C, Duan X, Tang W, Li Z. Precise quantitation of microRNA in a single cell with droplet digital PCR based on ligation reaction. *Anal Chem.* 2016; 88:11384–11389.
29. Wang P, Jing F, Li G, et al. Absolute quantification of lung cancer related microRNA by droplet digital PCR. *Biosens Bioelectron.* 2015;74:836–842.
30. Zhang H, Jenkins G, Zou Y, Zhu Z, Yang CJ. Massively parallel single-molecule and single-cell emulsion reverse transcription polymerase chain reaction using agarose droplet microfluidics. *Anal Chem.* 2012;84:3599–3606.
31. Ting DT, Wittner BS, Ligorio M, et al. Single-cell RNA sequencing identifies extracellular matrix gene expression by pancreatic circulating tumor cells. *Cell Rep.* 2014; 8:1905–1918.
32. White AK, VanInsberghe M, Petriv OI, et al. High-throughput microfluidic single-cell RT-qPCR. *Proc Natl Acad Sci U S A.* 2011;108:13999–14004.
33. Rajeswari PKP, Soderberg LM, Yacoub A, Leijon M, Svahn HA, Joensson HN. Multiple pathogen biomarker detection using an encoded bead array in droplet PCR. *J Microbiol Methods.* 2017;139:22–28.
34. Eastburn DJ, Sciambi A, Abate AR. Ultrahigh-throughput mammalian single-cell reverse-transcriptase polymerase chain reaction in microfluidic drops. *Anal Chem.* 2013;85:8016–8021.
35. Novak R, Zeng Y, Shuga J, et al. Single-cell multiplex gene detection and sequencing with microfluidically generated agarose emulsions. *Angew Chem Int Ed.* 2011;50: 390–395.
36. Mazutis L, Araghi AF, Miller OJ, et al. Droplet-based microfluidic systems for high-throughput single DNA molecule isothermal amplification and analysis. *Anal Chem.* 2009;81:4813–4821.
37. Berg JM, Tymoczko JL, Stryer L. *Biochemistry.* 5th ed. New York: W H Freeman; 2002.
38. Sims CE, Allbritton NL. Analysis of single mammalian cells on-chip. *Lab Chip.* 2007;7:423–440.
39. Eisenberg D, Marcotte EM, Xenarios I, Yeates TO. Protein function in the post-genomic era. *Nature.* 2000;405:823–826.
40. Shim J-U, Olguin LF, Whyte G, et al. Simultaneous determination of gene expression and enzymatic activity in individual bacterial cells in microdroplet compartments. *J Am Chem Soc.* 2009;131:15251–15256.
41. Huebner A, Srisa-Art M, Holt D, et al. Quantitative detection of protein expression in single cells using droplet microfluidics. *Chem Commun.* 2007;1218–1220.
42. Lee JS, Joung H-A, Kim M-G, Park CB. Graphene-based chemiluminescence resonance energy transfer for homogeneous immunoassay. *ACS Nano.* 2012;6:2978–2983.
43. Ma C, Fan R, Ahmad H, et al. A clinical microchip for evaluation of single immune cells reveals high functional heterogeneity in phenotypically similar T cells. *Nat Med.* 2011;17:738–743.
44. MacKenzie KR. Folding and stability of α-helical integral membrane proteins. *Chem Rev.* 2006;106:1931–1977.

45. Dallongeville S, Garnier N, Rolando C, Tokarski C. Proteins in art, archaeology, and paleontology: from detection to identification. *Chem Rev*. 2016;116:2–79.
46. Chan SD, Luedke G, Valer M, Buhlmann C, Preckel T. Cytometric analysis of protein expression and apoptosis in human primary cells with a novel microfluidic chip-based system. *Cytometry A*. 2003;55:119–125.
47. Huang B, Wu H, Bhaya D, et al. Counting low–copy number proteins in a single cell. *Science*. 2007;315:81–84.
48. Huebner A, Olguin LF, Bratton D, et al. Development of quantitative cell-based enzyme assays in microdroplets. *Anal Chem*. 2008;80:3890–3896.
49. Konry T, Dominguez-Villar M, Baecher-Allan C, Hafler DA, Yarmush ML. Droplet-based microfluidic platforms for single T cell secretion analysis of IL-10 cytokine. *Biosens Bioelectron*. 2011;26:2707–2710.
50. Martino C, Zagnoni M, Sandison ME, Chanasakulniyom M, Pitt AR, Cooper JM. Intracellular protein determination using droplet-based immunoassays. *Anal Chem*. 2011;83:5361–5368.
51. Chokkalingam V, Tel J, Wimmers F, et al. Probing cellular heterogeneity in cytokine-secreting immune cells using droplet-based microfluidics. *Lab Chip*. 2013;13: 4740–4744.
52. Mazutis L, Gilbert J, Ung WL, Weitz DA, Griffiths AD, Heyman JA. Single-cell analysis and sorting using droplet-based microfluidics. *Nat Protoc*. 2013;8:870–891.
53. Konry T, Smolina I, Yarmush JM, Irimia D, Yarmush ML. Ultrasensitive detection of low-abundance surface-marker protein using isothermal rolling circle amplification in microfluidic Nano-liter platform. *Small*. 2011;7:395–400.
54. Konry T, Golberg A, Yarmush M. Live single cell functional phenotyping in droplet nano-liter reactors. *Sci Rep*. 2013;3:3179.
55. Mei J, Leung NLC, Kwok RTK, Lam JWY, Tang BZ. Aggregation-induced emission: together we shine, united we soar! *Chem Rev*. 2015;115:11718–11940.
56. Feng G, Liu B. Aggregation-induced emission (AIE) dots: emerging theranostic nanolights. *Acc Chem Res*. 2018;51:1404–1414.
57. Lim X. The nanolight revolution is coming. *Nature*. 2016;531:26–28.
58. Jokerst JV, Gambhir SS. Molecular imaging with theranostic nanoparticles. *Acc Chem Res*. 2011;44:1050–1060.
59. Qin W, Ding D, Liu J, et al. Biocompatible nanoparticles with aggregation-induced emission characteristics as far-red/near-infrared fluorescent bioprobes for in vitro and in vivo imaging applications. *Adv Funct Mater*. 2012;22:771–779.
60. Geng J, Li K, Qin W, et al. Eccentric loading of fluorogen with aggregation-induced emission in PLGA matrix increases nanoparticle fluorescence quantum yield for targeted cellular imaging. *Small*. 2013;9:2012–2019.
61. Wang D, Qian J, Qin W, Qin A, Tang BZ, He S. Biocompatible and photostable AIE dots with red emission for in vivo two-photon bioimaging. *Sci Rep*. 2015;4:4279.
62. Guo B, Chen J, Chen N, et al. High-resolution 3D NIR-II photoacoustic imaging of cerebral and tumor vasculatures using conjugated polymer nanoparticles as contrast agent. *Adv Mater*. 2019;31:1808355.
63. Valencia PM, Farokhzad OC, Karnik R, Langer R. Microfluidic technologies for accelerating the clinical translation of nanoparticles. *Nat Nanotechnol*. 2012;7:623–629.
64. Li X, Zha M, Li Y, et al. Sub-10 nm aggregation-induced emission quantum dots assembled by microfluidics for enhanced tumor-targeting and reduced retention in the liver. *Angew Chem Int Ed*. 2020;59:21899–21903.
65. Zhang W, Gao N, Cui J, et al. AIE-doped poly(ionic liquid) photonic spheres: a single sphere-based customizable sensing platform for the discrimination of multi-analytes. *Chem Sci*. 2017;8:6281–6289.

66. Marie R, Kristensen A. Nanofluidic devices towards single DNA molecule sequence mapping. *J Biophotonics*. 2012;5:673–686.
67. Tegenfeldt JO, Prinz C, Cao H, et al. Micro- and nanofluidics for DNA analysis. *Anal Bioanal Chem*. 2004;378:1678–1692.
68. Ouyanga W, Hana J. Universal amplification-free molecular diagnostics by billion-fold hierarchical nanofluidic concentration. *Proc Natl Acad Sci U S A*. 2019;116: 16240–16249.
69. Kim SJ, Song Y-A, Han J. Nanofluidic concentration devices for biomolecules utilizing ion concentration polarization: theory, fabrication, and applications. *Chem Soc Rev*. 2010;39:912–922.
70. Napoli M, Eijkel JCT, Pennathur S. Nanofluidic technology for biomolecule applications: a critical review. *Lab Chip*. 2010;10:957–985.
71. Cipriany BR, Zhao R, Murphy PJ, et al. Single molecule epigenetic analysis in a nanofluidic channel. *Anal Chem*. 2010;82:2480–2487.
72. Dylla-Spears R, Townsend JE, Jen-Jacobson L, Sohn LL, Muller SJ. Single-molecule sequence detection via microfluidic planar extensional flow at a stagnation point. *Lab Chip*. 2010;10:1543–1549.
73. Das SK, Austin MD, Akana MC, Deshpande P, Cao H, Xiao M. Single molecule linear analysis of DNA in nano-channel labeled with sequence specific fluorescent probes. *Nucleic Acids Res*. 2010;38:e177.
74. Tegenfeldt JO, Prinz C, Cao H, et al. A single-molecule barcoding system using nano slits for DNA analysis. *Proc Natl Acad Sci U S A*. 2004;101:10979–10983.
75. Riehn R, Lu M, Wang Y-M, Lim SF, Cox EC, Austin RH. Restriction mapping in nanofluidic devices. *Proc Natl Acad Sci U S A*. 2005;102:10012–10016.
76. Jo K, Dhingra DM, Odijk T, et al. A single-molecule barcoding system using nanoslits for DNA analysis. *Proc Natl Acad Sci U S A*. 2007;104:2673–2678.
77. McCaffrey J, Sibert J, Zhang B, et al. CRISPR-CAS9 D10A nickase target-specific fluorescent labeling of double strand DNA for whole genome mapping and structural variation analysis. *Nucleic Acids Res*. 2016;44:e11.
78. Reisner W, Larsen NB, Silahtaroglu A, et al. Single-molecule denaturation mapping of DNA in nanofluidic channels. *Proc Natl Acad Sci U S A*. 2010;107:13294–13299.
79. Toprakcioglu Z, Challa PK, Morse DB, Knowles T. Attoliter protein nanogels from droplet nanofluidics for intracellular delivery. *Sci Adv*. 2020;6:eaay7952.
80. Wang C, Shi Y, Wang J, Pang J, Xia X-H. Ultrasensitive protein concentration detection on a micro/nanofluidic enrichment chip using fluorescence quenching. *ACS Appl Mater Interfaces*. 2015;7:6835–6841.
81. Teerapanich P, Pugnière M, Henriquet C, Lin Y-L, Chou C-F, Leïchlé T. Nanofluidic fluorescence microscopy (NFM) for real-time monitoring of protein binding kinetics and affinity studies. *Biosens Bioelectron*. 2017;88:25–33.
82. Frykholm K, Nyberg LK, Westerlund F. Exploring DNA-protein interactions on the single DNA molecule level using nanofluidic tools. *Integr Biol*. 2017;9:650–661.
83. Levy-Sakin M, Grunwald A, Kim S, et al. Towards single-molecule optical mapping of the epigenome. *ACS Nano*. 2014;8:14–26.
84. Lim SF, Karpusenko A, Sakon JJ, Hook JA, Lamar TA, Riehna R. DNA methylation profiling in nanochannels. *Biomicrofluidics*. 2011;5:034106.
85. Lim SF, Karpusenko A, Blumers AL, Streng DE, Riehn R. Chromatin modification mapping in nanochannels. *Biomicrofluidics*. 2013;7:064105.

CHAPTER TEN

Active matter dynamics in confined microfluidic environments

Rik Chakraborty, Arnab Maiti, Nikita Sharma, and Krishna Kanti Dey*
Discipline of Physics, Indian Institute of Technology Gandhinagar, Palaj, Gandhinagar, Gujarat, India
*Corresponding author: e-mail address: k.dey@iitgn.ac.in

Abstract

The field of active matter is a nascent area of research in soft condensed matter physics, which is drawing on the expertise of researchers from diverse disciplines. Small scale active particles—both inorganic and biological—display non-trivial emergent dynamics and interactions that could help us understand complex biological processes and phenomena. Recently, using microfluidic technologies, several research groups have performed important experimental and theoretical studies to understand the behavior of self-propelled particles and molecular active matter within confined environments—to glean a fundamental understanding of the cellular processes occurring under ultra-low Reynolds number conditions. In this chapter, we would like to review applications of microfluidics in active matter research, highlighting a few important theoretical and experimental investigations. We will conclude the discussion with a note on the future of this field mentioning a few open questions that are at the forefront of our minds.

Life involves orchestrated function of many biological processes where systems are driven away from equilibrium in harnessing energy from the environment and driving various metabolic activities. The cytoplasm of eukaryotic organisms is highly dynamic, where small molecules constantly move around by random thermal diffusion. In addition to the random motions of the macromolecules, the cytoplasmic matrix also gets constantly agitated by the self-propelled molecular motors[1] which play a critical role in intracellular transport and actuation of muscles.[2] Understanding the physics governing such active processes within the cytosolic environments has recently been one of the focus areas of investigation of soft matter physics. Several research groups have conducted important experimental and theoretical studies to understand behavior of self-propelled particles and

Progress in Molecular Biology and Translational Science, Volume 186
ISSN 1877-1173
https://doi.org/10.1016/bs.pmbts.2021.07.014

transport of active matter within confined environments, to glean a fundamental understanding of the cellular processes occurring under ultra-low Reynolds number conditions. These investigations helped in appreciating the nature of biochemical transformations and associated phenomena under complex crowded environments, interaction profiles among small molecules in motion and intracellular transport of materials driven by active diffusion and cytoplasmic advection.[3,4] Microfluidic setups have been found to offer useful platforms to investigate complex low Reynolds number phenomena, due to their ease of fabrication, ability to generate high quality data in relatively shorter periods of time and compact designs. Research results with active matter in microfluidic setup have so far demonstrated fascinating behavior of self-powered motors and molecules that are yet to be fully understood using known models and theories. Attempts have also been made to harness behavior of active materials within microfluidic setups in developing smart sensors and transducers. In this chapter, we would like to review application of microfluidics in active matter research, highlighting a few important theoretical and experimental investigations. We will close the discussion with a note on the future of this field mentioning a few open questions that are at the forefront of our minds.

Microfluidics refers to the study of fluid behavior and its precise manipulation within microscale confinement, where surface forces dominate over volumetric forces. Microfluidics have been successfully used in fabricating lab-on-a-chip devices—where many chemical and biochemical operations involving fluid transport and manipulation are integrated over a miniaturized and portable platform. Microfluidic devices offer many advantages compared to large-scale instruments and bulk analysis protocols including requirement of low sample volume, ease of operations, portability, rapid and accurate sample characterization, and opportunities for parallel experimentations.[5] These on-chip systems integrated with optical detection techniques have been extensively used in identification, separation and quantification of biomolecules, even at the single molecule level.[6,7] Lab-on-a-chip devices, also known as micro total analysis systems (μTAS), have found important applications in biological and biomedical engineering, especially in various protein and nucleic acid assays.[8] Although significant progress has been made in the development of highly efficient cellular analysis systems, usage of microfluidic devices in characterizing non-equilibrium systems and the dynamics of active fluids are yet to be vigorous. Brownian diffusion-based mixing of fluids and separation of molecules have been demonstrated within microfluidic chips (Fig. 1A), which are likely to find

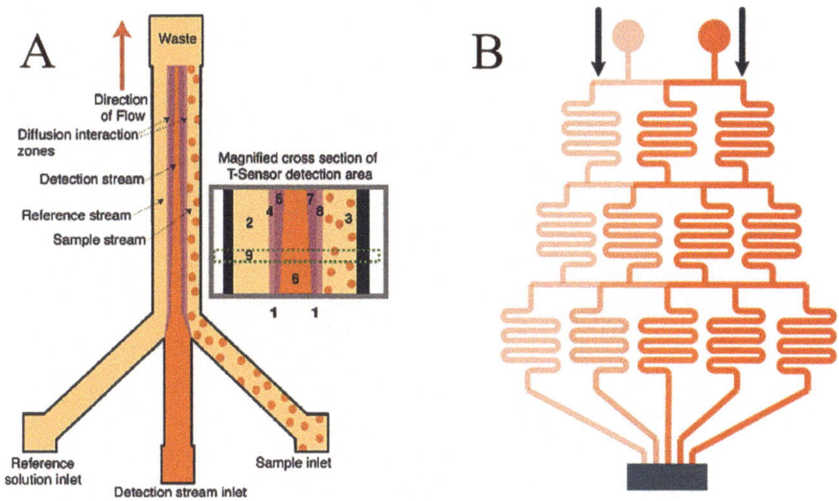

Fig. 1 (A) Schematic of diffusion-based separation of particles within a T-sensor network. Different solutions enter through the three inlets at the bottom and interact with each other until they reach the outlet. (B) Schematic of a microfluidic tree-shaped gradient generator. *Panel (A) image reproduced with permission from Weigl BH, Yager P. Microfluidic diffusion-based separation and detection.* Science. *1999;283(5400):346–7, Panel (B) the image is taken from Wang X, Liu Z, Pang Y. Concentration gradient generation methods based on microfluidic systems.* RSC Adv. *2017;7(48):29966–84 with permission.*

important applications in the development of smart sensors and reactors.[9,10] Microfluidic setups have also been found to be useful in generating stable concentration gradients of analytes (Fig. 1B), which facilitates the study of diffusiophoretic propulsion of particles under extremely controlled conditions.[11]

Formation of stable gradients or diffusion-based manipulation of particles are possible within microfluidic channels due to the laminar fluid flow within such confined geometries, where two or more layers of fluid flow next to each other without any mixing other than by diffusion of constituent molecules and components.[10] The nature of fluid flow (laminar or turbulent) within a channel is often characterized by a dimensionless parameter known as the Reynolds number, which is the ratio of inertial to viscous forces and is defined by the expression $R_e = \frac{\rho v D_h}{\mu}$.[12] Here ρ and μ are the density and viscosity of the liquid, D_h is the hydraulic diameter whose value depends on the cross-sectional geometry of the channel, and v is the velocity of the liquid.[13] In the case of microfluidic devices, the characteristic length D_h is very small which makes the Reynolds numbers less than

2000.[14] This results in laminar and predictable flows that can be manipulated very precisely—enabling controlled movement of molecules in liquids, and generation of stable chemical concentration gradients. It may be mentioned here that the Reynolds number for common microfluidic devices, with typical channel radii of 1–100 μm, and water as the working fluid, ranges within $\sim 10^{-6}$ and 10.[12] Blood flow within a medium-sized artery is laminar and corresponds to a typical Reynolds number value between 100 and 1000.[15] Swimming of microorganisms also occurs at a very low Reynolds number regime ($\sim 10^{-4}$),[16] which leads to their non-trivial locomotion patterns. The mass transport in a fluidic system may be further characterized by another dimensionless parameter known as the Péclet number, which is defined as $P_e = \frac{vD_h}{D}$. Here, D is the diffusion coefficient of the molecule being transported and P_e represents the ratio of advective to the diffusive transport in the system. For microfluidic systems, P_e becomes very small signifying a diffusion-dominant environment. Smaller dimensions of the microfluidic systems result in shorter diffusion times of molecules. This makes reaction times within microfluidic environments much quicker compared to that in the bulk systems.[13,14] Microfluidic approaches have found extensive applications in environmental remediation involving rapid detection and quantification of toxins and pollutants with minimal quantities of samples.[17] As mentioned earlier, the generation of diffusion-dominated flow and possibilities of precise flow manipulation also makes microfluidic devices ideal platforms to investigate and harness the properties of self-propelled micro and nanoscopic systems, where each molecule or constituent particle is capable of transducing energy from the environment and power its motion independently in liquids.

In a typical microfluidic device, the micro-channel network is connected to the external environment through inlet and outlet ports. The fluids are injected to the device using flow control arrangements such as syringe pumps, peristaltic pumps or pressure-driven pumps.[18] Although microfluidic chips may be fabricated using polymer, hybrid and composite materials or even papers,[19] for research with active matter, Polydimethylsiloxane (PDMS) microchannels have been mostly used so far following the reported fabrication protocols.[20,21] PDMS is inexpensive, flexible, and optically transparent material compatible with many optical methods of detection. It is also biocompatible and may be easily bonded over other surfaces.[21] The fabrication method begins with designing the microfluidic architecture using a computer aided design (CAD) software. The design is then printed over plastic films using UV opaque ink. The printed film is used as a photomask in UV-photolithography to generate a master pattern. To accomplish this, a thin layer of photoresist (such as SU-8) is spin-coated onto a clean silicon wafer. The photoresist is exposed to UV light through the photomask. In the case of a negative photoresist like SU-8, only the parts representing the channels are exposed to UV light and cured, while the other parts of the mold get protected by the opaque areas of the mask. A developing reagent is then used to dissolve the resist from the unexposed regions of the master, which serves as a mold with a photoresist replica of the patterns over it. The height of the microchannel to be fabricated is determined by the thickness of the photoresist deposited over the silicon wafer. To prevent irreversible binding of PDMS to the master, the latter is treated with fluorinated silane before a mixture of liquid PDMS and a cross-linking agent is poured into it. The PDMS is cured at 70 °C for an hour or so before it is peeled off the master, obtaining a replica of the designed microstructures on the PDMS block.[22–24] Fig. 2 is a schematic representation of the successive steps of PDMS channel fabrication.

Small holes are punched into the chip fabricated to make the inlets and outlets. The PDMS block is then treated with oxygen plasma to attach it irreversibly over other flat surfaces (such as glass).

PDMS consists of repeating -$OSi(CH_3)_2$- units where the -CH_3 groups make its surface hydrophobic. The hydrophobicity may result in poor

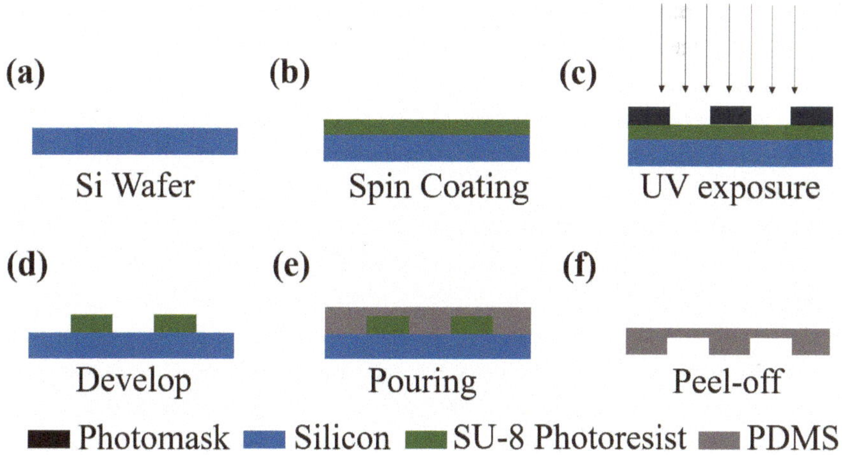

Fig. 2 Schematic representation of the steps in PDMS-based microfluidic channel fabrication process.

wettability of the channel surface with aqueous solvents, rendering its microchannels susceptible to air bubble trapping and prone to nonspecific adsorption to proteins and cells.[21] The hydrophobicity may be reduced by exposing the surface to an air plasma. The surface may be further functionalized with a layer of bovine serum albumin to minimize attachment of other proteins or enzymes over the microchannel surface.[25] For driving the fluid flow within PDMS microchannels, either the pressure-driven or electrokinetic technique is used. In pressure-driven flow (also called hydrodynamic flow) the volumetric flow rate is given by $Q = \frac{\Delta P}{R_H}$, where ΔP is the pressure drop across the channel and R_H is called the hydraulic resistance.[26] The pressure drop is usually created by applying a positive pressure at the inlet and opening the outlet to the atmospheric pressure. Flow control devices like syringe pumps however, in some occasions have been found to create fluctuations in aqueous microfluidic systems, which may perturb the flow of liquids and motion of suspended particles in the channel.[27] Nevertheless, syringe pumps are widely used in microfluidic measurements since these are easy to use and enable quick setup of the experiments. Finally, although the electrokinetic flow (generated due to the movement of charged particles in an electric field) based systems do not require any mechanical actuator for pumping fluids and have potential to be very useful for biological and chemical applications,[28–30] their use in active matter research has so far been limited.

The field of active matter primarily deals with colloidal assemblies of motors and molecules, where the constituent elements move autonomously in liquid media by harnessing energy from their environments. These systems differ from their equilibrium counterparts, where motion arises due to the application of external thermal, magnetic, or electric fields and the particles exhibit lock-step behavior rather than moving independently in liquids. As of now, little is known about the nature of physics that governs the behavior of self-propelled systems and their collective dynamics at the micro/nanoscales. Research on this front is being pursued to address the following grand challenges—(A) How can we design and understand the behavior of micro and nanoscopic active matter that are inherently driven out of equilibrium due to persistent energy consumption—resulting in invalidation of equilibrium concepts such as free energy, detailed balance, and time-reversal symmetry?[31] (B) How do we understand non-trivial dynamic interactions among molecules within biologically active systems and use such communication pathways in a variety of novel applications related to transport, assembly, and mixing?[32] It was realized that in a model non-equilibrium system, particles must have continuous supply of energy to sustain their motion. One of the simplest ways to achieve this is by making the particles catalytically active so that they could interact with their environment to generate the necessary mechanical energy. The studies reported by Whitesides,[33] Sen and Mallouk,[34] and Ozin[35] demonstrated the use of catalytic reactions as one of the most convenient means of powering small objects in liquids, analogous to ATP-driven propulsion of natural biomolecular motors over microtubular tracks. Soon after these demonstrations, a variety of active colloidal assemblies containing nano- and microscale catalytic motors have been designed, synthesized, and modeled to understand specific physics or to realize particular applications. These systems can be categorized not only on the basis of their size, shape, and propulsion mechanisms but also from the perspective of their potential applications.[36] Active systems could be biological, hybrid, or all synthetic, most of which display remarkable dynamics and offer model platforms to investigate controlled transfer of mass, energy and information between different parts of complex fluidic systems.[32]

Inorganic materials make up one of the most versatile classes of motors. The earliest nanomotors developed were cylindrical in shape.[36] Cylindrical motors are fabricated via electro-assisted deposition in pore templates to create uniform nanorods and are capable of self-propelling themselves in fluidic media based on different propulsion mechanisms.[36–38] Liu et al. fabricated bimetallic motors containing Cu and Pt segments that moved autonomously in dilute bromine and iodine solutions, acting as nanobatteries (Fig. 3A).[39] Rolled-up microtubes and conical microstructures have been explored by researchers as alternative designs for self-powered micro/nanomotors (Fig. 3B).[40] Solovev et al. developed a generalized fabrication technique to make rolled-up microtubes that allowed for further functionalization of the motor surface on the basis of its composition. This technique was used to make Au-Pt microtubes, which were found to move in hydrogen peroxide fuel following a bubble recoil mechanism.[41] Both uniformly and partially functionalized spherical particles have been used to achieve active propulsion in solutions. These particles catalyze chemical reactions on their surfaces and often propel themselves in solutions through various phoretic mechanisms.[42,43] Spherical Janus micromotors possess unique asymmetric structures and can integrate different functionalities on its two hemispherical surfaces.[44–46] Multifunctional motors using more than one method of propulsion, gives the systems another layer of complexity and endowing them with better controls over their propulsion and functionality.[47,48] Among

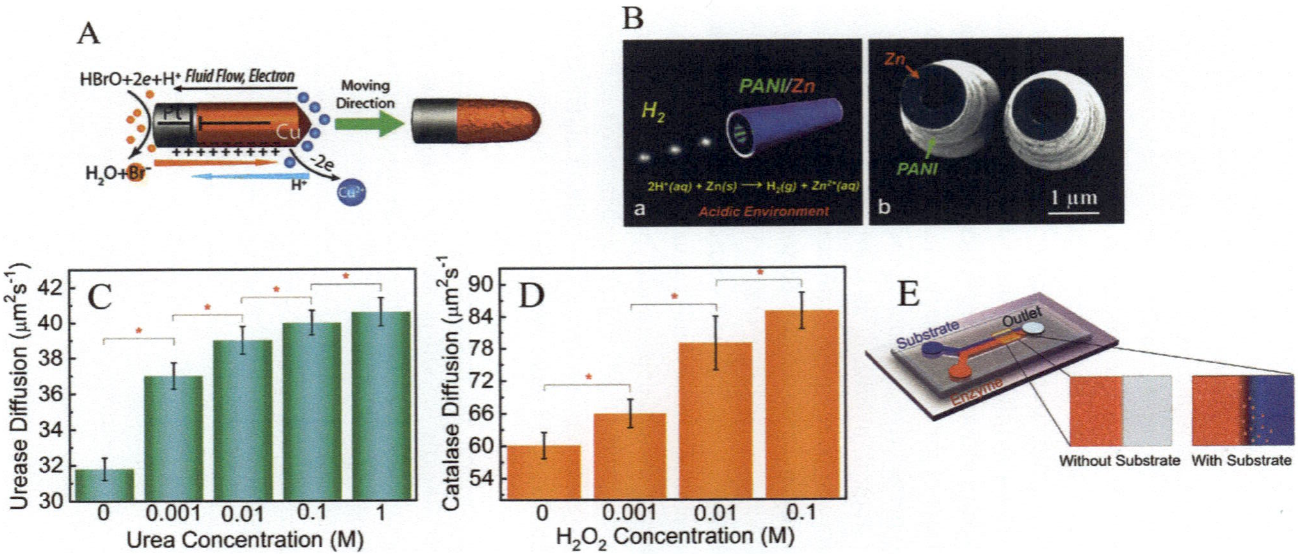

Fig. 3 (A) Self-electrophoretic Cu-Pt nanomotor moving in Br and I solutions, (B) Zinc-based micromotors propelled by hydrogen bubble recoil in acidic media. Diffusion of (C) urease and (D) catalase increased with increased substrate concentrations (urea and H_2O_2 respectively). (* indicates a significance value of $p < 0.05$. Here, the *p*-value, or probability value, is the number describing how likely it is that the experimental data would have occurred by random chance). (E) Schematic of the microfluidic channel used for enzyme chemotaxis experiments. *Panel (A) the image is reproduced from Liu R, Sen A. Autonomous nanomotor based on copper–platinum segmented nanobattery.* J Am Chem Soc. *2011;133(50):20064–7 with permission, Panel (B) the image is reproduced from Gao W, Uygun A, Wang J. Hydrogen-bubble-propelled zinc-based microrockets in strongly acidic media.* J Am Chem Soc. *2012;134(2):897–900 with permission, Panels (C–D) the data and the schematic are taken from Muddana HS, Sengupta S, Mallouk TE, Sen A, Butler PJ. Substrate catalysis enhances single-enzyme diffusion.* J Am Chem Soc. *2010;132(7):2110–1, Sengupta S, Dey KK, Muddana HS, Tabouillot T, Ibele ME, Butler PJ, et al. Enzyme Molecules as Nanomotors.* J Am Chem Soc. *2013;135(4):1406–14.*

various biological active matters, suspensions of motile bacteria often constitute ideal systems driven out of equilibrium on the micrometer scale, where each flagellated bacterium can efficiently convert chemical energy into propulsion using their cellular motors.[49,50] Going further down the length scale, Sen et al. demonstrated that active enzymes, while turning over substrates generate enough mechanical force to enhance their own diffusion in solution (Fig. 3C and D).[25,51–53] The diffusive movement of the enzymes increased with increasing substrate concentration. Moreover, in the presence of an imposed substrate concentration gradient, enzymes collectively moved toward areas of higher substrate concentrations, displaying a novel example of molecular chemotaxis (Fig. 3E).[25,52] Enzymes, being operative at very low Reynolds number conditions, have never been previously expected to display directional propulsion and energy transduction ability. The non-trivial collective migration of enzymes toward specific targets promised a variety of applications including targeted transport, sensing and delivery.

In recent years, microfluidic technology has become a useful tool in studying the behavior of active colloids and harnessing their properties for important applications. Researchers have taken recourse to microfluidic platforms and started investigating the motion of micro/nanomotors within confined diffusion-dominated environments to get a sense of the non-equilibrium physics that governs the particle behavior at lower length scales.[54–57] So far, nano/micromotors have been successfully used to deliver cargo at specific locations (Fig. 4B),[58,59] sort microobjects,[60] induce pumping,[61] mix fluids,[62] and concentrate samples[63] within microchannel network. Importantly, capillary microfluidic systems have also been used to fabricate multifunctional polymer or gas filled bubble micromotors (Fig. 4A).[64–67] A flow lithography integrated microfluidic spinning and spiraling system was developed by Wu et al. for the continuous generation of bioinspired helical micromotors.[68] The generation processes could be precisely tuned by adjusting the flow rates and the illuminating frequency of the system, thereby controlling the length, diameter, and pitch of the helical micromotors precisely. Baraban et al. used magnetic guidance techniques and investigated the velocity of catalytic Janus micromotors within microfluidic devices. Ultrathin magnetic films deposited over the surface of the motors helped in aligning their magnetic moments along the direction of propulsion leading to deterministic motion of the particles in the presence

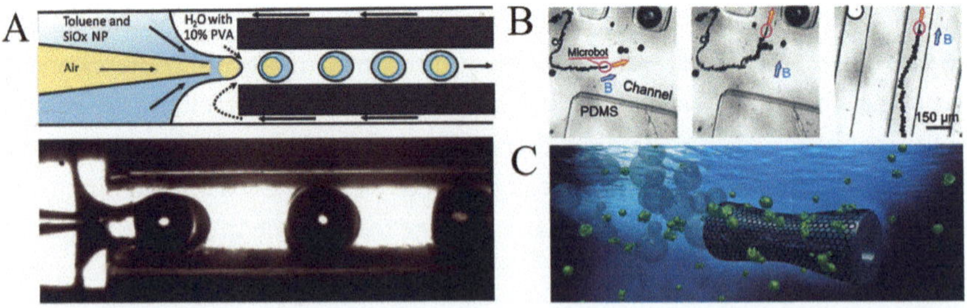

Fig. 4 (A) Generation of air-in-oil-in-water compound microbubbles using glass capillary microfluidic device. (B) Successive snapshots of a micromotor propulsion within the microchannels of a PDMS chip. The blue arrow represents the direction of the magnetic field. (C) Schematic of a graphene-based micromotor capable of removing toxic heavy metals from water. The micromotor may be subsequently recovered from the system and recycled. *Panel (A) image reproduced with permission from Adams LLA, Lee D, Mei Y, Weitz DA, Solovev AA. Nanoparticle-shelled catalytic bubble micromotor.* Adv Mater Interfaces. *2020;7(4):1901583, Panel (B) image reproduced with permission from Sanchez S, Solovev AA, Harazim SM, Schmidt OG. Microbots swimming in the flowing streams of microfluidic channels.* J Am Chem Soc. *2011;133(4):701–3, Panel (C) the image is taken from. Vilela D, Parmar J, Zeng Y, Zhao Y, Sánchez S. Graphene-based microbots for toxic heavy metal removal and recovery from water.* Nano Lett. *2016;16(4):2860–6.*

of external magnetic fields.[58] Zhao et al. showed that autonomous self-propelled millimeter sized capsules driven by the Marangoni effect, were able to sense pollutants within complex microchannel environments, label the sites and remove the contamination from the channel interior.[69] Graphene based micromotors have been also used to transport solid payloads within microfluidic chips[70] and remove pollutants and heavy metals from water (Fig. 4C).[71] Using microfluidic technology, Dong et al. showed the fabrication of substrate-free graphene oxide (GO)-based micromotors of different shapes and sizes that could self-propel in aqueous H_2O_2 solution, be controlled using external magnetic fields, and could remove molecules of antibiotics rapidly from a solution.[72] Dong et al. reported fabrication of graphene oxide-based helical micromotors with high throughput by the capillary microfluidics, which were able to move autonomously in a microchannel and used for drug delivery and environmental remediation applications.[73] Using appropriate magnetic field configurations and actuation time scales, Mandal et al. demonstrated that in a driven nanomotor system, it is possible to maneuver geometrically identical nanostructures in different directions, and subsequently position them at arbitrary locations with respect to each other.[74] This could be useful in applications that require remote and independent control over individual motors within confined microfluidic environments.

Using the microfluidic platforms several other important applications of self-powered systems have been demonstrated. To enhance fluid mixing within microchannels, Zizzari et al. used polyelectrolyte multilayer capsules (PMCs), embodying a catalytic polyoxometalate, as self-powered micromotors to generate chaotic flows by fuel-fed propulsions.[75] Autonomous micropumps, which are capable of creating fluid flows and imparting motion to nearby passive objects have also been realized in several investigations. Ibele et al. reported the fabrication of a bimetallic fluid pumping system that consisted of spatially defined palladium features over gold surfaces.[76] The direction of fluid flow in such a system was shown to be dependent on the choice of fuel. Light-driven TiO_2 based micropump system was developed by Hong et al., which offered an easy and controllable way of directing fluid within microfluidic channels.[77] Chang et al. showed a water-driven motile micropump, based on a ZnO/Ni/polystyrene Janus micromotor for achieving complex surface micropatterns.[78] The micromotor was driven by self-generated ion concentration gradient, while pumping away surrounding silica particles through diffusiophoretic mechanism. The resolution and efficiency of the micropatterning process could be regulated by controlling the diameters of the Janus micromotors. It was shown that diverse surface micropatterns might be fabricated through remote magnetic control of the motile micropump system.

A significant progress has been made in integrating biological active systems with microfluidic technology to underscore novel and scientifically important properties of materials. Dynamics of active bacterial suspensions have been widely investigated within microfluidic setups under different experimental conditions.[79–82] Mao et al. showed a microfluidic technique to measure bacterial chemotaxis within microchannels in which a gradient of chemoeffectors was established via diffusion between parallel streams of liquid having laminar flows.[83] Fluid mixing induced by random bacterial motility in porous microfluidic devices was studied by Singh et al.[84] Rapid estimation of the susceptibility of bacteria toward antibiotics, together with developing an understanding of the mechanism of antibiotic action was performed by Yang et al. using microfluidic devices without resorting to any high-resolution microscopy.[85] Active bacterial assemblies have also been reported to create turbulence within confined geometries, which may have important implications in material transport and mixing.[86,87] In a study reported by Kim and Breuer, bacterial motion was found to significantly influence the diffusion of a passive macromolecule dextran within a Y-shaped microfluidic channel (Fig. 5A).[88] It was observed that, in the presence of bacteria, there was a significant increase in the diffusion coefficient of

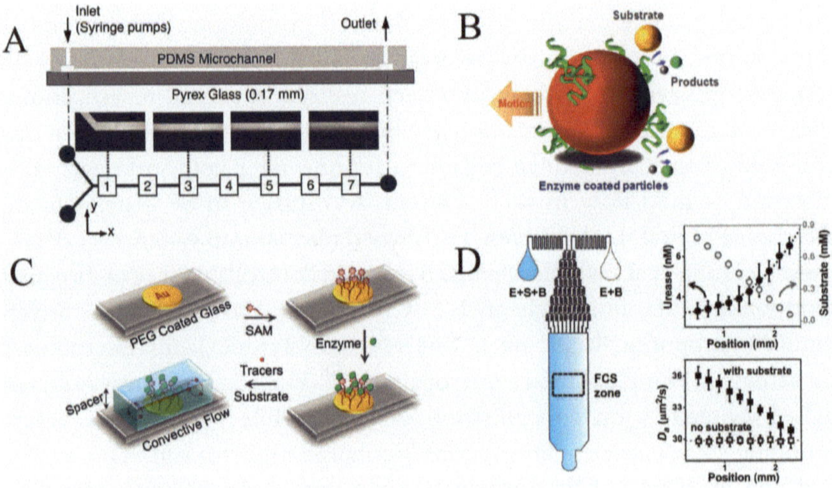

Fig. 5 (A) Schematic representation of the microfluidic setup used by Kim and Breuer to investigate the effect of bacterial motion on the diffusion of fluorescently labeled dextran. (B) Micromotors powered by enzyme catalysis, which show chemotaxis and move toward higher substrate concentration regions within a microfluidic channel. (C) Schematic of an enzyme powered micropump. (D) Design scheme of the microfluidic chip to observe anti-chemotaxis of active enzymes. The enzyme–substrate–buffer (E+S+B) enters one inlet, and the substrate-free enzyme solution in buffer (E+B) enters through the other, producing constant enzyme concentration across the channel but a linear gradient of its substrate. The graph at the top right corner shows urease concentration (solid circles) and calibrated urea concentration (open circles), plotted against position across the microchannel. The graph at the bottom right corner shows enzyme diffusion coefficient (D_a) extracted from FCS autocorrelation plotted against position in the channel, in the presence (solid squares) and absence (open squares) of substrate. *The images (A) and (D) are taken from Kim MJ, Breuer KS. Enhanced diffusion due to motile bacteria.* Phys Fluids. *2004;16(9):L78–81 and Jee A-Y, Dutta S, Cho Y-K, Tlusty T, Granick S. Enzyme leaps fuel antichemotaxis.* Proc Natl Acad Sci U S A. *2018;115(1):14–8, while (B) and (C) are taken from Dey KK, Zhao X, Tansi BM, Méndez-Ortiz WJ, Córdova-Figueroa UM, Golestanian R, et al. Micromotors Powered by Enzyme Catalysis.* Nano Lett. *2015;15(12):8311–5 and Sengupta S, Patra D, Ortiz-Rivera I, Agrawal A, Shklyaev S, Dey KK, et al. Self-powered enzyme micropumps.* Nat Chem. *2014;6(5):415–22 respectively.*

dextran which increased linearly with bacterial concentration. Moreover, with increased bacteria concentration, the nature of the diffusion was observed to undergo a cross-over from normal to superdiffusive behavior. Quantitatively, in normal diffusion, the mean squared displacement of a randomly moving particle varies linearly with time (MSD $\sim t$) whereas in superdiffusive motion the MSD varies nonlinearly with time (MSD $\sim t^{\alpha}$), with $\alpha > 1$. In biohybrid motor systems, biological and artificially imposed functionalities could be utilized jointly through a microrobotic device that can be controlled externally and directed toward specific targets.[89–91]

Microfluidic devices have been extensively used in characterizing the collective behavior of biological nanomotors. The group led by Sen investigated enzyme chemotaxis within model microfluidic networks with an aim to figure out the exact mechanism behind it.[25,52,92] Similar experimental setups have been used previously to study the chemotactic motions of neutrophils and bacteria.[93,94] Sen and his collaborators also demonstrated enzyme chemotaxis over paper based microfluidic devices, which were inexpensive and compatible with biological samples.[95] Using stimulated emission-depletion fluorescence correlation spectroscopy (STED-FCS), Jee et al. reported to observe enzyme migration toward lower substrate concentration and offered an alternative explanation of the observations reported by Sen (Fig. 5D).[96] Tethering active enzymes over large micron-scale particles and vesicles, enzyme powered hybrid motors were fabricated (Fig. 5B).[97–99] Micromotors powered by enzyme catalysis showed controlled chemotactic propulsion within microfluidic chips—offering a promising route to develop smart biocompatible payload carriers capable of functioning under complex physiological environments.[97]

Besides functioning as self-powered nanomotors and powering large micron-scale objects in liquids, active enzyme molecules, when immobilized on a surface, have also been shown to generate enough mechanical forces to cause directional fluid pumping (Fig. 5C).[100–102] The pumping velocity increased with increasing substrate concentration and reaction rate. These rechargeable devices can be triggered by the presence of specific analytes, which enabled the design of enzyme-based smart devices capable of acting both as sensors and pumps.

Apart from experimental investigations, there are many theoretical and computational studies, where non-trivial features of active matter dynamics have been analyzed within microfluidic setups.[103–106] The experimental and theoretical studies, however, are tightly coupled, with novel predictions from the modeling studies providing a springboard for new experimental explorations, and experimental breakthroughs yielding inspiration for the modeling effort. Thampi et al. proposed a microfluidic system that could harness mesoscale turbulence in an active assembly to generate useful work.[107] The group led by Saintillan recently analyzed the effective rheology of dilute suspensions of self-propelled slender particles confined between two infinite parallel plates and subject to a pressure-driven flow.[108] Their results showed that the effect of particle activity was strongest at low flow rates, where the pushers tend to decrease the suspension viscosity while the pullers tend to enhance it. Using both theory and experimental analysis, Das et al. showed the directed motion of spherical Janus micromotors, with their surfaces half coated with platinum, while they move close to the boundaries within a microfluidic channel.[109] The constrained in-plane swimming along the wall results from the active quenching of their Brownian rotation. Such prolonged directed transport is not dependent on any external fields or potentials and this strategy could be used to steer active colloids along arbitrary trajectories within confined microfluidic environments. Takatori et al. studied the use of an acoustic tweezer to confine self-propelled particles in two dimensions over distances large compared to their run lengths, which offered a novel way to monitor the motion and behavior of the swimmers operating under complex confined conditions.[110] A few studies have demonstrated that self-propelled particles exhibit many intriguing phenomena in confined spaces, such as accumulation at boundaries,[111–114] while some other studies have established that confinement of self-propelled particles via harmonic external fields may engender many useful features of active systems.[115–117]

There are other theoretical and computational investigations that sheds light into many novel and fascinating aspects of active matter dynamics operating within confinements. Using simulations, Hernandez-Ortiz et al. predicted large scale coherent vortex formation within populations of confined and hydrodynamically interacting swimming particles.[118] To explain the experimentally observed non-trivial velocity distribution of unicellular swimming algae *Euglena gracilis* within microfluidic channels, Romanczuk et al. developed a theory of active Brownian motion coupled with active fluctuations that provided important insights into the motion characteristics and dynamics patterns of the swimmers.[119] Moreover, using computational studies, it was predicted that incoherent behavior of light controlled swimmers can be rectified by microfluidic devices leading to systematic particle motion, mass transport and controlled pattern formation in fluids.[120] Based on mesoscopic particle-based simulations, a stochastic model was proposed to explain the features of colloidal chemotaxis, that has been experimentally observed within microfluidic devices, and compute the chemotactic force acting on the swimmers.[104] Recently, designs of self-propelled active sheets were proposed by Balazs et al. where self-generated flows affected both particle motion and shape of the swimmers and could manipulate propulsion, perform logic operation and execute multistage processes within fluid filled microchannels.[121] Several other theoretical and experimental studies have offered important insights into the novel and unexpected behavior of confined active matter systems, opening up possibilities to harness their dynamics for important technological applications.[108,122–124]

The study of micro and nanoscopic active matter is currently one of the most fascinating research areas in soft condensed matter physics. Unraveling, predicting, and controlling the behavior of active systems is a challenging endeavor, and requires collaborative efforts of researchers from multiple disciplines. Recent experimental and theoretical endeavors have offered insights into the exciting physics and unprecedented research opportunities in this area, which could lead to the development of smart devices and systems. Microfluidics provides a great opportunity to create devices with micro and nanoscopic active matter that would receive wide appreciation in fields ranging from healthcare to environmental remediation. Although numerous attempts have been made in integrating microfluidic technology with inorganic and hybrid micromotor systems, its usage in characterizing biological active matter is yet to be substantial. It is expected that in the coming years microfluidic technology would be of immense help in addressing the following questions: (a) What are the possible mechanisms of emergent dynamics in active systems over different length scales? (b) What are the ranges and magnitudes of reaction-induced force fields? (c) How are the collective and individual dynamics of active matter expected to change for media that are extremely crowded, inhomogeneous, and have complex rheological features? (d) Are there generic principles based on which active swimmers interact with their surroundings independent of their sizes and propulsion mechanisms? (e) Are there ways to control and harness non-equilibrium features of colloidal assemblies for useful purposes? Understanding the nature of physics governing active matter dynamics at smaller length scales would allow us to utilize the properties of nonequilibrium systems in a variety of novel applications related to transport, assembly, and mixing—especially under biologically relevant conditions. The detection and characterization of nonequilibrium effects in biological systems remain still an open question. It is anticipated that microfluidic technologies, when integrated with advanced imaging and data acquisition capabilities, would enable us to pursue sophisticated experiments with active materials under extremely defined environments. This would not only augment our fundamental understanding in various biological processes and phenomena but would also facilitate development of many advanced technologies related to healthcare and medicine based on functional micro/nanomotors.

References

1. Brangwynne CP, Koenderink GH, MacKintosh FC, Weitz DA. Cytoplasmic diffusion: molecular motors mix it up. *J Cell Biol*. 2008;183(4):583–587.
2. Hess H, Bachand GD. Biomolecular motors. *Mater Today*. 2005;8(12, Supplement 1): 22–29.
3. Brangwynne CP, Koenderink GH, MacKintosh FC, Weitz DA. Intracellular transport by active diffusion. *Trends Cell Biol*. 2009;19(9):423–427.
4. Mogre SS, Brown AI, Koslover EF. Getting around the cell: physical transport in the intracellular world. *Phys Biol*. 2020;17(6), 061003.
5. Panda S, Pyarajan S. Lab-on-chip devices for protein analysis. In: Li D, ed. *Encyclopedia of Microfluidics and Nanofluidics*. Boston, MA: Springer US; 2008:940–947.
6. Sin ML, Gao J, Liao JC, Wong PK. System integration—a major step toward lab on a chip. *J Biol Eng*. 2011;5(1):6.
7. Zhao Y, Chen D, Yue H, et al. Lab-on-a-chip technologies for single-molecule studies. *Lab Chip*. 2013;13(12):2183–2198.
8. Culbertson CT, Mickleburgh TG, Stewart-James SA, Sellens KA, Pressnall M. Micro total analysis systems: fundamental advances and biological applications. *Anal Chem*. 2014;86(1):95–118.
9. Fan J, Li S, Wu Z, Chen Z. Chapter 3—diffusion and mixing in microfluidic devices. In: Santos HA, Liu D, Zhang H, eds. *Microfluidics for Pharmaceutical Applications*. William Andrew Publishing; 2019:79–100.
10. Weigl BH, Yager P. Microfluidic diffusion-based separation and detection. *Science*. 1999;283(5400):346–347.
11. Wang X, Liu Z, Pang Y. Concentration gradient generation methods based on microfluidic systems. *RSC Adv*. 2017;7(48):29966–29984.
12. Squires TM, Quake SR. Microfluidics: fluid physics at the nanoliter scale. *Rev Mod Phys*. 2005;77(3):977–1026.
13. Beebe DJ, Mensing GA, Walker GM. Physics and applications of microfluidics in biology. *Annu Rev Biomed Eng*. 2002;4(1):261–286.
14. Convery N, Gadegaard N. 30 years of microfluidics. *Micro Nano Eng*. 2019;2:76–91.
15. Ku DN. Blood flow in arteries. *Annu Rev Fluid Mech*. 1997;29(1):399–434.
16. Wolgemuth CW. Collective swimming and the dynamics of bacterial turbulence. *Biophys J*. 2008;95(4):1564–1574.
17. Yew M, Ren Y, Koh KS, Sun C, Snape C. A review of state-of-the-art microfluidic technologies for environmental applications: detection and remediation. *Glob Chall*. 2019;3(1):1800060.
18. Zhang X, Xia K, Ji A. A portable plug-and-play syringe pump using passive valves for microfluidic applications. *Sens Actuators B*. 2020;304:127331.
19. Ren K, Zhou J, Wu H. Materials for microfluidic chip fabrication. *Acc Chem Res*. 2013;46(11):2396–2406.
20. Xia Y, Whitesides GM. Soft lithography. *Annu Rev Mater Sci*. 1998;28(1):153–184.

21. Sia SK, Whitesides GM. Microfluidic devices fabricated in poly(dimethylsiloxane) for biological studies. *Electrophoresis*. 2003;24(21):3563–3576.
22. Xia Y, Whitesides GM. Soft lithography. *Angew Chem Int Ed*. 1998;37(5):550–575.
23. McDonald JC, Duffy DC, Anderson JR, et al. Fabrication of microfluidic systems in poly(dimethylsiloxane). *Electrophoresis*. 2000;21(1):27–40.
24. Ji X, PKL C. Highly sensitive glucose sensor based on organic electrochemical transistor with modified gate electrode. In: Prickril B, Rasooly A, eds. *Biosensors and Biodetection*. Springer New York; 2017:205–216.
25. Dey KK, Das S, Poyton MF, et al. Chemotactic separation of enzymes. *ACS Nano*. 2014;8(12):11941–11949.
26. Oh KW, Lee K, Ahn B, Furlani EP. Design of pressure-driven microfluidic networks using electric circuit analogy. *Lab Chip*. 2012;12(3):515–545.
27. Li Z, Yi Mak S, Sauret A, Cheung SH. Syringe-pump-induced fluctuation in all-aqueous microfluidic system implications for flow rate accuracy. *Lab Chip*. 2014;14(4):744–749.
28. Chang C-C, Yang R-J. Electrokinetic mixing in microfluidic systems. *Microfluid Nanofluid*. 2007;3(5):501–525.
29. Horiuchi K, Dutta P. Electrokinetic flow control in microfluidic chips using a field-effect transistor. *Lab Chip*. 2006;6(6):714–723.
30. Cabrera CR, Yager P. Continuous concentration of bacteria in a microfluidic flow cell using electrokinetic techniques. *Electrophoresis*. 2001;22(2):355–362.
31. Gompper G, Winkler RG, Speck T, et al. The 2020 motile active matter roadmap. *J Phys Condens Matter*. 2020;32(19):193001.
32. Dey KK. Dynamic coupling at low Reynolds number. *Angew Chem Int Ed*. 2019;58(8):2208–2228.
33. Ismagilov RF, Schwartz A, Bowden N, Whitesides GM. Autonomous movement and self-assembly. *Angew Chem Int Ed*. 2002;41(4):652–654.
34. Paxton WF, Kistler KC, Olmeda CC, et al. Catalytic nanomotors: autonomous movement of striped nanorods. *J Am Chem Soc*. 2004;126(41):13424–13431.
35. Fournier-Bidoz S, Arsenault AC, Manners I, Ozin GA. Synthetic self-propelled nanorotors. *Chem Commun*. 2005;4:441–443.
36. Wong F, Dey KK, Sen A. Synthetic micro/nanomotors and pumps: fabrication and applications. *Annu Rev Mat Res*. 2016;46(1):407–432.
37. Gibbs J, Zhao Y. Catalytic nanomotors: fabrication, mechanism, and applications. *Front Mater Sci*. 2011;5(1):25–39.
38. Wang J. Template electrodeposition of catalytic nanomotors. *Faraday Discuss*. 2013;164:9–18.
39. Liu R, Sen A. Autonomous nanomotor based on copper–platinum segmented nanobattery. *J Am Chem Soc*. 2011;133(50):20064–20067.
40. Gao W, Uygun A, Wang J. Hydrogen-bubble-propelled zinc-based microrockets in strongly acidic media. *J Am Chem Soc*. 2012;134(2):897–900.
41. Solovev AA, Mei Y, Ureña EB, Huang G, Schmidt OG. Catalytic microtubular jet engines self-propelled by accumulated gas bubbles. *Small*. 2009;5(14):1688–1692.
42. Parmar J, Ma X, Katuri J, et al. Nano and micro architectures for self-propelled motors. *Sci Technol Adv Mater*. 2015;16(1), 014802.
43. Wang Q, Dong R, Wang C, et al. Glucose-fueled micromotors with highly efficient visible-light photocatalytic propulsion. *ACS Appl Mater Interfaces*. 2019;11(6):6201–6207.
44. Ye Y, Luan J, Wang M, et al. Fabrication of self-propelled micro- and nanomotors based on Janus structures. *Chem A Eur J*. 2019;25(37):8663–8680.
45. Pourrahimi AM, Pumera M. Multifunctional and self-propelled spherical Janus nano/micromotors: recent advances. *Nanoscale*. 2018;10(35):16398–16415.

46. Jurado-Sánchez B, Pacheco M, Maria-Hormigos R, Escarpa A. Perspectives on Janus micromotors: materials and applications. *Appl Mater Today*. 2017;9:407–418.
47. Gao W, Liu M, Liu L, Zhang H, Dong B, Li CY. One-step fabrication of multifunctional micromotors. *Nanoscale*. 2015;7(33):13918–13923.
48. Kochergin YS, Villa K, Novotný F, Plutnar J, Bojdys MJ, Pumera M. Multifunctional visible-light powered micromotors based on semiconducting sulfur- and nitrogen-containing donor–acceptor polymer. *Adv Funct Mater*. 2020;30(38):2002701.
49. Gachelin J, Rousselet A, Lindner A, Clement E. Collective motion in an active suspension of Escherichia coli bacteria. *New J Phys*. 2014;16(2), 025003.
50. Zhang HP, Be'er A, Florin E-L, Swinney HL. Collective motion and density fluctuations in bacterial colonies. *Proc Natl Acad Sci U S A*. 2010;107(31):13626–13630.
51. Muddana HS, Sengupta S, Mallouk TE, Sen A, Butler PJ. Substrate catalysis enhances single-enzyme diffusion. *J Am Chem Soc*. 2010;132(7):2110–2111.
52. Sengupta S, Dey KK, Muddana HS, et al. Enzyme molecules as nanomotors. *J Am Chem Soc*. 2013;135(4):1406–1414.
53. Sengupta S, Spiering MM, Dey KK, et al. DNA polymerase as a molecular motor and pump. *ACS Nano*. 2014;8(3):2410–2418.
54. Pumera M. Nanomaterials meet microfluidics. *Chem Commun*. 2011;47(20):5671–5680.
55. Wang J. Cargo-towing synthetic nanomachines: towards active transport in microchip devices. *Lab Chip*. 2012;12(11):1944–1950.
56. García M, Orozco J, Guix M, et al. Micromotor-based lab-on-chip immunoassays. *Nanoscale*. 2013;5(4):1325–1331.
57. Parmar J, Vilela D, Villa K, Wang J, Sánchez S. Micro- and nanomotors as active environmental microcleaners and sensors. *J Am Chem Soc*. 2018;140(30):9317–9331.
58. Baraban L, Makarov D, Streubel R, et al. Catalytic Janus motors on microfluidic chip: deterministic motion for targeted cargo delivery. *ACS Nano*. 2012;6(4):3383–3389.
59. Sanchez S, Solovev AA, Harazim SM, Schmidt OG. Microbots swimming in the flowing streams of microfluidic channels. *J Am Chem Soc*. 2011;133(4):701–703.
60. Burdick J, Laocharoensuk R, Wheat PM, Posner JD, Wang J. Synthetic nanomotors in microchannel networks: directional microchip motion and controlled manipulation of cargo. *J Am Chem Soc*. 2008;130(26):8164–8165.
61. Chang ST, Paunov VN, Petsev DN, Velev OD. Remotely powered self-propelling particles and micropumps based on miniature diodes. *Nat Mater*. 2007;6(3):235–240.
62. Chang ST, Beaumont E, Petsev DN, Velev OD. Remotely powered distributed microfluidic pumps and mixers based on miniature diodes. *Lab Chip*. 2007;8(1):117–124.
63. Restrepo-Pérez L, Soler L, Martínez-Cisneros C, Sánchez S, Schmidt OG. Biofunctionalized self-propelled micromotors as an alternative on-chip concentrating system. *Lab Chip*. 2014;14(16):2914–2917.
64. Zou M, Wang J, Yu Y, et al. Composite multifunctional micromotors from droplet microfluidics. *ACS Appl Mater Interfaces*. 2018;10(40):34618–34624.
65. Zhu H, Nawar S, Werner JG, et al. Hydrogel micromotors with catalyst-containing liquid core and shell. *J Phys Condens Matter*. 2019;31(21):214004.
66. Adams LLA, Lee D, Mei Y, Weitz DA, Solovev AA. Nanoparticle-shelled catalytic bubble micromotor. *Adv Mater Interfaces*. 2020;7(4):1901583.
67. Yu Y, Guo J, Zou M, Cai L, Zhao Y. Micromotors from microfluidics. *Chem Asian J*. 2019;14(14):2417–2430.
68. Yu Y, Shang L, Gao W, Zhao Z, Wang H, Zhao Y. Microfluidic lithography of bioinspired helical micromotors. *Angew Chem Int Ed*. 2017;56(40):12127–12131.
69. Zhao G, Pumera M. Marangoni self-propelled capsules in a maze: pollutants 'sense and act' in complex channel environments. *Lab Chip*. 2014;14(15):2818–2823.

70. Kumar S, Singh AK, Dasmahapatra AK, Mandal TK, Bandyopadhyay D. Graphene based multifunctional superbots. *Carbon*. 2015;89:31–40.
71. Vilela D, Parmar J, Zeng Y, Zhao Y, Sánchez S. Graphene-based microbots for toxic heavy metal removal and recovery from water. *Nano Lett*. 2016;16(4):2860–2866.
72. Dong Y, Yi C, Yang S, et al. A substrate-free graphene oxide-based micromotor for rapid adsorption of antibiotics. *Nanoscale*. 2019;11(10):4562–4570.
73. Dong Y, Wang L, Wang J, et al. Graphene-based helical micromotors constructed by "microscale liquid rope-coil effect" with microfluidics. *ACS Nano*. 2020;14 (12):16600–16613.
74. Mandal P, Chopra V, Ghosh A. Independent positioning of magnetic nanomotors. *ACS Nano*. 2015;9(5):4717–4725.
75. Zizzari A, Cesaria M, Bianco M, et al. Mixing enhancement induced by viscoelastic micromotors in microfluidic platforms. *Chem Eng J*. 2020;391:123572.
76. Ibele ME, Wang Y, Kline TR, Mallouk TE, Sen A. Hydrazine fuels for bimetallic catalytic microfluidic pumping. *J Am Chem Soc*. 2007;129(25):7762–7763.
77. Hong Y, Diaz M, Córdova-Figueroa UM, Sen A. Light-driven titanium-dioxide-based reversible microfireworks and micromotor/micropump systems. *Adv Funct Mater*. 2010;20(10):1568–1576.
78. Chang X, Chen C, Li J, et al. Motile micropump based on synthetic micromotors for dynamic micropatterning. *ACS Appl Mater Interfaces*. 2019;11(31):28507–28514.
79. Englert DL, Manson MD, Jayaraman A. Flow-based microfluidic device for quantifying bacterial chemotaxis in stable, competing gradients. *Appl Environ Microbiol*. 2009;75 (13):4557–4564.
80. Sipos O, Nagy K, Galajda P. Patterns of collective bacterial motion in microfluidic devices. *Chem Biochem Eng Q*. 2014;28(2):233–240.
81. Gurung JP, Gel M, Baker MAB. Microfluidic techniques for separation of bacterial cells via taxis. *Microb Cell*. 2020;7(3):66–79.
82. Männik J, Driessen R, Galajda P, Keymer JE, Dekker C. Bacterial growth and motility in sub-micron constrictions. *Proc Natl Acad Sci U S A*. 2009;106(35): 14861–14866.
83. Mao H, Cremer PS, Manson MD. A sensitive, versatile microfluidic assay for bacterial chemotaxis. *Proc Natl Acad Sci U S A*. 2003;100(9):5449–5454.
84. Singh R, Olson MS. Transverse mixing enhancement due to bacterial random motility in porous microfluidic devices. *Environ Sci Technol*. 2011;45(20):8780–8787.
85. Yang Y, Gupta K, Ekinci KL. All-electrical monitoring of bacterial antibiotic susceptibility in a microfluidic device. *Proc Natl Acad Sci U S A*. 2020;117(20):10639–10644.
86. Wensink HH, Dunkel J, Heidenreich S, et al. Meso-scale turbulence in living fluids. *Proc Natl Acad Sci U S A*. 2012;109(36):14308–14313.
87. Dunkel J, Heidenreich S, Drescher K, Wensink HH, Bär M, Goldstein RE. Fluid dynamics of bacterial turbulence. *Phys Rev Lett*. 2013;110(22):228102.
88. Kim MJ, Breuer KS. Enhanced diffusion due to motile bacteria. *Phys Fluids*. 2004;16 (9):L78–L81.
89. Schwarz L, Medina-Sánchez M, Schmidt OG. Hybrid BioMicromotors. *Appl Phys Rev*. 2017;4(3), 031301.
90. Sun M, Fan X, Meng X, et al. Magnetic biohybrid micromotors with high maneuverability for efficient drug loading and targeted drug delivery. *Nanoscale*. 2019;11 (39):18382–18392.
91. Xu H, Medina-Sánchez M, Magdanz V, Schwarz L, Hebenstreit F, Schmidt OG. Sperm-hybrid micromotor for targeted drug delivery. *ACS Nano*. 2018;12 (1):327–337.
92. Zhao X, Palacci H, Yadav V, et al. Substrate-driven chemotactic assembly in an enzyme cascade. *Nat Chem*. 2018;10(3):311–317.

93. Zhao W, Zhao H, Li M, Huang C. Microfluidic devices for neutrophil chemotaxis studies. *J Transl Med*. 2020;18(1):168.
94. Song J, Zhang Y, Zhang C, et al. A microfluidic device for studying chemotaxis mechanism of bacterial cancer targeting. *Sci Rep*. 2018;8(1):6394.
95. Ilacas GC, Basa A, Sen A, Gomez FA. Enzyme chemotaxis on paper-based devices. *Anal Sci*. 2018;34(1):115–119.
96. Jee A-Y, Dutta S, Cho Y-K, Tlusty T, Granick S. Enzyme leaps fuel antichemotaxis. *Proc Natl Acad Sci U S A*. 2018;115(1):14–18.
97. Dey KK, Zhao X, Tansi BM, et al. Micromotors powered by enzyme catalysis. *Nano Lett*. 2015;15(12):8311–8315.
98. Keller S, Teora SP, Hu GX, Nijemeisland M, Wilson DA. High-throughput design of biocompatible enzyme-based hydrogel microparticles with autonomous movement. *Angew Chem Int Ed*. 2018;57(31):9814–9817.
99. Somasundar A, Ghosh S, Mohajerani F, et al. Positive and negative chemotaxis of enzyme-coated liposome motors. *Nat Nanotechnol*. 2019;14(12):1129–1134.
100. Sengupta S, Patra D, Ortiz-Rivera I, et al. Self-powered enzyme micropumps. *Nat Chem*. 2014;6(5):415–422.
101. Ortiz-Rivera I, Shum H, Agrawal A, Sen A, Balazs AC. Convective flow reversal in self-powered enzyme micropumps. *Proc Natl Acad Sci U S A*. 2016;113(10):2585–2590.
102. Ortiz-Rivera I, Courtney TM, Sen A. Enzyme micropump-based inhibitor assays. *Adv Funct Mater*. 2016;26(13):2135–2142.
103. Cai S-C, Shen Y-X, Io C-W. The mesoscopic collective motion of self-propelling active particle suspension confined in two-dimensional micro-channel. *J Phys Condens Matter*. 2019;32(9), 095101.
104. Deprez L, de Buyl P. Passive and active colloidal chemotaxis in a microfluidic channel: mesoscopic and stochastic models. *Soft Matter*. 2017;13(19):3532–3543.
105. Wioland H, Lushi E, Goldstein RE. Directed collective motion of bacteria under channel confinement. *New J Phys*. 2016;18(7), 075002.
106. Woodhouse FG, Dunkel J. Active matter logic for autonomous microfluidics. *Nat Commun*. 2017;8(1):1–7.
107. Thampi SP, Doostmohammadi A, Shendruk TN, Golestanian R, Yeomans JM. Active micromachines: microfluidics powered by mesoscale turbulence. *Sci Adv*. 2016;2(7), e1501854.
108. Alonso-Matilla R, Ezhilan B, Saintillan D. Microfluidic rheology of active particle suspensions: kinetic theory. *Biomicrofluidics*. 2016;10(4), 043505.
109. Das S, Garg A, Campbell AI, et al. Boundaries can steer active Janus spheres. *Nat Commun*. 2015;6(1):8999.
110. Takatori SC, De Dier R, Vermant J, Brady JF. Acoustic trapping of active matter. *Nat Commun*. 2016;7(1):10694.
111. Sokolov A, Apodaca MM, Grzybowski BA, Aranson IS. Swimming bacteria power microscopic gears. *Proc Natl Acad Sci U S A*. 2010;107(3):969–974.
112. Berke AP, Turner L, Berg HC, Lauga E. Hydrodynamic attraction of swimming microorganisms by surfaces. *Phys Rev Lett*. 2008;101(3), 038102.
113. Fily Y, Baskaran A, Hagan MF. Dynamics of self-propelled particles under strong confinement. *Soft Matter*. 2014;10(30):5609–5617.
114. Yang X, Manning ML, Marchetti MC. Aggregation and segregation of confined active particles. *Soft Matter*. 2014;10(34):6477–6484.
115. Nash RW, Adhikari R, Tailleur J, Cates ME. Run-and-tumble particles with hydrodynamics: sedimentation, trapping, and upstream swimming. *Phys Rev Lett*. 2010;104(25):258101.
116. Wang Z, Chen H-Y, Sheng Y-J, Tsao H-K. Diffusion, sedimentation equilibrium, and harmonic trapping of run-and-tumble nanoswimmers. *Soft Matter*. 2014;10(18):3209–3217.

117. Tailleur J, Cates ME. Sedimentation, trapping, and rectification of dilute bacteria. *EPL Europhys Lett.* 2009;86(6):60002.
118. Hernandez-Ortiz JP, Stoltz CG, Graham MD. Transport and collective dynamics in suspensions of confined swimming particles. *Phys Rev Lett.* 2005;95(20):204501.
119. Romanczuk P, Romensky M, Scholz D, Lobaskin V, Schimansky-Geier L. Motion of Euglena gracilis: active fluctuations and velocity distribution. *Eur Phys J Spec Top.* 2015;224(7):1215–1229.
120. Stenhammar J, Wittkowski R, Marenduzzo D, Cates ME. Light-induced self-assembly of active rectification devices. *Sci Adv.* 2016;2(4), e1501850.
121. Laskar A, Shklyaev OE, Balazs AC. Designing self-propelled, chemically active sheets: wrappers, flappers, and creepers. *Sci Adv.* 2018;4(12), eaav1745.
122. Yu T, Athanassiadis AG, Popescu MN, et al. Microchannels with self-pumping walls. *ACS Nano.* 2020;14(10):13673–13680.
123. Bricard A, Caussin J-B, Das D, et al. Emergent vortices in populations of colloidal rollers. *Nat Commun.* 2015;6(1):7470.
124. Alonso S, Mikhailov AS. Towards active microfluidics: interface turbulence in thin liquid films with floating molecular machines. *Phys Rev E.* 2009;79(6), 061906.

CHAPTER ELEVEN

Paper-microfluidic signal-enhanced immunoassays

N. Sathishkumar and Bhushan J. Toley*
Department of Chemical Engineering, Indian Institute of Science, Bengaluru, KA, India
*Corresponding author: e-mail address: bhushan@iisc.ac.in

Contents

1. Introduction and the history of paper-based immunoassays 267
2. Classification of signal enhancement strategies used in paper-based immunoassays 271
   2.1 Nanomaterials based signal enhancement strategies 271
   2.2 Enzyme-based labels for signal enhancement 276
   2.3 Polymerization and hybrid labels-based signal enhancement strategies 280
3. SWOT analysis: Paper-based immunoassays 282
References 284

Abstract

Over the past decade, paper-based microfluidic devices have become popular for their simplicity and ability to conduct diagnostic tests at a low cost. An important class of diagnostic assays that paper-based analytical devices have been used for is immunoassays. The lateral flow immunoassay (LFIA), of which the home pregnancy test is the most prominent example, has been one of the most commercially successful membrane-based diagnostic tests. Yet, the analytical sensitivity of LFIAs is lower than the corresponding laboratory technique called ELISA (enzyme-linked immunoassay). As a consequence, traditional LFIAs fail to deliver on the promise of bedside diagnostic testing for many applications. Recognizing this shortcoming, several new developments have been made by researchers to enhance the sensitivity of membrane-based immunoassays. In this chapter, we present the various strategies that have been employed to this end. In the end, we present a brief SWOT analysis to guide future work in this area.

1. Introduction and the history of paper-based immunoassays

Immunoassays are used for qualitative or quantitative analysis of analytes, typically biomarkers, through the use of antibodies and antigens as biorecognition agents.[1] The analyte of interest, which could range from

Progress in Molecular Biology and Translational Science, Volume 186
ISSN 1877-1173
https://doi.org/10.1016/bs.pmbts.2021.07.015

small molecules to macromolecules, reacts with the biorecognition agent forming an immunocomplex, and by labeling the biorecognition agent, the analyte can be measured either qualitatively or quantitatively. Membrane-based immunoassays date back to the 1950s, deriving their roots from the latex agglutination assay.[2] The most significant developments in this field were the introduction of radio-immunoassays[3] followed by enzyme-linked immunosorbent assays (ELISAs). The development of membrane-based lateral flow immunoassays (LFIAs) started in the 1970s facilitated by advancements in various fields such as production and purification of monoclonal antibodies (most prominent biorecognition agent), manufacturing of nitrocellulose membranes, and development of reagent-dispensing equipment. LFIAs reached the commercial markets in the early 1990s and are now used for testing of many analytes, including disease biomarkers in humans and animals, environmental and agricultural contaminants, and biowarfare agents.[4]

A typical lateral flow test strip is shown in Fig. 1A. It is a one-step assay composed of overlapping membranes—a sample pad, a conjugate pad, a nitrocellulose test membrane, and a wicking pad, placed serially in this order and supported on a backing card. A sample containing the target is applied on the pretreated sample pad, which gets wicked into the conjugate release pad containing labeled antibodies specific to the target analyte, conjugated with a visual label such as a nanoparticle (NP) thus forming NP-analyte-antibody complexes. This complex then migrates along a porous membrane (usually composed of either nitrocellulose or polyvinylidene difluoride (PVDF)) and gets captured at the test line via a capture antibody. Unbound NP-analyte-antibody complexes bind to the control line via a different antibody. Recognition of the analyte results in an appropriate response on the test line, while a response on the control line indicates proper liquid flow through the membrane. The signal readout could be by the naked eye or using a reader. In general, LFIAs are simple, user-friendly, low-cost ($2–$12 per strip), and rapid (10–15 min) test devices. Most commercialized LFIAs provide a yes/no answer. Another type of one-step immunoassay is the dipstick immunoassay, mainly utilized in routine health checkups. A typical dipstick immunoassay is shown in Fig. 1B. Here, the conjugate pad and sample pad are eliminated, and the nitrocellulose strip is directly immersed into the test solution containing the NP-biomolecule conjugate, eliminating the need to dry-down the NP–biomolecule conjugate.[5] Note that highly inexpensive (<$1) urine dipsticks are commercially available to diagnose various health conditions,[6] but those are based on direct color-change assays, not immunoassays.

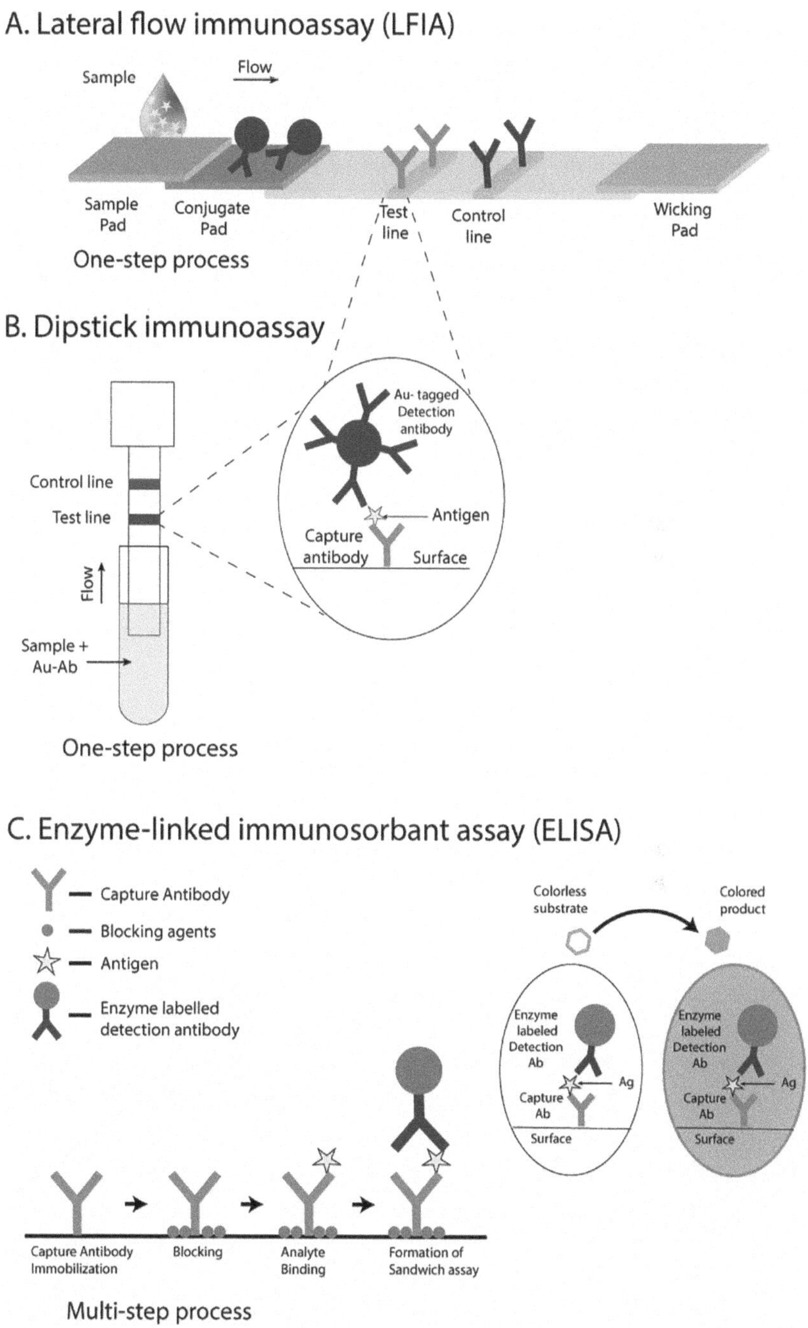

Fig. 1 Various immunoassay Schemes. (A) Lateral flow immunoassay (LFIA). (B) Dipstick immunoassay. (C) Enzyme-linked immunosorbent assay (ELISA).

Although simple and portable, both LFIAs and dipsticks suffer from poor sensitivity when compared to conventional laboratory techniques such as ELISA (enzyme-linked immunosorbent assay). ELISA enhances the sensitivity of detection by coupling the detection antibody to an enzyme, which catalyzes the conversion of a substrate into a colored product. However, it is a multistep assay not amenable to be conducted outside a laboratory. A schematic of the ELISA is shown in Fig. 1C. It is typically conducted in a well plate. The bottom surface of the well is coated with the capture antibody, and the remaining sites are blocked with a sacrificial protein. The sample containing the target analyte is added and incubated, and the surface is washed to remove nonspecifically bound molecules. This is followed by the addition of the enzyme-linked secondary antibody. After multiple washing steps, a colorless substrate is added, which is converted into a colored product, catalyzed by the enzyme. Newer ELISA-like techniques replace colorimetry by fluorescence, electrochemiluminescene (ECL), or chemiluminescence and substantially increase the sensitivity and multiplexing capability. However, these techniques require sophisticated detection instruments and trained personnel to conduct the test and thus are not compliant with low-resource settings.

Paper-based signal-enhanced immunoassays were introduced to bridge the above gap and enable conducting lab-quality immunoassays at the point-of-care. Note that in several texts, paper-based analytical devices are designated as μPADs (microfluidic paper analytical devices). Also, note the use of the term "paper" in this chapter refers to the use of any and all porous membranes. The development of paper-based immunoassays is an emerging field, but a plethora of ideas have emerged over the last decade. Primary innovations involve clever use of paper geometries to guide fluids in a programmed manner,[7] patterning paper to create isolated reaction zones as replacements for well plates,[8] use of cell phone imaging to replace spectrophotometers and other sophisticated instruments,[9] and incorporation of highly sensitive detection modalities like fluorescence into paper, coupled with low-cost readout instruments.[10] Increasingly, paper-based analytical devices are being envisioned as diagnostic devices that could satisfy WHO's ASSURED (Affordable, Sensitive, Specific, User-friendly, Rapid and robust, Equipment-free, and Delivered to the end-users) criteria for POC diagnostic devices.[11] Several excellent review articles are available in the literature explaining the various signal enhancement strategies for microfluidic devices.[12–16] In this chapter, we discuss signal enhancement techniques that have been used in paper-based immunoassays.

2. Classification of signal enhancement strategies used in paper-based immunoassays

In this chapter, we classify signal-enhancement strategies based on the type of labels used as (i) nanomaterials-based, (ii) enzyme-based, (iii) polymer-based and (iv) hybrid labels-based immunoassays.

2.1 Nanomaterials based signal enhancement strategies

Nanomaterials are widely used as labels in LFIA on account of their physical, chemical, thermal, and optical properties. Several nanomaterials such as colloidal gold, silver, copper, silica, carbon, and magnetic nanoparticles (MNPs) have been used as labels.

2.1.1 Colloidal gold nanomaterials-based assays

Colloidal gold is the most routinely used label for commercial LFIAs. The strategies used to enhance the signal of gold nanoparticle-based immunoassays are based on (i) modifying the size and shape of the particles, (ii) laser excitation, (iii) silver-based amplification, and (iv) gold-based amplification.

2.1.1.1 Signal enhancement using different sizes and shapes of AuNP

Gold nanoparticles used in traditional LFIAs vary from 30 to 40 nm in size and are spherical in shape. Improvement in the limit of detection is generally observed with an increase in the AuNP size, and it has been shown that the optimum AuNP size is 100 nm, beyond which steric hindrance occurs at the test zone.[17,18] AuNPs can be synthesized in various shapes such as nanospheres, nanorods, nanoshells and nanostars. Among these morphologies, gold nanostars exhibit better sensitivity than nanospheres and nanorods, especially when SERS is used as a detection method.[19]

2.1.1.2 Signal enhancement using laser excitation of AuNP

AuNPs exhibit localized surface plasmon resonance (LSPR), a phenomenon in which conduction band electrons in nanoparticles undergo collective oscillation when excited by the incident electromagnetic radiation, causing localized effects such as enhanced electromagnetic fields and heating. This principle led to the development of an enhancement technique named thermal contrast amplification (TCA).[20] In TCA, the AuNPs at the test and control line of the LFA strip were activated by irradiating with a 532 nm laser (Fig. 2A). The thermal contrast signal recorded by an IR camera showed a

32 × improvement in the sensitivity when compared with traditional LFIA. It is significant to note that the implementation of TCA with LFIA does not require any design change of the strip. Over the past decade, a flurry of methods has been developed by exploiting the laser excitation of plasmonic nanoparticles, including thermophotonic lock-in imaging (TPLI), photoacoustic (PA) signal enhancement and photothermal laser speckle imaging (PT-LSI).[16] All laser excitation-based amplification techniques are simple in the sense that they do not involve modification of the LFIA. However, they suffer from few disadvantages such as bulky instrumentation, need for an external reader and higher data acquisition time. In addition, they also need better optimization as the distribution of plasmonic nanomaterials is not uniform throughout the test zone.

2.1.1.3 Signal enhancement using silver amplification

Silver enhancement of gold nanoparticles (AuNPs) is a well-established method and is based on the reduction of silver ions to silver atoms by the reducing agent at the site of AuNPs, which act as development nuclei. The precipitation of silver atoms leads to the formation of large nanocrystals at or near the site of AuNPs, resulting in an enhanced signal. Introduced in the early 1980s, this method revolutionized the direct visualization of tissue sections under light microscopy[21] and was later adopted in many paper-based diagnostic devices.[22,23] Wada et al. introduced a fully automated immunochromatographic kit to detect influenza virus (Fig. 2B) using a silver-based amplification technique.[23] Recently, Fujifilm has developed a POC-compatible LFIA utilizing the silver enhancement technique for early detection of Tuberculosis.[24] One drawback of silver-based amplification is that it suffers from auto-nucleation and results in a high background signal. Controlling the reaction time of silver enhancement is crucial for lowering the background signal.

2.1.1.4 Signal enhancement using gold amplification

Similar to silver enhancement, gold enhancement of AuNP also works on the principle of reduction of Au ions at the site of AuNPs, resulting in an enhanced signal. Gold enhancement method for direct visualization of tissue sections under light microscopy was first introduced in 1991.[25] Like the silver amplification method, this principle was later adopted in many paper-based diagnostic devices.[26–28] Fu et al. utilized the gold enhancement method to detect PfHRP2 using an automated paper immunoassay.[27] Signal-enhanced dipstick immunoassays for the detection of two foodborne

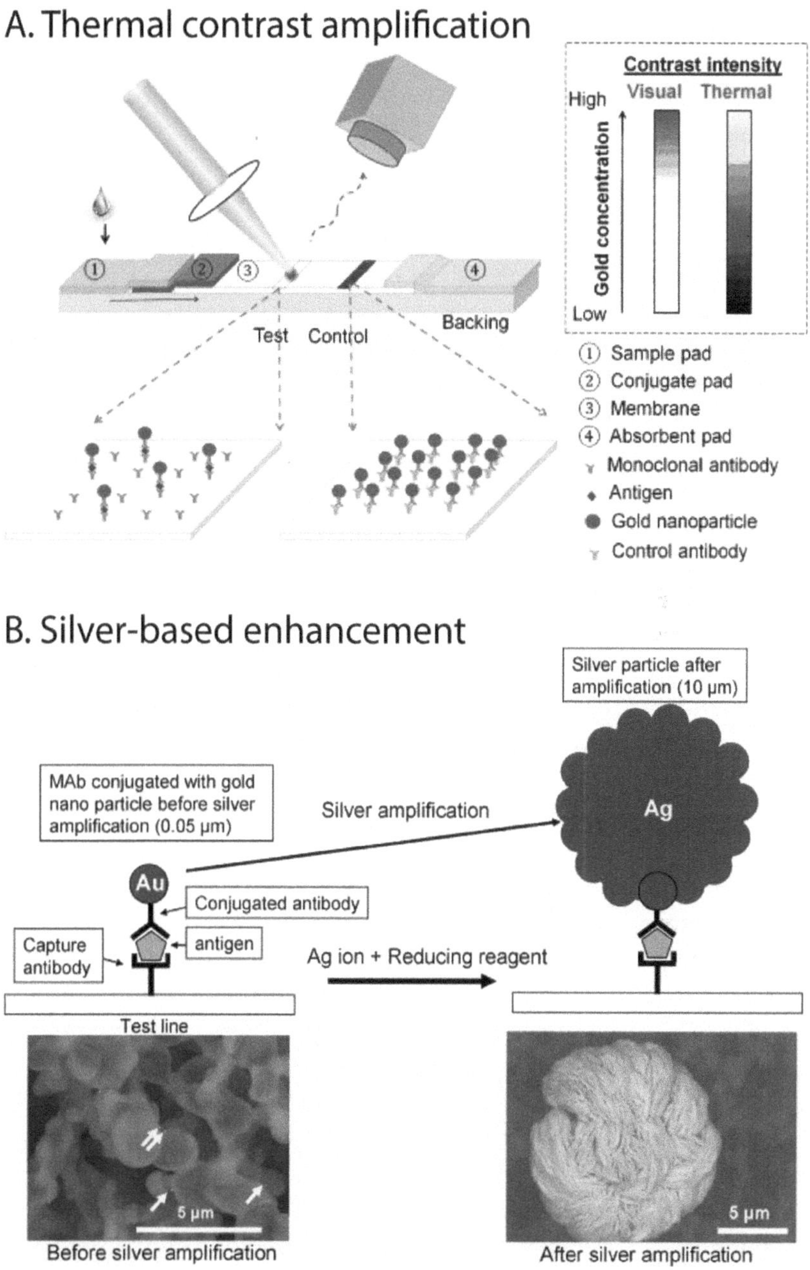

Fig. 2 Gold nanoparticle (AuNP)-based signal enhancement in paper-based immunoassay. (A) Concept of thermal contrast for LFIAs. (B) The principle of the silver-based amplification method. *Panel (A) adapted with permission from Qin Z, Chan WCW, Boulware DR, Akkin T, Butler EK, Bischof JC. Significantly improved analytical sensitivity of lateral flow immunoassays by using thermal contrast. Angew Chem Int Ed 2012;51 (18):4358–4361. doi:10.1002/anie.201200997, copyright 2012 John Wiley and Sons. Panel (B) reproduced from Wada A, Sakoda Y, Oyamada T, Kida H. Development of a highly sensitive immunochromatographic detection kit for H5 influenza virus hemagglutinin using silver amplification. Journal of Virological Methods 2011;178(1–2):82–86. doi:https://doi.org/10.1016/j.jviromet.2011.08.017, with permission from Elsevier.*

pathogens also utilized the gold enhancement method.[28] Gold enhancement shows better potential in signal amplification when compared to the silver enhancement method because auto-nucleation is minimal even after exposing the membrane for a long time (1 − 2 h).[26]

2.1.2 Carbon nanomaterials-based assays

Carbon-based nanomaterials have been used as alternative labels to AuNPs. A desirable feature is that they produce a black signal over a white background, increasing contrast and facilitating visual readout. The manufacturing of carbon nanomaterials requires lower-cost raw materials than gold or silver-based nanoparticles. Carbon nanomaterials, including carbon nanoparticles (CNPs), carbon nanotubes (CNTs) and graphene, can be used as labels.

Sensitivities of CNP-based assays are comparable to that of ELISA; for example, the limit of detection of an LFIA for circulating cathodic antigen (CCA) was found to be similar to the corresponding CCA-ELISA.[29] A multiplex lateral flow assay based on amorphous carbon nanoparticles (ACNP-LFA) showed an eightfold improvement in LOD when compared to a similar quantum dot-based assay for the detection of three Fusarium mycotoxins.[30] Compared to CNPs, carbon nanotubes (CNTs) have a higher specific surface area and can absorb many biomolecules per carbon volume. They also exhibit high electrical conductivity, and hence a simple resistance measurement across the membrane is enough for the quantification of the antigen. Both single-walled carbon nanotubes (SWCNT) and multiwalled carbon nanotubes (MWCNT) can be used as labels. 10-fold enhancement was observed using MWCNT when compared to conventional AuNP-based LFIA for the detection of methamphetamine.[31]

Graphene is a two-dimensional carbon nanomaterial with high specific surface area and conductivity. It is amenable to microfabrication techniques, making it a good label in paper-based analytical devices for POC usage. Wu et al. modified paper with graphene and silica nanoparticles to develop a multiplex electrochemical paper-based device. They successfully detected four different cancer biomarkers and achieved a sub-pg/mL limit of detection (LOD).[32] Owing to their low cost and unique material properties, carbon nanomaterials are increasingly becoming popular.

2.1.3 Magnetic nanoparticle-based assays

Magnetic nanoparticles (MNPs) are biocompatible, and their wide size range (few nm to few μm) makes them suitable for various biological applications. When MNPs are used as labels in LFIAs, they produce a brown precipitate with low background noise. Besides offering a visible signal, magnetic signals coming from MNPs can be detected by a reader. MNPs offer several advantages, including separation and preconcentration of various analytes using their magnetic property, stability of the magnetic signal over a long period of time, and the ability to measure signals over the entire depth of the membrane. Liu et al. developed a signal enhanced LFIA using MNPs and showed sensitivities comparable to that of a commercial electrochemiluminescence (ECLIA) assay for the detection of carcinoembryonic antigen (CEA).[33] MNPs are also commonly used to preconcentrate the analyte of interest and are combined with other materials such as AuNPs to enhance the sensitivity.

2.1.4 Quantum dot-based assays

Compared to colorimetric detection, fluorescent detection provides a superior signal to noise ratio, leading to higher sensitivity. While most of the traditional fluorescence-based labels suffer from severe limitations like photo-bleaching and instability, quantum dots (QDs) are highly stable, and they resist photobleaching. For a review of the use of QDs in sensing, please refer to Li et al.[34] Taranova et al. developed a multiplex immunochromatographic assay (ICA) using three types of QDs having different emission peaks for sensitive detection of antibodies against antibiotics in milk. The observed LODs were 80–200 times lower than ELISA while using the same set of antibodies.[35] Similarly, Hu et al. developed a LFIA using fluorescent nanospheres containing several CdSe/ZnS QDs to detect C-reaction protein (CRP), producing 380-fold enhancement than having one QD.[36] Pure QDs suffer from chemical and colloidal instability in the physiological environment, and to circumvent this limitation, QDs were either encapsulated or doped in polymers and other carriers.[37]

Quantum dot nanobeads (QBs) are polymer nanobeads synthesized by embedding numerous QDs into a polymer matrix. QBs possess higher fluorescent intensity than single QDs and hence used for signal amplification in paper-based devices.[38–40] QBs synthesized by embedding numerous CdSe/ZnS QDs, when used in the immunochromatographic assay (ICA), showed 10-fold improvement when compared with traditional AuNPs-based ICA for the detection of procalcitonin.[41]

2.1.5 Silica nanoparticle-based assays

Silica nanoparticles form a semi-conductive material that is biocompatible, has good water dispersibility and surface functionality. Silica nanoparticles can be used as a direct label owing to their ability to form intense signals. Zhu et al. have reported a novel immunochromatographic assay using colored silica nanoparticles as direct optical label for POC detection of clenbuterol.[42]

Silica nanoparticles are often used together with other materials for signal amplification in paper analytical devices,[43,44] especially with QDs, as they enhance their chemical stability. Zhao et al. reported the use of CdTe QDs-embedded silica nanoparticles for the detection of alpha-fetoprotein, showing a sensitivity improvement of 4.7-fold when compared with the nonsilica antibody-QD conjugate.[44] Xu et al. reported a highly sensitive lateral flow strip based on gold-nanoparticle decorated silica nanorods (AuNP-SiNRs) label for visual detection of proteins.[43] Silica nanoparticles were used as carriers and loaded with AuNPs and used as labels in an LFIA. While using rabbit IgG as a model analyte, they could lower the detection limit by 50 times compared to the traditional AuNP-based LFIA.

2.1.6 Upconverting nanoparticles (UCNPs)-based assays

Upconversion nanoparticles (UCNPs), a special type of nanomaterial doped with lanthanide ions, are emerging fluorescent labels for the biorecognition of various molecules. When excited at near-infrared wavelengths, UCNPs emit at higher frequencies in the visible light range resulting in large Stoke shifts, which are ideal for low-background fluorescence. UCNPs can be tuned to have a wide range of colors by varying the composition of the host material and lanthanide ions.[45] Among the traditional fluorescent labels, UCNPs possess unique properties such as photostability, low toxicity, low auto-fluorescence, tunable multicolor property and narrow emission spectrum.[46] UCNP based paper analytical devices have attracted attention for the detection of biomarkers with greater improvement in the sensitivity.[47–49] Lanthanide nanoparticles such as europium nanospheres have also been used to enhance the signal in paper-based devices. UCNPs have the potential to be used as a label for multiplexed, sensitive, and quantitative detection of biomolecules. However, their high production costs, complex preparation protocols and requirement of external readers limit their POC applications.

2.2 Enzyme-based labels for signal enhancement

Taking inspiration from the ELISA, the laboratory workhorse technique for conducting immunoassays, several researchers have adopted enzyme-based signal amplification strategies in paper-based immunoassays. The Whitesides group demonstrated the performance of ELISA in a 96-microzone plate fabricated in the paper, emulating the 96-well plate traditional ELISA.[8] Though the assay sensitivity was 10 times lower than traditional well plate-based ELISA, the ability to image the results on a flatbed scanner obviated the need to use expensive spectrophotometers. These devices are called P-ELISA devices. A schematic of a typical P-ELISA device is shown in Fig. 3. Nascimento et al. compared paper-based ELISA with traditional benchtop ELISA for the detection of IgG antibodies against Toxoplasma gondii.[50] Chen et al. developed an enzyme-based paper immunoassay in which SERS was used as the detection mechanism to detect mouse IgG.[51] Several other studies were conducted on P-ELISA.[52–55] While P-ELISA devices take the ELISA one step closer to be conducted in low resource settings, conducting the assay still requires multiple timed user steps.

In parallel to the development of the P-ELISA, the Yager group demonstrated two-dimensional paper networks (2DPN), which enables sequential delivery of multiple fluids over a detection zone enabling the automation of multistep immunoassays.[56] In a typical 2DPN device (Fig. 4A), the sample is added to the leg (A) along with a detection antibody, wash buffer to the leg (B), and a signal amplification reagent to the leg (C). To deploy 2DPN in remote settings, the reagents, especially enzymes, need to be dry stored for a long time. Ramachandran et al. described an automated 2DPN design where the reagents were dry-stored and performed an ELISA to detect PfHRP2 malarial antigen.[57] They reported that the device performed well after 5 months when stored at 45 °C, making the 2DPN ideal for low-resource settings. Huang et al. extended the use of 2DPN by demonstrating an integrated disposable device to detect both influenza A and B. The device had the capabilities to perform assay using a single activation step wherein the patient has to insert a nasal swab and close the device.[58]

A new type of sliding 3D microfluidic devices was introduced to replace pipetting steps with sliding operations, making them ideal for performing an ELISA at POC testing.[59] The sequential delivery was achieved by moving the strip containing the test zone through a series of paper zones where one reagent was delivered at a time. Han et al. introduced a more user-friendly version of sliding microfluidic device viz., 3D slip-PAD device where sequential delivery of multiple reagents was achieved with a single activation step.[60]

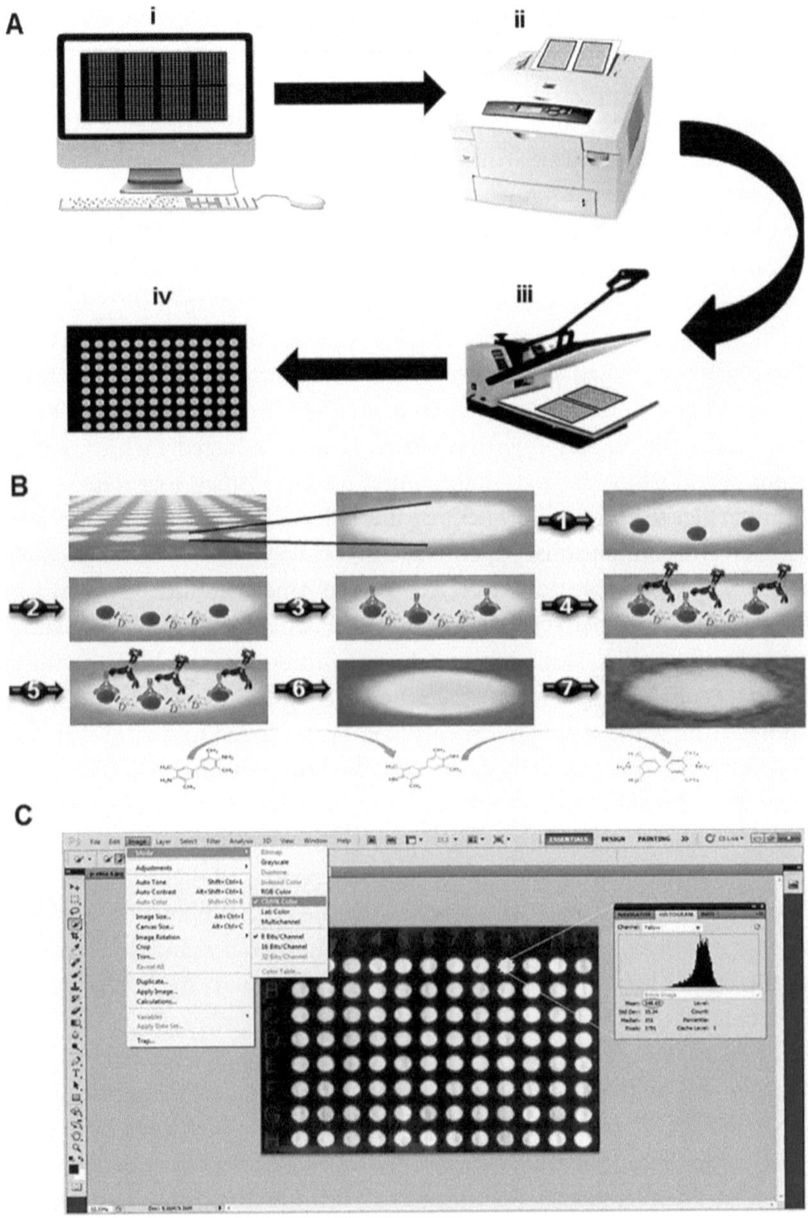

Fig. 3 Fabrication and operation of P-ELISA devices. (A) Steps involved in the fabrication of p-ELISA plates by wax printing. (B) Protocol for performing an immunoassay on a p-ELISA plate. (C) Digitalized image analysis on Adobe Photoshop. *Panel (A, B, and C) reproduced from Mazzu-Nascimento T, Morbioli GG, Milan LA, et al. Improved assessment of accuracy and performance indicators in paper-based ELISA. Anal Methods 2017;9 (18):2644–2653. doi:https://doi.org/10.1039/C7AY00505A, with permission from The Royal Society of Chemistry.*

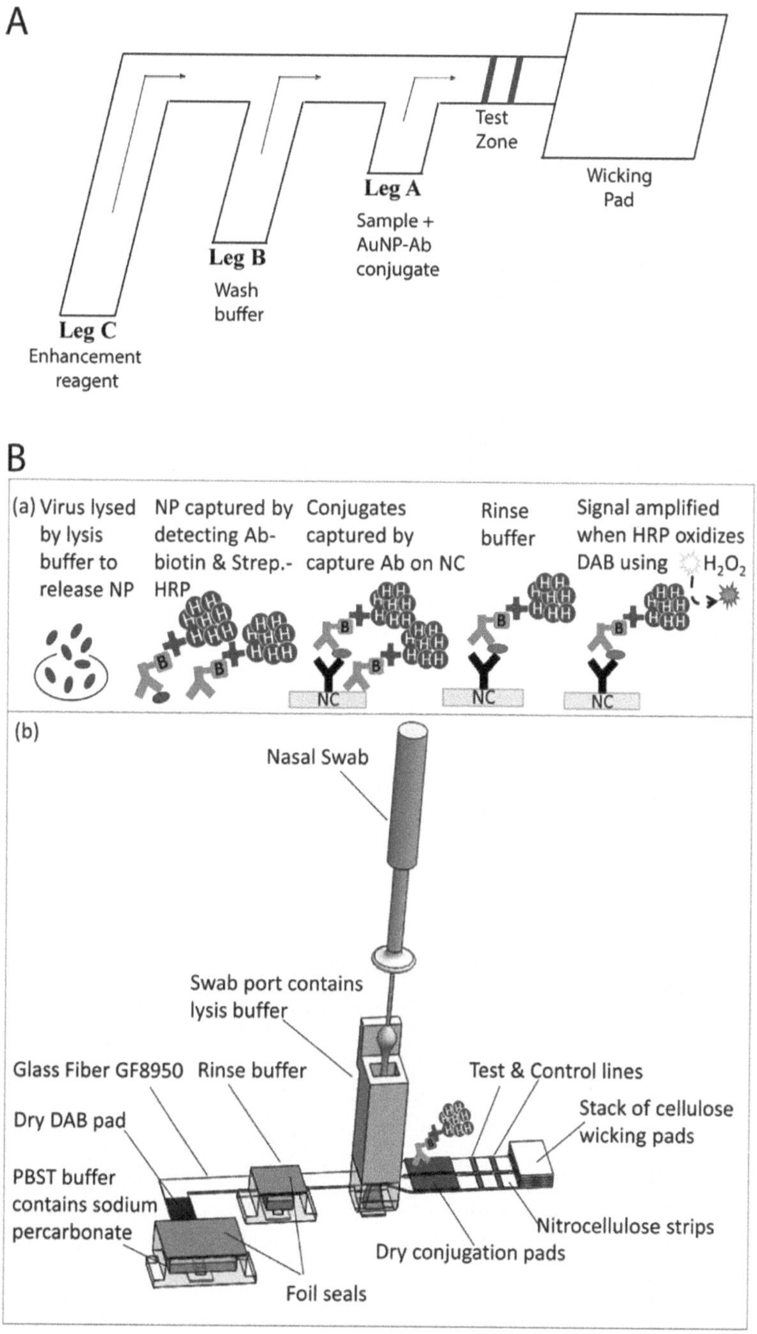

Fig. 4 Two-dimensional paper networks (2DPNs) for conducting signal-enhanced sandwich paper-based immunoassays. (A) Sequential delivery in a 2DPN. (B) 2DPN based automated single-step activation device for influenza detection. *Panel (B) reprinted with permission from Huang S, Abe K, Bennett S, et al. Disposable autonomous device for swab-to-result diagnosis of influenza. Anal Chem 2017;89(11):5776–5783. doi:10.1021/acs.analchem.6b04801. Copyright [2017] American Chemical Society.*

2.3 Polymerization and hybrid labels-based signal enhancement strategies

2.3.1 Polymerization-based signal enhancement strategies

The idea of polymerization-based signal enhancement technique for immunoassays was introduced in the late 2000s as an alternative strategy to nanoparticles and enzyme-based amplification.[61] In this technique, a polymerization initiator is labeled to the detection antibody. The antigen–antibody interactions localize the initiating molecules on the test line. When a monomer solution is added in the presence of light, polymerization-initiating radicals were produced, and a crosslinked hydrogel is formed provided the number of binding events exceeds a certain threshold.[62,63] Among these polymerization techniques, the most important method for paper-based immunoassays is photo-initiated free radical polymerization (Fig. 5A). Sikes and coworkers have successfully used this method for the detection of malarial antigen *Plasmodium falciparum* histidine-rich protein 2 (*Pf*HRP2).[64] They also developed a new visualization method (Fig. 5B) to detect the formation of the hydrogel using the color change of a pH indicator. They further compared their method with enzyme and nanomaterial-based signal enhancement for the detection of PfHRP2.[65] Results showed that the polymer-based enhancement was comparable with the silver enhancement technique, whereas the enzyme-based enhancement showed 20 times better sensitivity than polymer-based enhancement. In order to improve the sensitivity further, they encapsulated a large number of initiator molecules (Eosin-Y) into a liposome (Fig. 5C) to enhance the signal.[66] This approach led to a 30-fold improvement than the conventional polymer-based enhancement method. Another 10× improvement was observed while using continuous flow-through assays. The major limitation of polymer-based enhancement would be its insufficient sensitivity for early diagnosis of infectious diseases.[67] Moreover, photoinitiators like Eosin-Y suffers from poor stability and fluorescence quenching, which might cause high background signal. Nonspecific binding of liposomes to paper membranes might lead to high background signals.

2.3.2 Hybrid labels-based signal enhancement

This section deals with all the signal enhancement strategies where more than one label is used for detection. In most of these cases, AuNP is the common label. Sun et al. realized a dual signal enhancement strategy by employing gold-silver based bimetallic nanoparticles to detect carcinoembryonic antigen (CEA).[68] Surface modification of AuNPs with a thin coating of platinum nanoparticles results in high peroxidase-like catalytic activity for AuNPs while

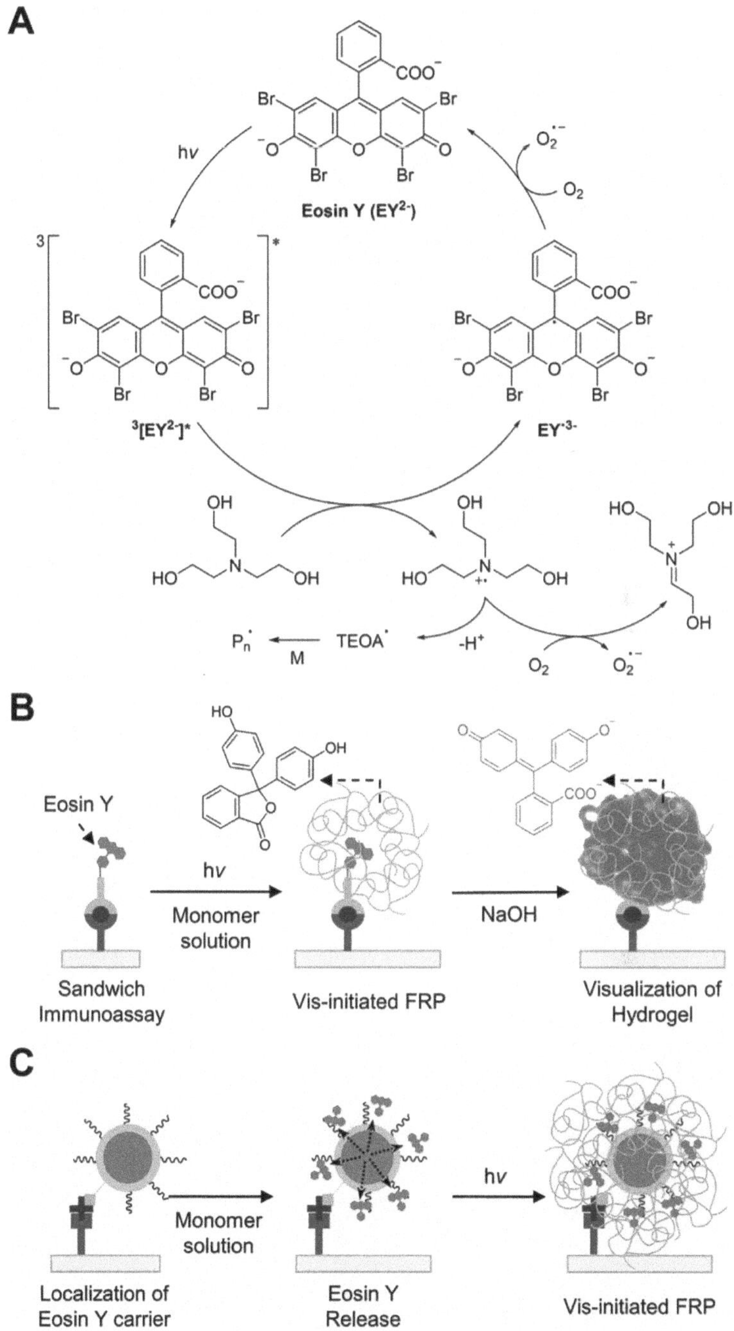

Fig. 5 Polymerization-based signal enhancement method in paper-based immunoassay. (A) Schematic account of photocatalytic initiation mechanism with eosin Y and triethanolamine (TEOA). (B) Vis-initiated free radical polymerization (FRP) for protein detection. (C) Sensitivity improvement obtained using eosin Y-loaded liposome. *Panel (A,B, and C) reproduced from Kim S, Sikes HD. Radical polymerization reactions for amplified biodetection signals. Polymer Chemistry 2020;11(8):1424–1444. doi:10.1039/c9py01801h, with permission from The Royal Society of Chemistry.*

retaining its plasmonic activity.[69] In addition to the visual readout, the peroxidase-like catalytic activity results in catalyzing the oxidation of colorless substrates to colored products, leading to two orders of improvement in the sensitivity. Similarly, a dual enhancement strategy can be realized in which enzymes such as horseradish peroxidase (HRP), alkaline phosphatase (ALP), etc., loaded on biofunctionalized AuNPs catalyze the oxidation of colorless substrate to colored product.[70,71]

MNPs are mostly used to preconcentrate the analyte of interest and are combined with other nanomaterials such as AuNPs to enhance the sensitivity. For example, MNPs were attached to a biotin-labeled antibody specific to the potato X virus and added to the sample to preconcentrate the analyte by a magnetic field.[72] The LFIA showed 32 times more sensitivity than the LFIA with MNPs alone. A dual enhancement strategy involving carbon nanotubes (CNT) and magnetic nanoparticle (MNP) was developed for the detection of proteins.[73] IgG was used as a model analyte in the LFIA for visual detection of protein in whole blood.

AuNPs-SERS tags find application in paper-based immunoassays due to their ultrasensitive and multiplexed detection capability. Hwang et al. developed a SERS-based LFIA that was three orders of magnitude more sensitive than the corresponding ELISA-based method.[74] Several reports for sensitive detection of numerous analytes on paper analytical devices using SERS have been published.[75–77] Making modified AuNPs with SERS tags is complex, and the reproducibility of SERS-based paper analytical devices remains the major drawback for widespread applications. In addition, the nonavailability of hand-held SERS detectors limits their applications in POC settings.

3. SWOT analysis: Paper-based immunoassays

Strengths: The obvious charm of using paper for conducting highly sensitive immunoassays is that paper is cheap, disposable, and therefore an ideal substrate for conducting immunoassays at the POC. If paper-based devices that match the sensitivities of ELISA could be developed and used in peripheral laboratories, it could radically reduce the cost and efficiency of diagnostic testing. However, a much-underappreciated strength of paper-based immunoassays is that they significantly reduced the assay time. Compared to ELISA well plates, small pores in paper membranes significantly reduce the time required for molecules to diffuse to surfaces and react. Paper-based immunoassays could thus reduce both the cost and speed of conducting sensitive immunoassays.

Weaknesses: One of the inherent challenges along the way of taking paper-based immunoassays to commercialization is the potential batch-to-batch variability in the structure and function of the paper membrane. The other major challenge is that the high surface area to volume ratio offered by paper also increases the possibility of nonspecific binding of molecules. Compared to ELISA well plates, it is significantly more challenging to wash the surface off of nonspecifically bound moieties. This increase background signals in paper-based signal-enhanced immunoassays and ultimately reduces sensitivity. Finally, dry storage of certain reagents could be more challenging on paper than in plastic tubes, which could be considered a weakness of paper-based platforms.

Opportunities: There are abundant opportunities for the development of paper-based immunoassays in today's rapidly changing world that is moving toward enabling individuals to conduct personalized health monitoring and diagnostics. An important opportunity in terms of application is that of rapid COVID-19 antigen detection tests. While LFIAs have been rolled out, they all suffer from poor sensitivity (compared to nucleic acid-based tests) because of the inherent limitations of traditional gold nanoparticle-based LFIAs. The plethora of modern techniques described in this chapter present opportunities to radically enhance the sensitivity of COVID-19 rapid antigen tests, which could enable confirmatory COVID-19 diagnostics at a doctor's office or even at home. The extension of this technology for the detection of other infectious diseases is obvious. In terms of the technology, there are two main directions of research that appear promising: (i) development of paper networks to program the movement of multiple fluids to conduct a multistep immunoassay, and (ii) development of improved labels that can enhance the sensitivity of detection in standard LFIA geometries using fluorescence, chemiluminescence, etc. There are pros and cons of each technique. The former increases the complexity of paper devices but enables visual detection without sophisticated instruments. The latter keeps the device geometry simple but requires some form of instrumentation for detection. Both approaches are promising and may be considered important opportunities in the path of commercialization of highly sensitive signal-enhanced immunoassays.

Threats: Like many other modern-day technologies that face this threat, paper-based signal-enhanced immunoassays also face the threat of never seeing the light at the end of the tunnel. This is happening too often in the field of technology development—exciting technologies are developed but perpetually remain at the proof-of-concept stage. This usually happens because of academic and entrepreneurial silos that prevail in several parts of the world. Academicians that develop such exciting new technologies are often not well equipped to take the technology beyond the lab. To mitigate this threat, commercial partners must be involved (as collaborators or partners) in the development of paper-based immunoassays early in the development process.

To conclude, if there is going to be a commercial device that can conduct a bench side immunoassay whose sensitivity matches that of ELISA, more than likely, it will be based on paper-based immunoassay technology. It is a promising new area that will benefit from the global forces that are pushing research and development toward frugal engineering.

References

1. Ju H, Lai G, Yan F. Introduction. In: *Immunosensing for Detection of Protein Biomarkers*. Elsevier; 2017:1–30. https://doi.org/10.1016/b978-0-08-101999-3.00001-3.
2. Plotz CM, Singer JM. The latex fixation test. I. Application to the serologic diagnosis of rheumatoid arthritis. *The American Jurnal of Medicine*. 1956;21(6):888–892.
3. Berson SA, Yalow RS. Quantitative aspects of the reaction between insulin and insulin-binding antibody. *J Clin Invest*. 1959;38(11):1996–2016. https://doi.org/10.1172/JCI103979.
4. Wong R, Tse H. In: Wong R, Tse H, eds. *Lateral Flow Immunoassay*. Totowa, NJ: Humana Press; 2009. https://doi.org/10.1007/978-1-59745-240-3.
5. de Puig H, Bosch I, Gehrke L, Hamad-Schifferli K. Challenges of the Nano–bio Interface in lateral flow and dipstick immunoassays. *Trends Biotechnol*. 2017. https://doi.org/10.1016/j.tibtech.2017.09.001.
6. Yamada K, Shibata H, Suzuki K, Citterio D. Toward practical application of paper-based microfluidics for medical diagnostics: State-of-the-art and challenges. *Lab Chip*. 2017; 17(7):1206–1249. https://doi.org/10.1039/C6LC01577H.
7. Toley BJ, Das D, Ganar KA, et al. Multidimensional paper networks: A new generation of low-cost pump-free microfluidic devices. *Journal of the Indian Institute of Science*. 2018;98(2):103–136. https://doi.org/10.1007/s41745-018-0077-1.
8. Cheng CM, Martinez AW, Gong J, et al. Paper-based elisa. *Angewandte Chemie - International Edition*. 2010;49(28):4771–4774. https://doi.org/10.1002/anie.201001005.
9. Lee S, Dell DO, Hohenstein J, Colt S, Mehta S. *NutriPhone: A Mobile Platform for Low-Cost Point-of-Care Quantification of Vitamin B 12 Concentrations*. Nature Publishing Group; 2016:1–8. https://doi.org/10.1038/srep28237.
10. Shah KG, Singh V, Kauffman PC, Abe K, Yager P. Mobile phone ratiometric imaging enables highly sensitive fluorescence lateral flow immunoassays without external optical filters. *Anal Chem*. 2018. https://doi.org/10.1021/acs.analchem.8b01241.
11. Kosack CS, Page A-L, Klatser PR. A guide to aid the selection of diagnostic tests. *Bull World Health Organ*. 2017;95(9):639–645. doi:https://doi.org/https://doi.org/10.2471/BLT.16.187468.
12. Giri B, Pandey B, Neupane B, Ligler FS. Signal amplification strategies for microfluidic immunoassays. *TrAC - Trends in Analytical Chemistry*. 2016;79:326–334. https://doi.org/10.1016/j.trac.2015.10.021.
13. Bishop JD, Hsieh HV, Gasperino DJ, Weigl BH. Sensitivity enhancement in lateral flow assays: A systems perspective. *Lab Chip*. 2019;19(15):2486–2499. https://doi.org/10.1039/c9lc00104b.

14. Liu L, Yang D, Liu G. Signal amplification strategies for paper-based analytical devices. *Biosens Bioelectron.* 2019;136(April):60–75. https://doi.org/10.1016/j.bios.2019.04.043.
15. Nguyen VT, Song S, Park S, Joo C. Recent advances in high-sensitivity detection methods for paper-based lateral-flow assay. *Biosens Bioelectron.* 2020;152(January):112015. https://doi.org/10.1016/j.bios.2020.112015.
16. Ye H, Liu Y, Zhan L, Liu Y, Qin Z. Signal amplification and quantification on lateral flow assays by laser excitation of plasmonic nanomaterials. *Theranostics.* 2020;10(10):4359–4373. https://doi.org/10.7150/thno.44298.
17. Li J, Duan H, Xu P, Huang X, Xiong Y. Effect of different-sized spherical gold nanoparticles grown layer by layer on the sensitivity of an immunochromatographic assay. *RSC Adv.* 2016;6(31):26178–26185. https://doi.org/10.1039/c6ra03695c.
18. Zhan L, Guo S, Song F, et al. The role of nanoparticle Design in Determining Analytical Performance of lateral flow immunoassays. *Nano Lett.* 2017;17(12):7207–7212. https://doi.org/10.1021/acs.nanolett.7b02302.
19. He S, Chua J, Tan EKM, Kah JCY. Optimizing the SERS enhancement of a facile gold nanostar immobilized paper-based SERS substrate. *RSC Adv.* 2017;7(27):16264–16272. https://doi.org/10.1039/C6RA28450G.
20. Qin Z, Chan WCW, Boulware DR, Akkin T, Butler EK, Bischof JC. Significantly improved analytical sensitivity of lateral flow immunoassays by using thermal contrast. *Angew Chem Int Ed.* 2012;51(18):4358–4361. https://doi.org/10.1002/anie.201200997.
21. Holgate CS, Jackson P, Cowen PN, Bird CC. Immunogold-silver staining: New method of immunostaining with enhanced sensitivity. *Journal of Histochemistry & Cytochemistry.* 1983;31(7):938–944. https://doi.org/10.1177/31.7.6189883.
22. Yeh CH, Hung CY, Chang TC, Lin HP, Lin YC. An immunoassay using antibody-gold nanoparticle conjugate, silver enhancement and flatbed scanner. *Microfluidics and Nanofluidics.* 2009;6(1):85–91. https://doi.org/10.1007/s10404-008-0298-0.
23. Wada A, Sakoda Y, Oyamada T, Kida H. Development of a highly sensitive immunochromatographic detection kit for H5 influenza virus hemagglutinin using silver amplification. *J Virol Methods.* 2011;178(1–2):82–86. https://doi.org/10.1016/j.jviromet.2011.08.017.
24. Broger T, Sossen B, Toit E, et al. Novel lipoarabinomannan point-of-care tuberculosis test for people with HIV: A diagnostic accuracy study. *Lancet Infect Dis.* 2019;19:852–861. https://doi.org/10.1016/S1473-3099(19)30001-5.
25. Morris RE, Ciraolo GM, Saelinger CB. Gold enhancement of gold-labeled probes: Gold-intensified staining technique (GIST). *Journal of Histochemistry and Cytochemistry.* 1991;39(11):1585–1591. https://doi.org/10.1177/39.11.1918934.
26. Lei KF, Butt YKC. Colorimetric immunoassay chip based on gold nanoparticles and gold enhancement. *Microfluidics and Nanofluidics.* 2010;8:131–137. https://doi.org/10.1007/s10404-009-0490-x.
27. Fu E, Liang T, Spicar-Mihalic P, Houghtaling J, Ramachandran S, Yager P. Two-dimensional paper network format that enables simple multistep assays for use in low-resource settings in the context of malaria antigen detection. *Anal Chem.* 2012. https://doi.org/10.1021/ac300689s.
28. Park J, Shin JH, Park JK. Pressed paper-based dipstick for detection of foodborne pathogens with multistep reactions. *Anal Chem.* 2016;88(7):3781–3788. https://doi.org/10.1021/acs.analchem.5b04743.
29. van Dam GJ, Wichers JH, Ferreira TMF, Ghati D, van Amerongen A, Deelder AM. Diagnosis of schistosomiasis by reagent strip test for detection of circulating cathodic antigen. *J Clin Microbiol.* 2004;42(12):5458 LP–5461. https://doi.org/10.1128/JCM.42.12.5458-5461.2004.
30. Zhang X, Yu X, Wen K, et al. Multiplex lateral flow immunoassays based on amorphous carbon nanoparticles for detecting three fusarium mycotoxins in maize. *J Agric Food Chem.* 2017;65(36):8063–8071. https://doi.org/10.1021/acs.jafc.7b02827.

31. Sun W, Hu X, Liu J, Zhang Y, Lu J, Zeng L. A novel multi-walled carbon nanotube-based antibody conjugate for quantitative and semi-quantitative lateral flow assays. *Biosci Biotechnol Biochem*. 2017;81(10):1874–1882. https://doi.org/10.1080/09168451.2017.1365590.
32. Wu Y, Xue P, Kang Y, Hui KM. Paper-based microfluidic electrochemical immunodevice integrated with nanobioprobes onto graphene film for ultrasensitive multiplexed detection of cancer biomarkers. *Anal Chem*. 2013. https://doi.org/10.1021/ac401445a.
33. Liu F, Zhang H, Wu Z, et al. Highly sensitive and selective lateral flow immunoassay based on magnetic nanoparticles for quantitative detection of carcinoembryonic antigen. *Talanta*. 2016;161:205–210. https://doi.org/10.1016/j.talanta.2016.08.048.
34. Li M, Chen T, Justin Gooding J, Liu J. Review of carbon and graphene quantum dots for sensing. *ACS Sensors*. 2019;4(7):1732–1748. https://doi.org/10.1021/acssensors.9b00514.
35. Taranova NA, Berlina AN, Zherdev AV, Dzantiev BB. "Traffic light" immunochromatographic test based on multicolor quantum dots for the simultaneous detection of several antibiotics in milk. *Biosens Bioelectron*. 2015. https://doi.org/10.1016/j.bios.2014.07.049.
36. Hu J, Zhang ZL, Wen CY, et al. Sensitive and quantitative detection of C-reaction protein based on immunofluorescent Nanospheres coupled with lateral flow test strip. *Anal Chem*. 2016. https://doi.org/10.1021/acs.analchem.6b01427.
37. Ouyang S, Zhang Z, He T, et al. An on-site, ultra-sensitive, quantitative sensing method for the determination of total aflatoxin in peanut and rice based on quantum dot nanobeads strip. *Toxins*. 2017. https://doi.org/10.3390/toxins9040137.
38. Li X, Li W, Yang Q, et al. Rapid and quantitative detection of prostate specific antigen with a quantum dot nanobeads-based immunochromatography test strip. In: *ACS Applied Materials and Interfaces*; 2014. https://doi.org/10.1021/am5012782.
39. Duan H, Huang X, Shao Y, Zheng L, Guo L, Xiong Y. Size-dependent Immunochromatographic assay with quantum dot Nanobeads for sensitive and quantitative detection of Ochratoxin a in corn. *Anal Chem*. 2017. https://doi.org/10.1021/acs.analchem.7b00869.
40. Tsai TT, Huang CY, Chen CACF, et al. Diagnosis of tuberculosis using colorimetric gold nanoparticles on a paper-based analytical device. *ACS Sensors*. 2017;2(9):1345–1354. https://doi.org/10.1021/acssensors.7b00450.
41. Zhou S, Peng Y, Hu J, et al. Quantum dot nanobead-based immunochromatographic assay for the quantitative detection of the procalcitonin antigen in serum samples. *Microchem J*. 2020;159. https://doi.org/10.1016/j.microc.2020.105533.
42. Zhu C, Zhao G, Dou W. Immunochromatographic assay using brightly colored silica nanoparticles as visible label for point-of-care detection of clenbuterol. *Sens Actuators B*. 2018. https://doi.org/10.1016/j.snb.2018.03.085.
43. Xu H, Chen J, Birrenkott J, et al. Gold-nanoparticle-decorated silica nanorods for sensitive visual detection of proteins. *Anal Chem*. 2014;86(15):7351–7359. https://doi.org/10.1021/ac502249f.
44. Zhao M, Li H, Liu W, Chu W, Chen Y. Paper-based laser induced fluorescence immunodevice combining with CdTe embedded silica nanoparticles signal enhancement strategy. *Sens Actuators B*. 2017;242:87–94. https://doi.org/10.1016/j.snb.2016.11.004.
45. Zijlmans HJMAA, Bonnet J, Burton J, et al. Detection of cell and tissue surface antigens using up-converting phosphors: A new reporter technology. *Anal Biochem*. 1999. https://doi.org/10.1006/abio.1998.2965.
46. Jin B, Wang S, Lin M, et al. Upconversion nanoparticles based FRET aptasensor for rapid and ultrasenstive bacteria detection. *Biosens Bioelectron*. 2017. https://doi.org/10.1016/j.bios.2016.10.029.

47. He M, Li Z, Ge Y, Liu Z. Portable Upconversion nanoparticles-based paper device for field testing of drug abuse. *Anal Chem.* 2016. https://doi.org/10.1021/acs.analchem.5b04863.
48. Zhao Y, Liu X, Wang X, et al. Development and evaluation of an up-converting phosphor technology-based lateral flow assay for rapid and quantitative detection of aflatoxin B1 in crops. *Talanta.* 2016. https://doi.org/10.1016/j.talanta.2016.08.058.
49. You M, Lin M, Gong Y, et al. Household fluorescent lateral flow strip platform for sensitive and quantitative prognosis of heart failure using dual-color Upconversion nanoparticles. *ACS Nano.* 2017. https://doi.org/10.1021/acsnano.7b02466.
50. Mazzu-Nascimento T, Morbioli GG, Milan LA, et al. Improved assessment of accuracy and performance indicators in paper-based ELISA. *Anal Methods.* 2017;9(18):2644–2653. https://doi.org/10.1039/C7AY00505A.
51. Chen Y, Cheng H, Tram K, et al. A paper-based surface-enhanced resonance Raman spectroscopic (SERRS) immunoassay using magnetic separation and enzyme-catalyzed reaction. *Analyst.* 2013;138(9):2624. https://doi.org/10.1039/c3an36647b.
52. Khan MS, Pande T, van de Ven TGMM. Qualitative and quantitative detection of T7 bacteriophages using paper based sandwich ELISA. *Colloids Surf B Biointerfaces.* 2015; 132:264–270. https://doi.org/10.1016/j.colsurfb.2015.05.028.
53. Wu D, Zhang J, Xu F, et al. A paper-based microfluidic dot-ELISA system with smartphone for the detection of influenza a. *Microfluidics and Nanofluidics.* 2017;21 (3). https://doi.org/10.1007/s10404-017-1879-6.
54. Chi J, Gao B, Sun M, et al. Patterned photonic nitrocellulose for Pseudopaper ELISA. *Anal Chem.* 2017. https://doi.org/10.1021/acs.analchem.7b01732.
55. Ortega GA, Pérez-Rodríguez S, Reguera E. Magnetic paper-based ELISA for IgM-dengue detection. *RSC Adv.* 2017. https://doi.org/10.1039/c6ra25992h.
56. Fu E, Kauffman P, Lutz B, Yager P. Chemical signal amplification in two-dimensional paper networks. *Sens Actuators B.* 2010;149(1):325–328. https://doi.org/10.1016/j.snb.2010.06.024.
57. Ramachandran S, Fu E, Lutz B, Yager P. Long-term dry storage of an enzyme-based reagent system for ELISA in point-of-care devices. *Analyst.* 2014;139(6):1456–1462. https://doi.org/10.1039/C3AN02296J.
58. Huang S, Abe K, Bennett S, et al. Disposable autonomous device for swab-to-result diagnosis of influenza. *Anal Chem.* 2017;89(11):5776–5783. https://doi.org/10.1021/acs.analchem.6b04801.
59. Liu XY, Cheng CM, Martinez AW, et al. A portable microfluidic paper-based device for elisa. In: *Proceedings of the IEEE International Conference on Micro Electro Mechanical Systems (MEMS)*; 2011:75–78. https://doi.org/10.1109/MEMSYS.2011.5734365.
60. Han KN, Choi JS, Kwon J. Three-dimensional paper-based slip device for one-step point-of-care testing. *Sci Rep.* 2016;6(May):1–7. https://doi.org/10.1038/srep25710.
61. Sikes HD, Hansen RR, Johnson LM, et al. Using polymeric materials to generate an amplified response to molecular recognition events. *Nat Mater.* 2008;7(1):52–56. https://doi.org/10.1038/nmat2042.
62. Sikes HD, Jenison R, Bowman CN. Antigen detection using polymerization-based amplification. *Lab Chip.* 2009;9(5):653–656. https://doi.org/10.1039/b816198d.
63. Kaastrup K, Sikes HD. Polymerization-based signal amplification under ambient conditions with thirty-five second reaction times. *Lab Chip.* 2012;12(20):4055–4058. https://doi.org/10.1039/c2lc40584a.
64. Badu-Tawiah AK, Lathwal S, Kaastrup K, et al. Polymerization-based signal amplification for paper-based immunoassays. *Lab Chip.* 2015;15(3):655–659. https://doi.org/10.1039/C4LC01239A.

65. Lathwal S, Sikes HD. Assessment of colorimetric amplification methods in a paper-based immunoassay for diagnosis of malaria. *Lab Chip*. 2016;16(8):1374–1382. https://doi.org/10.1039/C6LC00058D.
66. Kim S, Sikes HD. Liposome-enhanced polymerization-based signal amplification for highly sensitive naked-eye biodetection in paper-based sensors. *ACS Applied Materials and Interfaces*. 2019;11(31):28469–28477. https://doi.org/10.1021/acsami.9b08125.
67. Kim S, Sikes HD. Radical polymerization reactions for amplified biodetection signals. *Polymer Chemistry*. 2020;11(8):1424–1444. https://doi.org/10.1039/c9py01801h.
68. Sun G, Ding YN, Ma C, et al. Paper-based electrochemical immunosensor for carcinoembryonic antigen based on three dimensional flower-like gold electrode and gold-silver bimetallic nanoparticles. *Electrochim Acta*. 2014;147:650–656. https://doi.org/10.1016/j.electacta.2014.09.149.
69. Gao Z, Ye H, Tang D, et al. Platinum-decorated gold nanoparticles with dual functionalities for ultrasensitive colorimetric in vitro diagnostics. *Nano Lett*. 2017. https://doi.org/10.1021/acs.nanolett.7b02385.
70. Merkoc A, Parolo C, de la Escosura-Muniz A. Enhanced lateral flow immunoassay using gold nanoparticles loaded with enzymes. *Biosens Bioelectron*. 2013;40:412–416. https://doi.org/10.1016/j.bios.2012.06.049.
71. Cho IH, Bhunia A, Irudayaraj J. Rapid pathogen detection by lateral-flow immunochromatographic assay with gold nanoparticle-assisted enzyme signal amplification. *Int J Food Microbiol*. 2015;206:60–66. https://doi.org/10.1016/j.ijfoodmicro.2015.04.032.
72. Razo SC, Panferov VG, Safenkova IV, Varitsev YA, Zherdev AV, Dzantiev BB. Double-enhanced lateral flow immunoassay for potato virus X based on a combination of magnetic and gold nanoparticles. *Anal Chim Acta*. 2018;1007:50–60. https://doi.org/10.1016/j.aca.2017.12.023.
73. Huang Y, Wen Y, Baryeh K, et al. Magnetized carbon nanotubes for visual detection of proteins directly in whole blood. *Anal Chim Acta*. 2017. https://doi.org/10.1016/j.aca.2017.09.025.
74. Hwang J, Lee S, Choo J. Application of a SERS-based lateral flow immunoassay strip for the rapid and sensitive detection of staphylococcal enterotoxin B. *Nanoscale*. 2016;8(22):11418–11425. https://doi.org/10.1039/c5nr07243c.
75. Rong Z, Xiao R, Xing S, et al. SERS-based lateral flow assay for quantitative detection of C-reactive protein as an early bio-indicator of a radiation-induced inflammatory response in nonhuman primates. *Analyst*. 2018;143(9):2115–2121. https://doi.org/10.1039/c8an00160j.
76. Zeng N, You Y, Xie L, Zhang H, Ye L. Optik a new imaged-based quantitative reader for the gold immunochromatographic assay. *Optik - International Journal for Light and Electron Optics*. 2018;152:92–99. https://doi.org/10.1016/j.ijleo.2017.09.109.
77. Luo H, Lin X, Peng Z, Song M, Jin L. Rapid and sensitive detection of bisphenol a based on self-assembly. *Micromachines*. 2020;11(1). https://doi.org/10.3390/mi11010041.

CHAPTER TWELVE

Challenges and opportunities in micro/nanofluidic and lab-on-a-chip

Nidhi Verma and Alok Pandya*
Department of Engineering and Physical Sciences, Institute of Advanced Research, Gandhinagar, Gujarat, India
*Corresponding author: e-mail address: alokpandya20@gmail.com

Contents

1. Understand micro/nanofluidic (MF/NF) 289
2. Understand lab-on-a-chip 292
3. Challenges in designing micro/nanofluidic 293
4. Challenges in designing lab-on-a-chip device 296
5. Opportunities in micro/nanofluidic and lab-on-a-chip devices 299
Acknowledgments 301
References 301

Abstract

Fluidic systems are prevalent in many areas of science due to its advantage in miniaturization, development of unique tools for diseases diagnosis and biomolecule separation. In the chapter, we will describe some of the key features of microfluidic/nanofluidic (MF/NF) and lab-on-a-chip system in diverse field over the past years. In addition, we will highlight the major challenges for the microfluidic/nanofluidic and lab-on-a-chip system. All-purpose and universal micro/nanofluidic platforms that can perform multiplexed assays on real biological samples are in high demand. However, the adoption of novel microfluidic devices has been carried out at a slow pace due to translation gap in development of new devices to realization into commercialization. By addressing the challenges of system integration, low-cost technology availability, rapid regulatory approval, and clinical acceptance, a pipeline of promising microdevice technologies can be developed.

1. Understand micro/nanofluidic (MF/NF)

In recent years, microfluidic platforms or technologies have emerged as powerful tools in a broad range of fields, from chemistry, biological sciences and engineering. In fact, nowadays, it is recognized for rapid

Progress in Molecular Biology and Translational Science, Volume 186
ISSN 1877-1173
https://doi.org/10.1016/bs.pmbts.2021.07.016

experiments on small-sample volumes effectively. Microfluidics refers to a science of fluid handling platform which is designed and developed with manipulation of small volumes of fluids having a network of channels in the dimensions of 1–100 μm. Fundamentally, it is a subject of fluid mechanics, where describing the physics of fluids at the micro length scales in microfluidic systems. The commonly used methods for driving flows in microchannels are pressure gradients, pressure-driven flow and externally applied electric fields. Along with, acoustic, capillary, and magnetic forces have also been used to move fluids in microchannels. Nowadays, fabricating microelectronic structures and photolithography is the most common method used for microfluidic design. Since the last 30 years, this technology has been established for printed circuit boards, integrated circuits, microscale gas chromatography, ink-jet printers, fluid engineering, analytical chemistry, synthetic chemistry, and life sciences.[1–3] Most importantly, microfluidics enables the manipulation of small volumes of fluid in a precise and effective manner which is crucial for the preparation and delivery of small synthetic molecules or sensing applications.[4]

Geometrical length scale constraints of microfluidic/nanofluidic platforms remain consistent but the design and methods used to achieve dependent on the targeted application such separation and detection of biomolecule. Normally, microfluidic flows are closed channels in the range of 10 μm × 10 μm which maintain robust fluid flow. Various kinds of microfluidic platform such as open microfluidics, droplet-based microfluidics, paper-based microfluidics, continuous-flow microfluidics, digital microfluidics, particle detection microfluidics, microfluidic-assisted magnetophoretic and associated techniques have matured over the years for different applications. In open microfluidics, the types include open-channel microfluidics, railed based microfluidics, and thread-based microfluidics which has drawbacks including contamination, evaporation, and confined flow rate. Continuous flow microfluidics is a liquid flow through narrow channels or porous media which significantly accelerates the fluid flow in capillary elements with computer-aided design automation approaches.[5] Microfluidics using droplets manipulates discrete concentrations of fluids in immiscible phases with low Reynolds numbers and laminar flow regimes. It allows for easy handling of small volumes of fluids (litre to femtolitre) for better mixing, encapsulation, sorting, and sensing, and is well suited for high-throughput experiments.[6] As an alternative to the above, analogy of digital microelectronics, referred to as digital microfluidics, is introduced to the use of electrocapillary forces to move droplets on a digital track. In digital microfluidics, electrowetting on dielectric (EWOD) is a popular actuation process, and several lab-on-a-chip applications have been demonstrated. Other droplet manipulation techniques such as magnetic force, surface acoustic waves, optoelectrowetting, and mechanical actuators are becoming more common. Paper-based microfluidics works on the phenomenon of capillary action in porous media.[7] Paper-based microfluidic devices are growing immensely for portable, affordable, and easy to use biomedical diagnostic systems.[8] Current applications include portable glucose detection, environmental testing, COVID 19 detection, HIV detection and many more with hopes of reaching rural areas or low resource settings. Particle detection microfluidics is the area which has seen significant efforts in academic some commercial. It is used to detect small fluid-particles about 1 μm in diameter, where electrical signals are produced, when a weakly-conducting fluid passed through a 100 μm diameter small pore. Particle detection microfluidics associated technique microfluidic resistive pulse sensing (MRPS) is used to measure the size and count of erythrocytes as well as leukocytes for standard blood analysis. In recent years, microfluidics has emerged with the integration of microfluidic devices with magnetophoretic and referred as microfluidic-assisted magnetophoretic or the migration of particles by a magnetic field. Example like, milk exhibit paramagnetism and therefore, milk can be flowed before packaging, through channels with magnetic gradients to purify the metal contaminants. Similarly, more such applications of microfluidic-assisted magnetophoretic are targeted toward cell separation.

In the past few years, use of nanomaterials has significantly brought many prospects for scientists and engineers for technology development. And nanofluid is one of the amazing consequences of such development. Nanofluids are suspending nanoparticles in the range of 1–100 nm typically made of metals, oxides, carbides, or carbon nanotubes with base fluid such as water, oil, and ethylene glycol, etc. which provide dramatic improvements in the thermal properties of host fluids in comparison to traditional heat transfer fluids. Nanofluids is known for potential use in many applications including microelectronics, pharmaceutical processes, hybrid powered engines, fuel cells, engine cooling/vehicle thermal management, domestic refrigerator, chiller and heat exchanger. As base fluids, a number of liquids such as water, ethylene glycol, and oils have been used. Stabilization can be difficult, but ongoing research shows that it is possible. Clove buds were used to build a bio-based, environmentally friendly system for covalently functionalized multi-walled carbon nanotubes (MWCNTs). Several studies have demonstrated the benefits of nanofluids in biomedical methods in a variety of areas, including diagnosis, drug delivery, imaging, and antibacterial activities. Magnetic nanofluid systems, for example, are useful for drug delivery, hyperthermia, and differential diagnosis. Nowadays, researchers are exploring nanofluid production and development of biofluid (biological fluid such as blood).[9]

Nanofluids are less researched in the field of biological sciences. A lot of research is pursued using microfluidic. Today, the relevance of microfluidics is significant by a range of features that leads to system miniaturization.[10,11] The microfluidics toolset has influenced many areas of science so here we will see how advanced microfluidics such as the lab on chip have been explored. Microfluidics lab-on-a-chip technologies have significantly impacted the research in clinical diagnostics outside of laboratories, and in particular, at the point-of-care (POC).[12,13]

2. Understand lab-on-a-chip

A lab-on-a-chip (LOC) is a miniaturized device that integrates into a single chip which is capable of carrying out separations and detections with high resolution.[14] This miniaturization also offers advantages, like cost efficiency, diagnostic, and sensitivity for the analysis such as DNA sequencing and biochemical detection. In fact, microfluidic manufacturing methods such as laser machining, lithography, xurography have enabled lab-on-a-chip devices with a smaller footprint at a reduced cost especially for the POC application. The microPCR, a lab-on-a-chip integrated technology used to amplify DNA 10 times faster than with conventional PCR systems. Nowadays, lab on chip microfluidic emerged in molecular biology for DNA/RNA amplification and detection like ELVESYS developed the world's fastest qPCR system—FASTGENE. In fact, researchers are working on sequence genomes thousands of times faster using lab on chip microfluidics especially nanopore technologies. Proteomics is also a field where lab on a chip make a significant contribution. This advancement provides the opportunity to perform shorten protein analysis with integrating the extraction from the cell and separation by electrophoresis in same chip. Protein crystallization is also an area where researchers are exploring the use of lab on chip microfluidic technologies. Due to the same typical size of cell as microchannel in microfluidic, research is also focused on cell biology. Lab on a chip shows the ability to control cells at the single-cell level in seconds. It also includes applications such as micro patch clamp, control of stem cell differentiation, high-speed flow cytometry and cell sorting. Similarly, nowadays, extensive research has been focused on using lab-on-a-chip microfluidics and highly parallelized micro chemical reactors. However, there are limitations of lab-on-a-chip technologies compared to classic technologies like industrialization and standardization. It increases the signal/noise ratio which provides poorer results. Ethical permission is also a major concern before adapting such technology.

3. Challenges in designing micro/nanofluidic

As described earlier, interest in microfluidic technologies is evident in fields such as molecular biology, genomics, proteomics, cell biology, single cell analysis, chemical engineering, drug discovery, high-throughput screening, and diagnostic.[15,16] And it is because it is miniaturized and performs rapid, sensitive and efficient experiments on small-sample volumes almost reduced to the micrometer scale. As evident, microfluidics platforms have amazingly influenced many areas of science and impacted well as needed for an hour.[17] But various challenges in design and development of microfluidic make attempts difficult for commercialization. Therefore, further we discuss the challenges in industrialization of lab-on-a-chip technologies.

Materials used for the development of microfluidic is crucial. In all, PDMS is widely used although having limitations and not completely suited for mass production of microfluidic channels. Afterwards, glass, silicon and ceramic substrates were used as robust material.[18] However, it requires costly sophisticated fabrication processes. Interestingly, for nanofluidic devices, silicon and glass have emerged as important materials which offer high aspect ratio channels, integrated electrodes, and e-beam patterned nanoscale features.[19] Recently, 3D printers have been applied to microfluidic device fabrication used for over the past 5 years. 3D printed microfluidic is still in an early stage. Additionally, it also requires printer hardware and software control and printing material selections.

The inclusion of unconventional *boundary conditions at micrometric scale*, the regulation of physicochemical phenomena such as adsorption and desorption in unusual situations, the driving of single molecules through nanochannels, and the use of short turn-over times for chemical measurements are all critical parts of microfluidic design. There are two main types of boundary conditions, i.e., no-slip and slip boundary conditions for the flow velocity which needs to be tailored hydrodynamically.[20]

Modular components: The concept of mass-producing modular parts that could be assembled into more complex structures and interchanged at will became popular during the industrial revolution (and much earlier with the printing press). Parts could be swapped between systems, and changes to the larger system could be made without re-engineering unchanged components, according to the rationale. Of course, interchangeable parts necessitate standardized interconnections (e.g., screws and bolts with defined geometries and threading). During the microelectronics revolution, this paradigm was taken to a whole new stage, with modularity being applied at different levels. Capacitors, resistors, transistors, sensors, microprocessors, and other circuit elements on circuit boards are abstracted to "black boxes" with specified input and output responses and seamlessly connected and interchanged to perform logical operations, with standardized wiring on the breadboard connecting components. Microfluidic devices have also been subjected to the same paradigm. For continuous flow microfluidics, analogous microfluidic "breadboards" with embedded fluidic "wiring" and fluidic circuit elements have been produced[21] and are commercially available,[22] but variations in how electrons and fluids containing reagents behave have restricted. First, despite the fact that metal wiring has negligible resistance and capacitance, either high fluidic resistances (small channels) or massive dead volumes (large channels) become issues inside the breadboard, resulting in reagents, particles, or cells becoming lost between components and/or large pressure drops accumulating. Second, while the flow of electrons in a wire has no effect on the function of logical operations, the arrangement of fluid parcels in a flow that is deformed or disrupted as it passes through larger (i.e., higher Reynolds number) interconnects can have a significant impact on downstream function. Nonetheless, modular design has benefited both single-phase and multiphase microfluidic platforms, but these approaches currently involve device-specific modifications for assembly into a functional system.[23,24] These modules, unlike conventional electronic circuits, are not mounted on a breadboard, and the engineer must perform custom calculations for each part and link in the device (e.g., droplet splitting to equally sized smaller droplets requires balanced downstream resistances).[3] The level of modularity and abstraction used in the design of monolithic integrated circuits, where custom software allows the assembly of components and fast prediction of the entire system's outputs, may be useful in the future design of monolithic microfluidic devices. Circuit designers can iterate and predict the behavior of an integrated circuit using software like Spice or Modern Electronic Design Automation without having to know specific semiconductor physics for each transistor.[25] The disadvantage is that the designer is restricted to a small number of well-known transistor geometries and design rules. Software packages that abstract difficult-to-simulate fluid dynamics to obtain the input-output response for a subset of fluid channel geometries are still in their early stages, but they are progressing toward similar functionality for pre-computed fluidic circuit elements.

The challenges of microfluidics in biomedical applications are particularly for measurement devices where biomolecular interactions can lead to inaccurate results. Advancement is driven by the focusing on synergy between fundamental materials development and new microfluidic capability.

The *interplay of complex materials*, such as plastics, adhesives, and other materials can alter assay function. Becton-Dickinson is used a wide-scale recall of blood tubes due to a plasticizer leachate which affects the lead tests. If we add the additional complexities of a system employing pumps and valves, myriad permutations, configuration of a device, and Murphy's Law can show up. It must be achieved with aggressive cost-of-goods targets which reflective of the cost pressures on diagnostics tests.[26]

Multiphase flow is the flow activity of molecules and chemical reactions in micrometer and sub micrometer-sized networks is known as multiphase flow. Multiphase flows can improve and expand the efficiency of single-phase microfluidic systems in a variety of ways. By introducing a second, immiscible fluid stream, the long diffusion times and wide dispersion bands associated with single-phase pressure-driven flow can be reduced. When two or more partially or not miscible fluids come into contact and are exposed to a pressure gradient, multiphase flows form. Fluid pathways on a chip must frequently be changed between different reaction, separation, and detection phases in integrated micro reaction solutions. A number of active valves for fluid control have been produced previously in polymers using soft-lithographic techniques[27,28] and in the MEMS group based on metals and silicon.[29] Most of these solutions, on the other hand, present significant challenges in terms of chemical and thermal compatibility, fabrication expense, and ease of integration for many microchemical systems.

Is the value of miniaturization transferable? Miniaturization has long-term cost advantages in all fields of technology due to lower material and energy consumption; however, when scaling down, critical new capabilities emerge. The reduction in operating scale allows for the use of special micro or nanoscale physics, which can help researchers solve problems that were previously thought to be insurmountable. When downscaling bulk gold or silver to nanoscale particles, this concept is nicely demonstrated in nanotechnology. Since surface electron density dominates behavior, optical absorption and emission properties change drastically during this transition.[30,31] In microfluidics, exploitation of scale-dependent effects is frequently crucial for eventual impact and application. Low Reynolds number activity with relatively high velocities and peclet numbers (which measure the ratio of advective to diffusive transport rates) enables reagent flow streams to be shaped and particle, droplet, and cell motion to be manipulated without turbulence. or diffusive blurring, which is almost difficult to do with aqueous solutions in larger channels. Furthermore, in contrast to interfacial forces, relatively small viscous (capillary number, Ca) or inertial (Weber number, We) forces allow for accurate "dripping" of uniformly sized drops. Small volumes of these "micro-droplets," or valved chambers containing single cells or molecules, allow discretization. Using microfluidic plumbing to valve and route flows from one location to another—directly scaling down larger-scale fluid handling—has, on the other hand, had some benefit but little usefulness.

4. Challenges in designing lab-on-a-chip device

With a plethora of ingenious devices and individual components,[2–32] lab-on-a-chip (LOC) technology has advanced from the research lab to clinical practice. In comparison to their conventional lab-based counterparts, complex assays can now be performed in a single instrument, quickly and with high sensitivity. However, many researchers are thriving hard to overcome the challenge of technology transfer, increased due to endpoint detection sensitivities, added device functionality and lower resolution manufacturing processes from the micro to the nanoscale. Moreover, challenge like integration, standardization, cost effective. System integration, world-to-chip interface, fluid delivery, application programmability, and automation are thus major integration problems for the LOC researcher. The four major obstacles to the transformation of LOC systems into clinical practice must be addressed as shown in Fig. 1.

1. *System integration*: Integrating LOC systems into POC architectures that fully reproduce the features available in lab settings. To operate such a system, ancillary equipment such as fluidic pumps, high current power supplies, and signal acquisition devices are needed. It is also significantly larger than the lab-on-a-chip systems themselves, typically led to

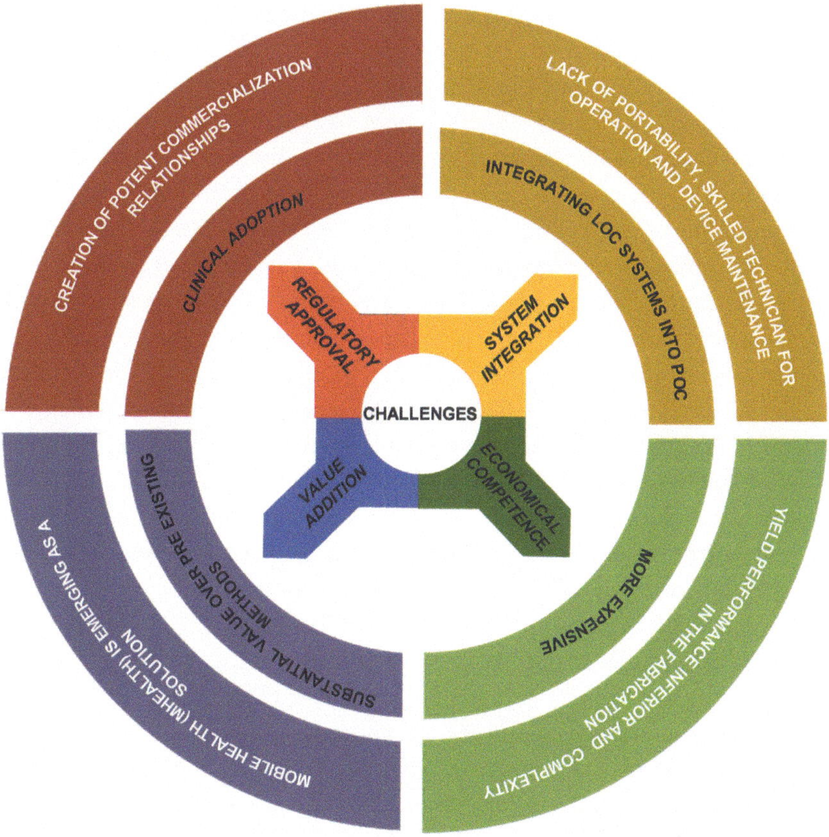

Fig. 1 Key challenges for the translation of lab on chip systems to clinical practice.

chip-in-a-lab not lab on chip.[33,34] It also pertains to user training and standardization protocols, portability, and skilled technician. Simplifying the lab-to-chip interface is also required major consideration for the introduction of buffers and reagents to the device. The disposal of contaminated devices and potentially biohazardous contents majorly need to be focused. Moreover, the adaptability of LOC with proper fluid delivery, programmability and automation is also crucial for development of LOC. The use of a bespoke photodetector and data processing device with a camera and the computing capacity of a cell phone, which not only streamlined the overall design but also reduced the cost through part reduction. It can be made to simplify the user AP interface and the management of data. It is anticipated that such functionality and interconnectivity of LOC devices will become more need of an hour in the future for lab-on-a-chip device development.[35]

2. *Economical competence*: In terms of cost and efficiency, current LOC systems fall short of central and remote laboratories. In general, POC devices are more costly and produce results that are inferior to those obtained in a laboratory environment. To realize their full potential, these modern LOC devices must achieve both cost reduction and improved efficiency as compared to traditional laboratory methods. Researchers focus on proving a hypothesis in one dimensional manner without addressing the issues such as feasibility, viability, with the field of lab-on-a-chip being no exception. Many lab on chip systems require advanced and labor-intensive manufacturing techniques to fabricate individual components and the entire product. Dealing with a variety of issues related to cost-effective manufacturing and scalability, such as reducing the number of components, increasing robustness, reducing assembly steps, and specifying appropriate materials are the key requirement of making LO based POC and expected from researchers. Currently the research groups are using poly (dimethylsiloxane) (PDMS) for the fabrication of microfluidic systems despite the alternative materials, such as polymers, paper, silicon, glass, etc. However, due to the difficulty in scaling up mass production and the comparatively high cost compared to alternatives such as polymers, PDMS has been limited in its commercial application. More importantly, native PDMS' inherent material properties can cause evaporation, leeching, and absorption of a fluid sample. Therefore, it is important for the research community to focus on the up-scaled manufacturing of PDMS or look for alternative fabrication materials of choice for lab-on-a-chip applications, such as low-cost polymers.
3. *Value addition*: New content must be developed for LOC systems that adds significant value to the target application and is currently inaccessible at central and remote labs. In order for a clinical population to embrace emerging technology, the device must have significant benefit over current methods. A new POC-based lab-on-a-chip invention that not only overcomes the limitations of existing methodologies/devices in terms of efficiency, cost, and usability, but also performs one or two orders of magnitude better in delivering a solution of value to the end user, must be qualified as potential for commercialization. Biomarker measurements, for example, play a critical role in clinical decision making and would benefit from a lab-on-a-chip platform to provide a result quickly and in a compact manner, although the same would not be true for disease genetic testing. Lab on chip applications, in particular, can be

incorporated with mobile health (mHealth), which is gaining traction as a solution to issues such as high disease burden, limited resources, and inaccessibility.

4. *Regulatory approval and clinical adoption*: Technology developers need to develop strategies for streamlining regulatory approval route to clinical adoption. Translational research programs are heavily invested and their research efforts remain largely decoupled from real-world clinical practice. As a result, the acceptance of groundbreaking medical technology by regulatory agencies is at an all-time low. Device production in academia, national laboratories, and industry is typically a complex and time-consuming process that proceeds in a linear, inefficient fashion, wasting valuable time and money from venture capital and public funding sources. In the absence of evidence supporting the use of the candidate technology for an intended diagnostic objective, a range of assumptions are often made prior to product approval. Invalidity of the intended target, lack of a demand for its use, or improbability of its acceptance as part of clinical practice are all examples of misleading facts. As a result, promising medical microdevice innovations can be made possible by identifying unmet clinical needs, forming world-class clinical partnerships, developing standard testing protocols and operating procedures, defining stringent user specifications, and establishing powerful commercialization relationships. Early in the development process, clinical and consumer needs must be identified, and efforts to identify the competitive environment and regulatory pathways for these new test modalities are crucial. To resolve obstacles to regulatory approval, commercialization, and clinical implementation, technology developers must also tap into multidisciplinary expertise through close clinical partnerships. Device manufacturers should be involved in developing standardized research procedures, ensuring access to essential clinical samples, and exchanging knowledge about consumer criteria for new diagnostic tests in nontraditional healthcare settings.

5. Opportunities in micro/nanofluidic and lab-on-a-chip devices

While microfluidics offers many benefits to the researcher, it is not a panacea. Many features of microfluidic systems that were once considered essential to their use or acceptance are now less convincing as traditional technologies advance and refine. The automation of microfluidic design may be an interesting solution to many of the issues listed above. Currently, most laboratories need a considerable amount of handcrafting, hands-on experience, and scientific expertise to construct usable microfluidic systems. While this hasn't stopped progress, it has resulted in many microfluidic systems performing below expectations. Electrical engineers can teach microfluidic researchers a lot about auto-mating architecture in this regard. As a result, software that abstracts microfluidic components and then integrates them in a way that enables them to operate predictably is in high demand. Many of the early predictions about accuracy, usefulness, and functionality have been realized thanks to microfluidic systems. However, their true potential as facilitating methods in chemical and biological experimentation has yet to be realized. We conclude with two recommendations to aid future efforts in the region. To begin, consider the final product. Understand how the basic and applied research you're doing now will affect something people will want to use in 1–5 years, while also taking into account other technologies in progress. Second, look at an idea, a physical phenomenon, or a design paradigm that is so new that no one knows what to do about it. Since no one is aware of it or has considered it in conjunction with other variables, it has the ability to have a wide-ranging effect.

It is becoming increasingly evident that research into lab-on-a-chip technology has grasped a tipping point, which must be converted into functional consumer goods. This task would necessitate a change in researchers' mindsets to interact more with their industrial counterparts in order to gain a deeper understanding functionality, cost, sensitivity, and device complexity, as well as how this will translate to affordable mass manufacturing. This is a significant stumbling block in the commercialization of technology. Ironically, the reductionist approach might be required in stripping lab-on-a-chip technology down to its bare essentials in order to reduce chip-to-chip variations by reducing the number of variables that negatively impact the final device's efficacy. This method may be at odds with recent developments in the field, in which researchers seek to demonstrate novel functionalities at the cost of increased complexity and, as a result, a reduced potential for mass production. It's debatable if there is a single "killer application" for lab-on-a-chip/microfluidic technology, but by carefully considering the commercialization aspect, an end-user needs, and the added value the technology offers for a specific application. Although the race to collect vast quantities of health data is still in its early stages (e.g., through large pharmaceutical/medical firms, insurance companies, or governments), the uses of these high-quality and high-dimensional clinical datasets will play a significant role in medicine in the coming years as direct to consumer research becomes more common. As a result, product developers must form strong clinical partnerships to define disease targets, oversee the development of standardized testing protocols, and address unmet clinical needs in order for the new LOC technology to have the greatest social and economic effect possible.

Acknowledgments

We gratefully acknowledge Gujarat State Biotechnology Mission (GSBTM Project ID: AJJV48) and SHODH: ScHeme of Developing High Quality Research Scholarship (Ref No: 201901440009) for financial support.

References

1. Wang J, Nuñovero N, Zhan C, et al. Microscale gas chromatography with microsensor array detection: challenges and prospects. In: *Paper Presented at: Multidisciplinary Digital Publishing Institute Proceedings*; 2017.
2. Whitesides G. The lab finally comes to the chip! *Lab Chip*. 2014;14(17):3125–3126.
3. Xi H-D, Zheng H, Guo W, et al. Active droplet sorting in microfluidics: a review. *Lab Chip*. 2017;17(5):751–771.
4. Dixit CK, Kaushik AK, Kaushik A. *Microfluidics for Biologists*. Springer; 2016.
5. Antfolk M, Laurell T. Continuous flow microfluidic separation and processing of rare cells and bioparticles found in blood—a review. *Anal Chim Acta*. 2017;965:9–35.
6. Feng H, Zheng T, Li M, et al. Droplet-based microfluidics systems in biomedical applications. *Electrophoresis*. 2019;40(11):1580–1590.
7. Gong MM, Sinton D. Turning the page: advancing paper-based microfluidics for broad diagnostic application. *Chem Rev*. 2017;117(12):8447–8480.
8. Carrell C, Kava A, Nguyen M, et al. Beyond the lateral flow assay: a review of paper-based microfluidics. *Microelectron Eng*. 2019;206:45–54.
9. Jama M, Singh T, Gamaleldin SM, et al. Critical review on nanofluids: preparation, characterization, and applications. *J Nanomater*. 2016;2016.
10. Wang C, Liu M, Wang Z, Li S, Deng Y, He N. Point-of-care diagnostics for infectious diseases: from methods to devices. *Nano Today*. 2021;37:101092.
11. Kumar S, Nehra M, Khurana S, et al. Aspects of point-of-care diagnostics for personalized health wellness. *Int J Nanomedicine*. 2021;16:383.
12. Sachdeva S, Davis RW, Saha AK. Microfluidic point-of-care testing: commercial landscape and future directions. *Front Bioeng Biotechnol*. 2021;8:1537.
13. Hin S, Lopez-Jimena B, Bakheit M, et al. Fully automated point-of-care differential diagnosis of acute febrile illness. *PLoS Negl Trop Dis*. 2021;15(2), e0009177.
14. Reyes DR, van Heeren H, Guha S, et al. Accelerating innovation and commercialization through standardization of microfluidic-based medical devices. *Lab Chip*. 2021; 21(1):9–21.
15. Vera D, García-Díaz M, Torras N, Álvarez M, Villa R, Martinez E. Engineering tissue barrier models on hydrogel microfluidic platforms. *ACS Appl Mater Interfaces*. 2021;13:13920–13933.
16. Kim SK, Kim YH, Park S, Cho S-W. Organoid engineering with microfluidics and biomaterials for liver, lung disease, and cancer modeling. *Acta Biomater*. 2021.
17. Dai B, Yin C, Wu J, et al. A flux-adaptable pump-free microfluidics-based self-contained platform for multiplex cancer biomarker detection. *Lab Chip*. 2021;21(1):143–153.

18. Elvira KS, i Solvas XC, Wootton RC, Demello AJ. The past, present and potential for microfluidic reactor technology in chemical synthesis. *Nat Chem*. 2013;5(11):905.
19. Levy SL, Craighead HG. DNA manipulation, sorting, and mapping in nanofluidic systems. *Chem Soc Rev*. 2010;39(3):1133–1152.
20. Nagayama G, Cheng P. Effects of interface wettability on microscale flow by molecular dynamics simulation. *Int J Heat Mass Transf*. 2004;47(3):501–513.
21. Bhargava KC, Thompson B, Malmstadt N. Discrete elements for 3D microfluidics. *Proc Natl Acad Sci U S A*. 2014;111(42):15013–15018.
22. Zhou P, Young L, Lynch K. The LabMatrix™ system. *J Assoc Lab Autom*. 2003; 8(5):5–10.
23. Mazutis L, Araghi AF, Miller OJ, et al. Droplet-based microfluidic systems for high-throughput single DNA molecule isothermal amplification and analysis. *Anal Chem*. 2009;81(12):4813–4821.
24. Niu X, de Mello AJ. Building droplet-based microfluidic systems for biological analysis. *Biochem Soc Trans*. 2012;40(4):615–623.
25. Nagel LW. SPICE-simulation program with integrated circuit emphasis. In: *Memo No. ERL-M382*. Berkeley: Electronics Research Laboratory, Univ. of California; 1973.
26. Hou X, Zhang YS, Trujillo-de Santiago G, et al. Interplay between materials and microfluidics. *Nat Rev Mater*. 2017;2(5):1–15.
27. Dunne A, Francis W, Delaney C, Florea L, Diamond D. *Stimuli-Controlled Fluid Control and Microvehicle Movement in Microfluidic Channels*. Proceedings, MDPI; 2017.
28. Rose MA, Bowen JJ, Morin SA. Emergent soft lithographic tools for the fabrication of functional polymeric microstructures. *ChemPhysChem*. 2019;20(7):909–925.
29. Franssila S, Tuomikoski S. MEMS lithography. In: *Handbook of Silicon Based MEMS Materials and Technologies*. Elsevier; 2020:399–416.
30. Maier SA. Plasmonic field enhancement and SERS in the effective mode volume picture. *Opt Express*. 2006;14(5):1957–1964.
31. Psaltis D, Quake SR, Yang C. Developing optofluidic technology through the fusion of microfluidics and optics. *Nature*. 2006;442(7101):381–386.
32. Tymm C, Zhou J, Tadimety A, Burklund A, Zhang JX. Scalable COVID-19 detection enabled by lab-on-chip biosensors. *Cell Mol Bioeng*. 2020;13(4):313–329.
33. Mohammed MI, Haswell S, Gibson I. Lab-on-a-chip or chip-in-a-lab: challenges of commercialization lost in translation. *Procedia Technol*. 2015;20:54–59.
34. Chin CD, Linder V, Sia SK. Commercialization of microfluidic point-of-care diagnostic devices. *Lab Chip*. 2012;12(12):2118–2134.
35. Qamar AZ, Shamsi MH. Desktop fabrication of lab-on-chip devices on flexible substrates: a brief review. *Micromachines*. 2020;11(2):126.

Index

Note: Page numbers followed by "*f*" indicate figures and "*t*" indicate tables.

A

Abstraction, 46
Active matter, in confined microfluidic environments, 245–260
Active systems, 250–251
Additive manufacturing, 21*f*, 22, 33–39
  advantages and disadvantages, 38–39, 40–42*t*
AdnaTes, 194
Affinity chromatography, 67
Agarose gel droplets, 223
Agglutination technique, 162
Aggregation-caused quenching (ACQ), 228–230
Aggregation induced emission (AIE), 228–231, 230*f*
AIEgen-doped poly(ionic liquid) (PIL) photonic spheres, 231
AIEgens, 230–231
AIE quantum dots, 230–231, 231*f*
Alkaline phosphatase (ALP), 280–282
α-fetoprotein (AFP), 180–181
Amperometry-based detection, 72
ASSURED criteria, 111, 164–166
Asthma, 196
Augmented Reality (AR) app, 182–183
Au-Pt microtubes, 251–253
AutoCAD, 4, 204–205
Autonomous micropumps, 255
Autoradiography, 73

B

Bacterial motion, on diffusion of fluorescently labeled dextran, 255–257, 256*f*
Barcode, 234–235
Bed volume, 116–117
Bimetallic motors, 251–253
Bio-functional nanoslit, 237–238, 237*f*
Biomarkers, 197
Biomolecules, paper-based detection, 181–182, 182*f*
Biotin-avidin bridging, 227–228
Biotinylated proteins, 67–68
Blood plasma separation, 137, 140*f*
Brownian diffusion, 246–247
Brownian motion, 258–259
Brownian motor, 93–94
Bulk micromachining, 22–24, 26–27, 40–42*t*

C

Cancer, 193–194
Cancer diagnostics, μPADs, 137, 141–142*f*
Cancer theranostics, 193–194, 193*f*
Capacitive coupled contactless conductivity detection (C4D), 73
Capillary electrophoresis (CE), 63, 66, 91–92
Capillary flow rate, paper, 115–116
Capillary isoelectric focusing (CIEF), 63, 66
Capillary isotachophoresis (CITP), 63
Capillary microfluidic systems, 253–255
Capillary zone electrophoresis (CZE), 63, 66
Carbon nanomaterials-based assays, 274
Carbon nanoparticles (CNPs), 274
Carbon nanotubes (CNTs), 274
Carcinoembryonic antigen (CEA), 180–181
  gold-silver based bimetallic nanoparticles, 280–282
Cardiovascular diseases, 194–195
  microfluidic-based detection, 194–195
Cellulose fiber filter paper, 115–116
Cellulose paper, 160–161
cfDNA extraction, 100*f*
Channel architecture, 46–47
Chemiluminescence, 71–72
Chemiluminescence-based assay, 177
Chemstrip1 Micral test strips, 164
Cholesterol, amperometric detection, 181
Chromatography, 63, 64–65*t*
Circulating cathodic antigen (CCA), 274
Circulating tumor cells (CTCs), 194, 221–222
Click-chemistry, 61–62

Clinical diagnostics, paper-based analytical devices, 180–181
Colloidal gold nanomaterials-based assays
signal enhancement using
gold amplification, 272–274, 273*f*
laser excitation, 271–272
silver amplification, 272
sizes and shapes, 271
Color homogeneity, 176
Colorimetric paper-based assay, 174–176, 175*f*
quantification in, 176–177
Computer-aided design (CAD), 33–34, 248–249
Conductivity-based detection, 72
Conformity Assessment Procedure, 147
Continuous flow (CF) microchip, 9
Continuous flow microfluidics, 290–291
Continuous-flow polymerase chain reaction (CF-PCR) devices, 9
Continuous inkjet (CIJ) inkjet printing, 172–173
Conventional fluorophores, 228–230
COVID-19 antibody test strips, 166
COVID, μPAD test, 133–136
COVID-19 rapid antigen tests, 283
C-reaction protein (CRP), CdSe/ZnS QDs in detection of, 275
CRISPR/Cas9, 234–235
Cu-Pt nanomotor, 251–253, 252*f*
Cylindrical motors, 251–253

D

Darcy's law, 114*f*, 115, 161
Deoxyribonucleic acid (DNA), 219–221
Design for manufacturing and assembly (DFMA), 48
Design thinking, 16, 43–50
stages of, 44, 45*f*
define, 46–47
empathize, 44–45
ideate, 47–48
prototyping, 48–50
testing, 50
Detectors, 177
Diabetes, microfluidic-based detection, 195–196, 195*f*
Dielectrophoresis, 91–92
Diffusion, 18
Digital fabrication, 16
Digital light projection (DLP) printer, 34–35, 34*f*
Dipstick immunoassay, 268, 269*f*
Dipstick paper biosensors, 164
Direct ink writing (DIW), 36, 37*f*
Disease diagnostics, 10–11
DNA interaction study, nanofluidic systems for, 235–239
DNA polymerase, 219–221
DNA-protein interactions, 238–239
DNA/RNA detection and separation, 8–9
detection techniques, 94–95, 95*f*
glass based inorganic material, 97–98
paper based microfluidics, 99–103, 101*f*
polydimethylsiloxane (PDMS) based, 96–97
silica based inorganic substrate in, 98–99
future prospects, 103–104, 104*f*
materials used, 87*f*
separation techniques
electrokinetic based, 92–93
electrophoresis, 90–92
entropic based, 87–90
microfluidic and nanofluidic, 87, 88*f*
Ratchet mechanism, 93–94
Double-sided sticking method, 125
Droplet-based microfluidics, 290–291
Droplet nanofluidic platform, 235–237
Droplets-based single cell analysis, 207, 208*f*
Dropon-demand (DOD) inkjet printing, 172–173
Drug analysis, paper-based analytical devices, 181
Drug discovery, 9–10
Dual signal enhancement strategy, 280–282
Dynamic light scattering (DLS), 73

E

Ebola virus, paper-based detection, 137, 138–139*f*
Eectro-assisted deposition, 251–253
Electrocapillary forces, 290–291
Electrochemical detection, 11–12
paper microfluidics, 177–178
protein, 72–73

Electrochemical paper-based device, 169
Electrochemiluminescence assay, 178, 275
Electrokinetic flow, 249–250
Electrokinetics, 92
Electrokinetic separation, nucleic acids, 92–93
Electron-beam lithography (EBL), 22–24
Electrophoresis, nucleic acid separation, 90–92
Electrospray ionization mass spectrometry, 7–8
Electrowetting on dielectric (EWOD) technique, 163, 290–291
ELISAs. *See* Enzyme-linked immunosorbent assays (ELISAs)
Embossing, 31, 33, 40–42*t*
Emission quenching technique, 235–237
Empathetic understanding, 44–45
Energy harvesting, paper-based microfluidic devices, 137, 143*f*
Entropic trapping, 87–90, 88*f*
Enzyme-based labels, for signal enhancement, 276–279
Enzyme-linked immunosorbent assays (ELISAs), 267–268, 269*f*, 270, 276–277
Enzymes, 251–253
Eosin-Y, 280
ePADs, 6–7
Epithelial cell adhesion molecule (EpCAM), 227–228
EPithelial ImmunoSPOT (EPISPOT), 194
Escherichia coli, 205–206
  single cell analysis, 208–209
  smartphone-based detection, 182
Europium nanospheres, 276
Extended-release GI luminal delivery systems, 10
Extrusion-based 3D printing, 36, 37*f*, 40–42*t*
Ezrin, 197

F

Fabrication techniques, 21–22, 21*f*
  additive manufacturing, 33–39
  advantages and limitations, 39, 40–42*t*
  attainable channel dimensions, 39, 43*f*
  formative manufacturing, 28–33
  subtractive manufacturing, 22–28
Fast technology analysis (FTA) cards, 161–162
Femtosecond direct laser writing (DLW), 23*f*, 25
Femtosecond laser direct writing, 23*f*, 25, 40–42*t*
Field fractionation, 68
Filter papers, 112
Filtration isolation nucleic acid (FINA) technology, 162
FLASH technique, 117–121, 122–124*f*
Flexographically printed technique, 117–121, 122–124*f*
Flexographic printing, 171, 172*f*
Flow-cytometry, 226–227
Flow profile, in microchannel, 19, 19*f*
Fluid control, 47
Fluid flow, 247–248, 255
Fluidic devices, future prospects, 103–104, 104*f*
Fluidic resistance, 18–19
Fluorescein-labeled dog serum albumin (FITC-DSA) protein, 235–237
Fluorescence, 218–219
Fluorescence activated cell sorting, 204
Fluorescence-based protein detection method, 71
Fluorescence imaging/scanning, 71
Fluorescent based PCR method, 232–233
Fluorescent detection, 177
Fluorescent video microscopy, 88–89
Fluorosurfactants, 207
Focused ion beam milling (FIB), 23*f*, 24, 40–42*t*
Food quality monitoring, μPADs, 137
Formative manufacturing, 21*f*, 22, 28–33
  advantages and disadvantages, 32–33, 40–42*t*
Form factor, 47
Frequency resonance energy transfer (FRET), 233–234
Functional components, 47
Fused-deposition modeling (FDM), 36

G

Gadolinium thiocyanate method, 162
Gel electrophoresis (GE), 67, 91–92
Glass, 22–24

Glass capillary microfluidic device, 253–255, 254*f*
Glass chip, 97–98
Glucose-stimulated insulin secretion (GSIS) assays, 196
Gold amplification, of AuNP, 272–274
Gold-nanoparticle decorated silica nanorods (AuNP-SiNRs), 276
Gold nanoparticles (AuNPs), 164–166, 175–176
  signal enhancement
    gold amplification, 272–274, 273*f*
    laser excitation, 271–272
    silver amplification, 272
    sizes and shapes, 271
Graphene, 274
  micromotors, 253–255, 254*f*
Graphene oxide (GO)-based micromotors, 253–255
Gravity separation method, 88–89

H

Hagen-Poiseuille law, 161
Hard photolithography mask, 28–30
Heart-on-a-chip platforms, 195
Helical micromotors, 253–255
Hexaphenylsilole (HPS), 228–230
Horseradish peroxidase (HRP), 280–282
Hot embossing, 31
Hybridization chain reaction (HCR), 219–221
Hybrid labels-based signal enhancement, 280–282
Hydrolase, 180–181
Hydrophobicity, polydimethylsiloxane (PDMS) microchannels, 249–250
Hyper branched rolling circle amplification (HRCA) technique, 224–225

I

Ideate, 47–48
ImageJ, 205–206
Imbibition, 161
Immunoassays, 10–11, 267–268
Immunochromatographic assay (ICA), 275
Immunofluorescence, 225–226
Impedance-based detection system, 72
Influenza A, detection of, 137, 138–139*f*
Injection molding, 31–32, 40–42*t*, 206–207
Inkjet 3D printing, 35–36, 36*f*, 40–42*t*
Inkjet etching, 169–170, 170*f*
Inkjet printing, 117–121, 122–124*f*, 172–173
Ink stamping, 125, 173–174, 174*f*
Instrument-free readout devices, 180
Interfacial forces, 20
International Organization for Standardization (ISO), 147
In vitro transcription/translation (IVT), 224–225
Islet perifusion systems, 196

K

Keto Diastix Reagent Test Strip, 164

L

Lab-on-a-chip (LOC) devices, 246–247
  challenges in designing, 296–299
    economical competence, 298
    regulatory approval and clinical adoption, 299
    system integration, 296
    value addition, 298
  opportunities in, 299–301
Laminar flows, 17–18
  fluidic resistance for, 18–19
Lanthanide nanoparticles, 276
Laser ablation/engraving, 25, 26*f*, 40–42*t*, 206–207
Laser capture microdissection, 204
Laser excitation, of AuNP, 271–272
Laser treatment, paper-based microfluidic devices, 117–121, 122–124*f*
Lateral flow assay (LFA), 164–166, 165*f*
Lateral flow immunoassays (LFIAs), 267–268, 269*f*
Layered paper and tape technique, 117–121, 122–124*f*
LFIAs. *See* Lateral flow immunoassays (LFIAs)
Ligand Foot-printing (LiF), 72
Ligation-based droplet digital polymerase chain reaction (dd-PCR) method, 219–221, 220*f*
Light-assisted 3D printing, 34–35, 34*f*, 40–42*t*

Light-induced polymerization amplification, 176
Lithic reduction, 21–22
Litmus paper, 110
Localized surface plasmon resonance (LSPR), 271–272
Lucas-Washburn equation, 114, 161
Luer-lock, 49

M

Magnetic nanoparticle-based assays, 274–275
Magnetic nanoparticles (MNPs), 282
Magneto-resistive sensors, 73
Marangoni effect, 253–255
Mass spectrometry (MS), 72
Materials, choice of, 46
mChip assay, 10–11
Medi-Test Combi 11, 164
Methyl-binding-domain (MBD) protein, 238–239, 238*f*
Micellar electrokinetic capillary chromatography (MEKC), 63, 66
Micro-channel network, 248–249
Microchip gel electrophoresis (MGE), 67–68
Microchip gel electrophoresis with electrochemical detection (MGE-ED), 91–92
Micro droplets, 227, 228*f*
Micro-electromechanical system (MEMS), 16
Microfluidic chip electrophoresis (MCE), 62–67
Microfluidic devices, 248–249, 257
Microfluidic plumbing, 295–296
Microfluidic resistive pulse sensing (MRPS), 290–291
Microfluidics chip, design and fabrication, 204–205, 205*f*
Micromanipulator-assisted cell picking, 204
Micromilling, 26, 27*f*, 40–42*t*
Micro/nanofluidics, 1–2, 246–247, 289–290
  applications, 2–3, 3*f*
    disease diagnostics and personalized medical, 10–11
    DNA/RNA detection and separation, 8–9
    drug discovery, screening and delivery, 9–10
    protein detection and separation, 7–8
  biomedical applications
    aggregation induced emission (AIE), 228–231
    nucleic acid sensing and analysis, 219–225
    protein sensing and analysis, 225–228
  challenges in designing, 293–296
    interplay of complex materials, 295
    materials, 293
    miniaturization, 295–296
    modular components, 294–295
    multiphase flow, 295
  definition, 60
  design and fabrication, 4
  opportunities in, 299–301
  paper based, 4–7
  types of, 3–4
MicroPAD (μPAD), 4–5, 99–101, 167
  applications, 133–136
  fabrication techniques, 117–121, 122–124*f*
  flow management in, 126
  paper-based microfluidic device as, 112, 113*f*
  production of, 142
MicroPCR, 292–293
MicroRNAs (miRNAs), electrokinetic separation, 92–93
Micro total analysis systems (μTAS). *See* Lab-on-a-chip (LOC) devices
Mie scattering detection, 178–179
Miniaturization, 73
Mobile-based clinical microscopy, 10–11
Modular design, 49–50
Molecular chemotaxis, 251–253, 252*f*
Multidimensional device fabrication method, 118*t*
Multifunctional motors, 251–253
Multiphase flows, 295
Multiple-jet (MJ) printers, 35–36, 36*f*
Multiplex protein detection, 60, 60*f*
Multistix 10SG reagent strips, 164
Multi-walled carbon nanotubes (MWCNTs), 274, 291–292

μPAD. *See* MicroPAD (μPAD)
Mycobacterium tuberculosis, detection of, 137, 138–139*f*

N

Nanofluidics, 68
biomedical applications
DNA interaction, 235–239
nucleic acid sequencing and sensing, 232–235
protein binding kinetics, 235–239
paper, 183
Nanofluids, 291–292
Nanoimprint lithography (NIL), 22–24, 31
Nanomaterials
signal enhancement in paper-based immunoassays, 271–276
carbon nanomaterials, 274
colloidal gold nanomaterials, 271–274
magnetic nanoparticles (MNPs), 274–275
quantum dots (QDs), 275
silica nanoparticles, 275–276
upconversion nanoparticles (UCNPs), 276
Nanopore sensing, 7–8
Nano-RCA droplets, 227–228, 229*f*
Neutrophils, 196
Next-generation sequencing, 195–196
Nicking endonuclease, 234–235, 235*f*
Nick labeling strategy, 234, 235*f*
Nitrocellulose, 268
Nitrocellulose paper, 161
Nominal pore size, 116
Non-gel sieving-CE, 63
Nucleic acid amplification test, 98–99, 134–135*t*, 137, 138–139*f*
Nucleic acids, 87
detection
glass based inorganic material, 97–98
paper based microfluidics, 99–103, 101*f*
polydimethylsiloxane (PDMS) based, 96–97
silica based inorganic substrate in, 98–99
integrated paper-fluidic molecular diagnostic chip for detection, 137, 138–139*f*
sensing and analysis
microfluidic systems for, 219–225
nanofluidic systems for, 232–235
separation
electrokinetic based, 92–93
electrophoresis, 90–92
entropic based, 87–90
Ratchet mechanism in, 93–94

O

On Chip amplification (On Chip-HDA), 97–98
On-chip diffusion immunoassay, 10–11
On-Chip RP-HPLC, 63
Open microfluidics, 290–291
Optical detection, 11–12
Optical mapping technology, 232–233
Organs-on-a-chip, 195
Oxidase, 175
Oxygen plasma cleaner, 204–205

P

Paper, 4–5, 160
chemistry of, 112–113
invention of, 110
physics of, 113–117, 114*f*
as sample pretreatment platform in clinical diagnosis, 161–163
Paper-based immunoassays, 270
history of, 267–270
signal enhancement strategies
enzyme-based labels for, 276–279
hybrid labels-based, 280–282
nanomaterials based, 271–276
polymerization-based, 280, 281*f*
SWOT analysis, 282–284
Paper-based microfluidic devices, 4–7, 148, 290–291
advantages, 111–112
applications, 6, 133–141, 133–135*t*
in biomolecule detection, 181–182, 182*f*
capillary flow rate, 115–116
challenges and research, 184
in clinical diagnostics, 180–181
competitive advantages, 145
competitive study, 144*f*

detection techniques
chemiluminescence, 177
colorimetry, 174–176, 175*f*
electrochemical method, 177–178
electrochemiluminescence, 178
fluorescent detection, 177
in drug analysis, 181
fabrication techniques, 117–125, 118–120*t*
flexographic printing, 171, 172*f*
inkjet etching, 169–170, 170*f*
inkjet printing, 172–173
ink stamping, 173–174, 174*f*
photolithography, 168–169, 168*f*
plasma treatment, 173
plotting method, 169
screen-printing, 169
wax dipping, 171–172
wax printing, 170–171, 171*f*
fluid flow control and liquid handling, 125–133
chemical processes, 129–130*f*, 130–132
geometrical approach, 126–130, 127–128*f*
mechanical methods, 131–132*f*, 132–133
instrument-free readout devices, 180
limitations and future works, 142–147
market analysis, 142–145
in nucleic acid detection, 99–103, 101*f*
principle and prospect, 112
regulatory challenges, 147
smartphone based applications, 182–183
smartphone-based devices, 178–179
Paper chromatography, 110
Paper continuous flow microfluidic devices (p-CMF), 163
Paper digital microfluidic devices (p-DMF), 163
Paper microfluidic diagnostic devices, 163–164
Paraffin patterned filter paper, 110
PDMS. *See* Polydimethylsiloxane (PDMS)
Péclet Number, 247–248
P-ELISA device, 276–277, 278*f*
Peroxidase, 175
Personalized medicine, 192–193
applications, 10–11
cancer theranostics, 193–194, 193*f*
cardiovascular diseases, detection of, 194–195
diabetes, detection of, 195–196, 195*f*
pulmonary disease, 196–197
Photoacoustic (PA) signal enhancement, 271–272
Photodynamic therapy (PDT), 228–230
Photo-initiated free radical polymerization, 280
Photolithography, 22–24, 168–169, 168*f*
Photomasks, 28–30
Photoresist, 248–249
Photothermal laser speckle imaging (PT-LSI), 271–272
Physical blocking of pores, 117–121
PIBEX chip, 100*f*
Planner device fabrication method, 118*t*
Plasma treatment, paper microfluidic devices, 117–121, 122–124*f*, 173
Plotting method, 169
Point-of-care (POC) devices, 111
for infectious disease diagnostics, 136*f*
Poly(ethylene glycol) diacrylate (PEGDA), 35
Polydimethylsiloxane (PDMS), 4, 298
microchannels, 248–250, 249*f*
in nucleic acid detection, 96–97
paper-based device fabrication, 117–121, 122–124*f*
replica molding with, 16, 28–30, 29*f*
self-locating feature on, 48–49, 49*f*
Polyelectrolyte multilayer capsules (PMCs), 255
Polymerase chain reaction (PCR), 219–221
Polymerization-based signal enhancement, 280, 281*f*
Polymer-lipid hybrid nanoparticles (P/LNPs), 10
Polyvinylidene difluoride (PVDF), 268
Pore size, paper, 116
Porosity, 116–117
Post-transitionally modified (PTM) proteins, 67
Pregnancy test kit, 166
Pressure-driven flow, 249–250

Pressure driven liquid chromatography microfluidic device, 66–67
Print, cut and laminate (PCL) technique, 91–92
Printed circuit technology, 117–121, 122–124*f*
Proteinase K, 223
Proteins, 61
  binding kinetics, nanofluidic systems for, 235–239
  crystallization, 292–293
  detection and separation, 7–8
  detection system, 62, 63*f*
    classification of, 70*f*
    electrochemical, 72–73
    mass spectrometry, 72
    spectrophotometric method, 70–72
  sensing and analysis, microfluidic systems for, 225–228
  separation systems, 63–68
Proteomics, 292–293
Prototyping, 48–50
Protran membranes, 112–113
Pulmonary disease, detection and therapy, 196–197

Q

Qualitative assay, colorimetry, 176
Quantum dot-based assays, 275
Quantum dot nanobeads, 275
Quantum dots (QDs), 275
Quartz, 22–24
Quick Response Code-based immunosensors, 182–183
Quorum sensing-based synthetic oscillator, 210

R

Ratcheted electrophoresis, 93–94
Ratchet mechanism, 93–94
Reactive oxygen species (ROS), 228–230
Recombinase polymerase amplification (RPA), 9
Replica molding, PDMS, 16, 28–30, 29*f*
Resistive pulse sensing technique (RPS), 96–97
Restriction reaction approach, 232–233
Reverse transcription polymerase chain reaction (RT-PCR), 221–223
Reynolds number, 17–18, 247–248, 290–291
Ribonucleic acid (RNA), 219–221
Richard's equation, 114*f*, 115
Rolled-up microtubes, 251–253
Roller embossing, 31
Rolling circle amplification (RCA), 219–221, 227–228
RT-PCR test, 148–149

S

Salmonella, smartphone-based detection, 182
SARS, μPAD test, 133–136
Screen-printing, paper-based devices, 169
Self-locating features, 48–49, 49*f*
Semiquantitative assay, colorimetry, 176
Shear stress, 19*f*, 20
Signal enhancement strategies, in paper-based immunoassays
  enzyme-based labels for, 276–279
  hybrid labels-based, 280–282
  nanomaterials based, 271–276
    carbon nanomaterials, 274
    colloidal gold nanomaterials, 271–274
    magnetic nanoparticles (MNPs), 274–275
    quantum dots (QDs), 275
    silica nanoparticles, 275–276
    upconversion nanoparticles (UCNPs), 276
  polymerization-based, 280, 281*f*
Silica nanoparticle-based assays, 275–276
Silicon, 22–24
Silicon based lithography, 88–89
Silicon chip, DNA/RNA detection, 98–99
Silver amplification, of gold nanoparticles (AuNPs), 272
Silver nanoparticles (AgNPs), 175–176
Single cell analysis, 204
  droplet-based method, 207, 208*f*
  microfluidic platform for, 222, 223*f*
  microfluidics devices for, 208–210, 209*f*
Single point fluorescence measurement, 71
Single-walled carbon nanotubes (SWCNT), 274
Small interference RNAs (siRNAs), 10

Smartphone-based devices, 178–179
Smartphones, 11–12
Sodium dodecyl sulfate (SDS), 223
Sodium dodecyl sulfate-polyacrylamide gel electrophoresis (SDS-PAGE), 63, 66
Soft lithography, 16, 28
  advantages and limitations, 40–42*t*
Soft photolithography mask, 28–30
Spectrophotometric method, 70–72
Spherical Janus micromotors, 257–258
Stamp embossing, 31
Stimulated emission-depletion fluorescence correlation spectroscopy (STED-FCS), 257
Strand displacement amplification (SDA), 219–221
Streptavidin, 237–238
Substrates, 95
Subtractive manufacturing, 21–28, 21*f*
  advantages and disadvantages, 26–28, 40–42*t*
SU-8 master molds, 30–31
Surface enhanced Raman scattering (SERS), 92–93
Surface plasmon resonance (SPR), 175–176
Surfactants, 207
SWOT analysis, paper-based immunoassays, 282–284
SYBR Green cyanine dye, 221–222
Syringe pumps, 249–250

T

TelomeScan, 194
Test, in design thinking, 50
Tetraphenylethylene (TPE), 228–230
Thermal contrast amplification (TCA), 271–272
Thermophotonic lock-in imaging (TPLI), 271–272
Thermoplastic polymers, 31–32
Thiol-OEG-biotin conjugates, 237–238
2D-μPAD, 110
2D shaping/cutting, papers, 117–121, 122–124*f*
3D slip-PAD device, 277
3D paper-based microfluidic devices, 128–130
3D-printed embossing molds, 31
3D printers, 293
3D printing, 16
  advantages, 38–39
  extrusion-based, 36, 37*f*
  inkjet, 35–36, 36*f*
  light-assisted, 34–35, 34*f*
3D wax printing, 117–121, 122–124*f*
Throughput screening, 9–10
Time lapse fluorescent microscope (TLFM), 205–206
Tissue engineering, μPADs, 137, 141–142*f*
Total chemical analysis system (TAS), 16
Tracking, cell, 210
Transferases, 180–181
Trapping, cell, 210
Triggerable Mutually Amplified Signal (TMAS) probe, 92–93
T4 RNA ligase 2, 219–221
Tumor organoids, 194
Tween 20, 205–206
Two-dimensional paper networks (2DPN), 277, 279*f*

U

Unisys 11,001, 164
Upconversion nanoparticles (UCNPs), 276
Upconverting nanoparticles (UCNPs)-based assays, 276
UV embossing, 31
UV-photolithography, 248–249

V

Veterinary diagnostics, μPADs, 137

W

Wax dipping, 170–172
Wax printing, 170–171, 171*f*
Wax screen printing, 117–121, 122–124*f*
Whatman chromatographic paper, 170, 184
Whatman 903 paper, 102–103

X

Xurography, 25, 27*f*

FSC
www.fsc.org
MIX
Papier | Fördert
gute Waldnutzung
FSC® C083411

Kolibri 360 GmbH
Ferdinand-Jühlke-Straße 7
99095 Erfurt, Deutschland
produktsicherheit@kolibri360.de